World Survey of Nonferrous Smelters

World Survey of Nonferrous Smelters

Proceedings of Symposium on World Survey of Nonferrous Smelters, sponsored by The Pyrometallurgical Committee of TMS, at the TMS-AIME Annual Meeting in Phoenix, January 25-29, 1988.

Edited by:

John C. Taylor
Jan H. Reimers and Associates, Inc.
221 Lakeshore Road East
Oakville, Ontario, Canada L6J 1H7

and

Dr. Heinrich R. Traulsen
c/o Lurgi Gmbh
Gervinusstrasse 17/19, Postfach 11 12 31
D-6000 Frankfurt am Main 1
West Germany

A Publication of  The Metallurgical Society, Inc.

A Publication of The Metallurgical Society, Inc.
420 Commonwealth Drive
Warrendale, Pennsylvania 15086
(412) 776-9000

Printed in the United States of America.
Library of Congress Catalogue Number 87-42883
ISBN NUMBER 0-87339-026-1

TABLE OF CONTENTS

FOREWORD

At the annual meeting in 1984, the Pyrometallurgical Committee of The Metallurgical Society initiated a survey of selected non-ferrous smelters to determine whether or not they would be willing to supply, for publication, average operating data for their pyrometallurgical process steps. Response from the industry to the Committee's preliminary inquiry was encouraging, and detailed questionnaires were sent to the major operating companies around the world in the latter part of 1985 and early 1986. The responses from those companies which replied to the questionnaire are recorded in this publication in tabular form, according to geographic region and primary metal.

Data for over ninety non-ferrous pyrometallurgical operations are presented. However, as operators know, smelters are dynamic and process variables change. These changes will occur from day to day, and indeed from hour to hour, consequently the data reported should be considered typical for the particular plant and must be viewed by the reader in this context. For more specific or detailed information, we would suggest that the reader contact the respondent directly; the address of each is included, along with the name of the individual who arranged for the completion of the questionnaire. In response to the concern expressed by some in the industry the Committee took care to structure the questionnaire in a manner which avoided sensitive areas or data from which confidential information could be calculated. This point should be taken into account when working with the data presented here or in making further inquiries of the respondents.

Where practical, simplified process flow diagrams have been included to further describe the various plants and the interaction of the various sections of the plant and process streams. In addition, the Committee has selected papers from various sources which, in conjunction with the diagrams and specific references, will provide additional background and explanation of the main processes for which data are tabulated. In cases where the paper was presented prior to the fall of 1986, the tabulated data will be the more current. The reader will appreciate that it would be unreasonable to ask the volunteers serving on the Task Force to compile a complete metallurgical text. It was thought, however, that some papers describing particular plants should be included to provide an explanation of the data presented, particularly for those readers not closely involved with the industry.

The majority of companies approached were very cooperative in providing the information requested. The Committee appreciates the time and effort taken by these companies in support of this project. Other companies declined to participate for various reasons, and the absence of their data is regretted. Nevertheless, the Committee believes that, in this publication, the major non-ferrous operations around the world have been summarized and this will provide a valuable reference to those directly and indirectly involved in our industry. To the best of our knowledge, with the exception of the Committee's publication entitled: "Copper and Nickel Converters" edited by Robert E. Johnson in 1979, there has not been a comprehensive publication of operating data for the non-ferrous smelting industry. The more recent TMS publication entitled: "The Electrorefining and Winning of Copper", edited by J. E. Hoffmann et al, provides additional operating details on fire refining, anode casting, and anode copper composition which supplements the copper smelter data presented here.

The Editors and the Pyrometallurgical Committee of TMS express their appreciation to the members of the Task Force who spent considerable time and effort in soliciting answers to the questionnaire, analyzing the data, and preparing this volume. Without the hard work of such volunteers, this project would never have been carried through to completion. The strong support of Lurgi GmbH through the use of their facilities is gratefully acknowledged by the Committee. In addition, particular thanks go to Ms. L. Dunleavy of Jan H. Reimers and Associates and to the TMS publications staff who made final publication possible.

Heinrich R. Traulsen
Norddeutsche Affinerie Aktiengesellschaft
Hamburg, F. R. GERMANY

John C. Taylor
Jan H. Reimers and Associates
Oakville, Ontario, CANADA

THE PYROMETALLURGICAL COMMITTEE - TMS SURVEY TASK FORCE

CO-ORDINATOR
John C. Taylor
Consulting Engineer and
Advisor to the Committee

AFRICA

L. J. Hanschar - Consulting Metallurgist
ZCCM Ltd., Zambia

AUSTRALIA

B. S. Andrews - Vice President, Technological Resources
CRA Limited, Melbourne, Australia

CANADA, CENTRAL AND SOUTH AMERICA

D. G. Pannell - Manager of Production
Noranda Inc., Noranda, Quebec, Canada

CHILE

C. M. Diaz - Section Head, Pyrometallurgy
Inco Ltd., Mississauga, Ontario, Canada

EUROPE

H. R. Traulsen - Senior Manager - Metallurgy Department
Lurgi GmbH, Frankfurt am Main, West Germany

JAPAN

H. Asao - Manager, Metallurgy Department
Mitsubishi Metal Corporation, Tokyo, Japan

SCANDINAVIA

G. Lindkvist - General Manager - Technology
Boliden Metall AB, Skelleftehamn, Sweden

UNITED KINGDOM

J. G. Eacott - Consulting Engineer
Graham Eacott & Associates, Toronto, Ontario, Canada

UNITED STATES

W. J. Chen - Smelter Superintendent
Phelps Dodge, Hidalgo, New Mexico, U.S.A.

D. B. George - Technical Superintendent
Kennecott Minerals Co., Utah Copper Division
Magna, Utah, U.S.A.

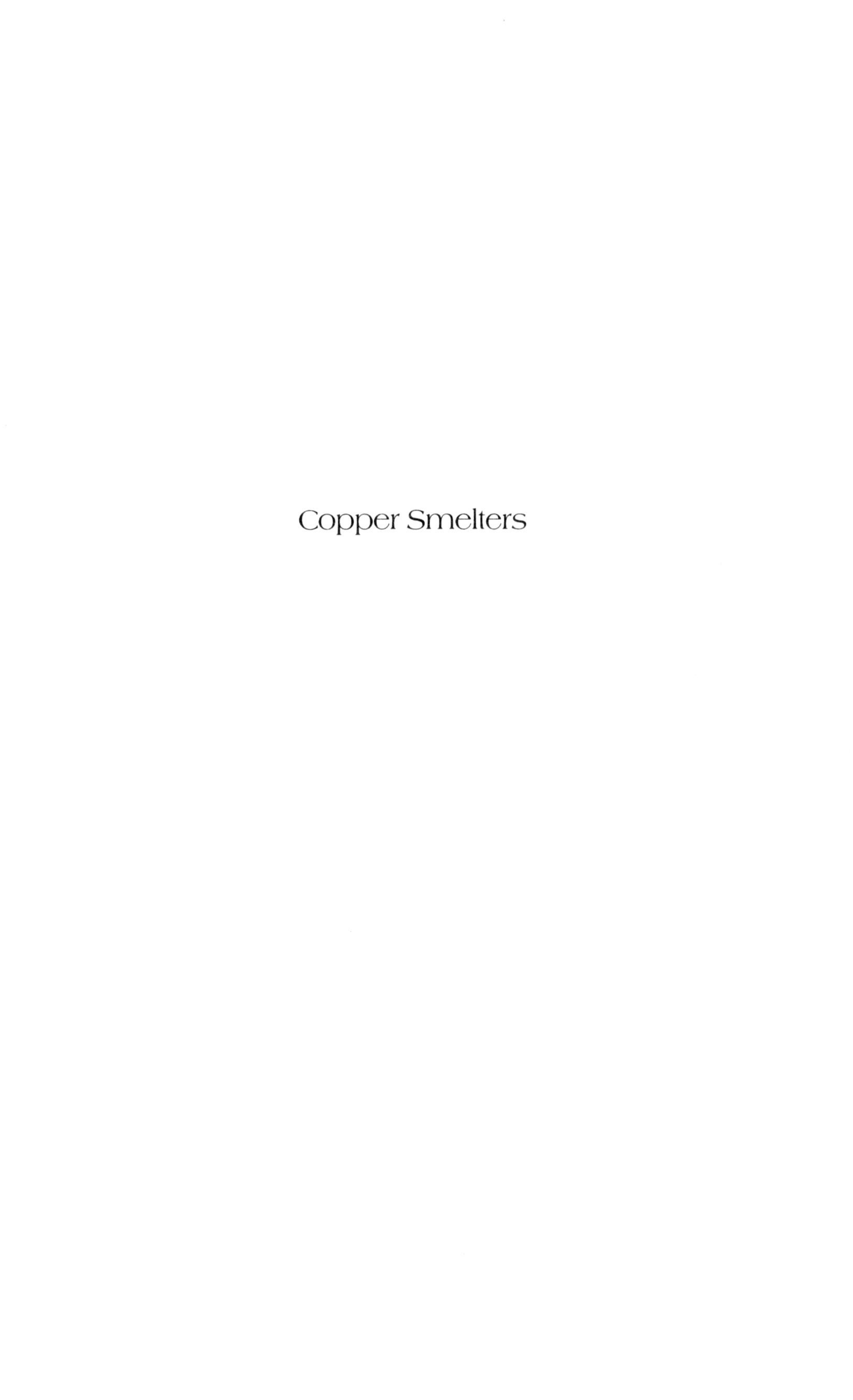

Copper Smelters

A SURVEY OF WORLD

COPPER SMELTERS

D.G. Pannell, Manager of Production

NORANDA MINERALS INC. - HORNE DIVISION
P.O. BOX 4000, ROUYN-NORANDA, QUEBEC
J9X 5B6

ABSTRACT

Data summarizing the operations and equipment of 47 copper smelters have been tabulated and reviewed. Information has been included on all standard aspects of smelter operation from feed preparation and analysis through to casting and sulphur fixation. The review attempts to compare and group the varied processes and where possible note any trends or standardization that may be discerned. The questionnaires, on which this paper is based, were completed by the respective smelter personnel during 1986 and early 1987.

INTRODUCTION

The survey presents data for a total of 47 copper smelters ranging in nature from the large, predominantly concentrate consuming, Japanese Sagonoseki to the smaller but complex polymetallic and secondary smelters of Europe such as Hoboken, Belgium, and Luenen, Germany. The smelters surveyed represent together a total annual production capacity of 5.1 million metric tons from 21 different countries. 50% of the smelters have an annual copper production capacity falling in the 80,000 to 110,000 MTPY range.

Although the number of smelters and countries represented in this survey is considerable the list is far from complete. With the exception of Poland and Yugoslavia, entries are missing from the entire communist block (Russia, China, Rumania, Bulgaria) as well as Peru, Brazil, India, and Zaire. The Chuquicamata smelter of Chile, as well as White Pine, San Manuel, Southwire, El Paso, Hayden and Chemetco smelters of the U.S.A. are also significant omissions. Although no further discussions will be made concerning these omissions, some additional general references have been provided to assist in accessing a still wider range of information (1) (2) (3).

The data presented here have been assembled from questionnaires completed during 1986 and early 1987. The information has not been updated since the completed questionnaires were received even though it subsequently became evident that some of it was out of date, e.g. the closure of the Douglas smelter. The survey can thus be viewed as a snapshot of part of the copper smelting industry during a particular period of time.

When reviewing and comparing data it has not been the intention to criticize the techniques used by any particular operation, indeed it may be assumed that smelters have modified their operations to suit local conditions of feed analyses and manpower, supply, energy and capital costs. It is fascinating to observe how operations have both adapted to these widely varying conditions and yet in many cases maintained a certain degree of conformity with other smelters. This variety and quantity of data, which will be further analysed by many of those interested in the copper industry, is available because of the cooperation of the participating companies and their staff. Their contribution to this survey is gratefully acknowledged.

FEED

Of the thirty nine smelters reporting feed analyses two main groups can be identified. Twenty nine smelters are treating feed averaging 25% copper, typical of a chalcopyrite concentrate, and a second group of seven smelters, Africa, Chile and La Caridad,Mexico, treat higher grade concentrates averaging 34% copper, typical of bornite and chalcocite. The three exceptions are Flin Flon treating copper zinc concentrates, Brixlegg consuming a 65% copper secondary feed and Mugal-Goektas treating the lowest grade feed at 17.8% copper.

FURNACE FLUX

Two distinct types are reported. Silica flux averaging 83% silica but generally varying from 70% to 95% silica which is used by the majority of smelters. The second, a lime based flux, is used by seven smelters. This is typically 52% lime, except at the Quimigal blast furnace where a 92% limestone is used. Of these seven smelters it is interesting to note that five are treating higher grade concentrates. The three Polish smelters treat essentially self-fluxing concentrates containing 18% silica.

CONVERTER FLUX

Converter silica fluxes are on the whole of a similar composition to furnace fluxes; however, several smelters report a slightly richer silica content for their converter fluxes. Use of a lime based flux is limited to the two Mitsubishi furnaces at Naoshima and Kidd Creek. Comparison with the silica analyses listed by Johnson et al. (1), is possible for twenty-four smelters common to both surveys. Of these ten appear to have increased and eleven decreased the silica grade of their converter fluxes.

FEED PREPARATION - DRYER

Only nine of the smelters do not dry feed in some way before the principal smelting furnace, of these four are complex or secondary smelters. The rotary dryer, by far the most common form of dryer, is installed in a total of twenty plants and the rotary-flash dryer in a further nine plants. The once ubiquitous roaster is represented as now being in use in only eight plants and virtually all of these roasters are of the modern fluid bed type.

70% of dryers achieve product moistness of 0.5% or less. The remaining 30% of the dryers which deliver a concentrate product of 5 to 10% moisture are, with one exception, associated with the more traditional reverberatory furnace operations.

Heavy oil is used to fire dryers at fourteen smelters and natural gas at eleven. The few remaining smelters use either coal (three), producer gas (two) or light oil (two). The Ashio smelter in Japan uses shredded tires as a source of fuel in addition to heavy oil.

It is perhaps surprising that more plants are not using low grade heat or recycled fuels for what is essentially a requirement for relatively low grade heat. In this respect Ashio, mentioned above, Harjavalta using anode furnace gases, and the Roennskaer Kaldo plant, using steam, report movement in this direction although Onahama have earlier reported the use of shredded tires for reverberatory furnace fuel (4).

PRIMARY SMELTING

Reverberatory furnaces still comprise the largest class of smelting vessel there being a total of thirty-one reported in this study. The flash furnace, Outokumpu and Inco, is the next most common class of vessel with a total of sixteen units; however the combined tonnage of copper produced from these vessels, 2 million tonnes annually, exceeds that produced from the thirty-one reverberatory furnaces, 1.7 million tonnes. Thirteen shaft or blast furnaces are reported with a combined annual copper production of 0.5 million tonnes, three electric furnaces; 0.35 million tonnes, four Noranda reactors; 0.32 million tonnes, two Mitsubishi continuous units; 0.18 million tonnes, and one Kaldo furnace.

Oxygen enriched air is used in the primary furnaces of about half of the smelters surveyed and, of these smelters, the average level of enrichment is to 50% oxygen. Within the different types of smelting vessels there is little variation in oxygen enrichment, 96-98% for the Inco flash; 42-43% for the Mitsubishi furnaces; and 34% for the Noranda reactors. Apart from the few Outokumpu flash furnaces operating without oxygen, the level of enrichment varies from 27 to 85%, the average level of enrichment being 42%. Only three smelters report oxygen enrichment of reverberatory furnace burners, Nkana, Noranda, oxy-fuel burners, and Palabora.

In spite of the fact that the majority of primary smelting processes can be classified into relatively few types a comparison of these processes is extremely complex and in fact becomes a detailed discussion of the individual factors influencing each plant and location. These are best revealed by reference to the original entry for each smelter. The various factors influencing furnace operations have been reviewed by Verney (5), Andersonn (6), Traulsen (7) and Eacott (8) amongst others. The great variation can be clearly seen by reference to Table I which lists smelters in ascending order of matte grade along with matte and slag chemical data. Some interesting general trends can nevertheless be discerned.

As is well known an increase in the matte grade causes slag losses from the primary furnace to increase at a proportionally greater rate; for example, for an average furnace operating in the 30% to 50% copper in matte range, an increase in matte grade of 10%, e.g.: 35% to 45% Cu, causes slag losses to increase by 0.1% copper. For furnaces in the 50% to 70% matte grade range a similar increase in the matte grade causes the slag analysis to increase by 1.6% copper. This is obviously a reflection on the type of processes predominant in the differing matte grade ranges as well as the inherent chemistry. The comparison is however interesting in that it gives some measure of the price to be paid for using the oxidation of sulphur as a source of heat in the primary smelting vessel. It also underlines in practical terms how far we remain from the oft-stated goal (8) (9) (10) (11) of producing copper from concentrates in a single vessel whilst at the same time producing discardable slags.

The general trend noted in the previous paragraph is subject to some major qualifications. Table I lists the smelters in ascending order of matte grade and, all other other things being equal, one would expect slag losses to fall into the same ascending order. It is evident that smelters, whose slag contains significant quantities of lime, are often able to improve the copper analysis of their slag beyond the average reported for that matte grade. Mt. Isa, Mufilira, Legnica, Glocow 1, and Luenen are clearly examples of this phenomenon. The three last examples cited are of course also blast furnace operations.

The wide variation of iron and sulphur analyses between mattes of a similar copper content also reflects local conditions and constituents such as lead and zinc matte contents.

TABLE I - SMELTER PRIMARY FURNACE MATTE AND SLAG ANALYSIS ARRANGED IN ASCENDING ORDER OF MATTE COPPER ANALYSIS

SMELTER	PRIMARY FCE MATTE			PRIMARY FURNACE SLAG					SEPARATE SLAG CLEANING STAGE	
	Cu	Fe	S	Cu	Fe	SiO_2	Fe/ SiO_2	CaO	Primary Furnace	Converter
Mugel-Goektas	19	41	25	0.6	61.8	27.7	2.2			
Horne 1	32	31	27	0.6	33.0	35.0	0.9	0.5		
Cananea	34	32	29	0.6	19.0	37.0	0.5	11.0		flotation
Gaspe	35	25	26	0.5	32.5	39.0	0.8	5.0		
Flin Flon	36	28	24	0.7	33.2	37.8	0.9		fuming	
Inspiration	37	34	25	0.8	33.0	38.0	0.9	8.0		
Onahama	38	28	26	0.5	36.0	31.0	1.2	3.0		
Bor	40	33	25	0.5	37.0	37.0	1.0	5.7		
Naoshima 1	43	26	26	0.5	30.9	34.5	0.9	9.6		flotation
Mt. Isa	44	27	26	0.4	35.8	36.6	1.0	8.7		flotation
Inco	45	26	25	0.6	40.0	33.0	1.2			Ni circuit
Ray	45	28	24	0.6	30.0	40.0	0.8	2.0		flotation
Douglas	45	25	25	0.7	34.0	39.0	0.9	3.0		
Boliden Cu	45	19	23	1.0	33.0	33.0	1.0		fuming	
Boliden Kaldo	45	19	19	1.3	32.4	31.3	1.0		fuming	
Ventanas	47	23	27	0.9	36.5	33.0	1.1	5.5		
Hoboken	48	9	15	0.6	22.0	29.0	0.8	13.0		separate fce
Chino	48	25	23	0.7	44.0	33.0	1.3			
Palabora	49	24	24	0.6	32.5	33.0	1.0			
Kembla	50	16	21	0.5	35.0	30.0	1.2	1.5		
Quimigal	50	29	21	0.5	38.0	32.0	1.2	10.0		
Isabel	50	22	24	0.6	36.0	34.3	1.0	1.7		flotation
Luanshya	50	21	25	0.6	27.0	40.0	0.7	7.0		
Palpote	50	24	26	0.8	32.0	35.0	0.9	6.7		
Tsumeb	50	15	23	0.9	22.0	38.0	0.6	13.0		Pb smelter
Ashio	50	22	23	1.5	33.0	28.2	1.2		electric	flotation
Mufilira	55	18	23	0.5	27.0	37.0	0.7	13.0		
Huelva	57	17	21	2.0	41.0	29.0	1.4	0.1	electric	electric
Kosaka	58	12	23						electric	flotation
Toyo	58	14	23	0.9	37.4	33.0	1.1	1.5	electric	flotation
Nkana	59	16	23	1.1	22.0	43.6	0.5	7.8		
Sagonoseki	60	14	22	1.0	40.0	33.0	1.2	2.0	electric	flotation
La Caridad	60	16	23	1.3	35.5	28.4	1.3		electric	electric
Hidalgo	60	16	23	1.5	42.0	30.0	1.4	1.3	electric	electric
Onsan	60	14	22	1.6	41.7	30.0	1.4		electric	electric
Tamano	61	15	22	0.5	37.8	32.0	1.2	4.2	fnce elctrde	flotation
Legnica	62	11	3	0.5		38.0		15.0		
Hamburg	62	15	22	1.3	39.5	31.0	1.3	4.5	electric	
Glocow 1	63	10		0.5		39.0	0.0	17.0		
Horne 2	65	7	20	4.6	40.0	22.0	1.8	2.0	flotation	
Naoshma 2	66	10	21						electric	
Kidd	68	7	22						electric	
Outokmpu	69	8	24	2.5	40.0	27.5	1.5	2.3	flotation	flotation
Utah	73	4	21	7.0	42.2	22.8	1.9		flotation	
Luenen	75	5	2	1.0	45.0	20.0	2.3	15.0		
Glocow 2	99			14.5	5.5	30.0	0.2	12.0	electric	electric

N.B.: Where a smelter has supplied an analysis range the average of the high and low values has been assumed.

CONVERTING

The Peirce-Smith converter remains by far the most common copper matte converting vessel. Forty plants surveyed, representing 128 installed vessels, continue to use this process now virtually unchanged in some eighty-five years. Eighteen Hoboken converters are installed in a further five plants and Mitsubishi continuous converters in two plants.

The trend towards increasing matte grades, noted by Johnson et al. (1), continues. The average matte grade of converters surveyed here is 53% compared to approximately 46% of the Johnson survey completed eight years earlier. The potential advantages of varying matte grades were recently discussed in papers by Partelpoeg (12) and Persson (13). The grade of converter flux however, as previously noted, does not appear to have undergone any parallel upgrading.

Slag grades reported in the survey vary widely bearing only a passing relationship to matte grade and other listed parameters. It is likely that both operating techniques and the difficulty of consistently sampling a two-phase system have more influence on the apparent converter slag quality than any other factors. Several operations logically produce a converter slag that is best suited to reducing total metal losses in their slag cleaning stage, thus slag volume is minimized, often to the detriment of classical slag quality.

Oxygen enrichment of converting air is practiced in sixteen of the smelters surveyed. The two Mitsubishi converters at Naoshima and Kidd Creek report the highest consistent levels of enrichment at 30% oxygen compared to an average of 24.7% oxygen for the remaining fourteen plants. Only Inco among Peirce-Smith operators reports enrichment up to 30%.

A comparison of the data listed by Johnson (1) and the present survey would indicate that there has been a modest increase, particularly in Japan, in the practice of enriching converting air. Six smelters, Palabora, Bor, Naoshima, Tomano, Kosaka and Toyo, included in the previous survey as using no oxygen enrichment are now doing so. Only two smelters reported in this survey, Douglas and Onahama, have apparently ceased using oxygen enrichment. As would be expected oxygen enrichment is associated with higher matte grades, the average matte copper analysis of smelters practicing oxygen enrichment is 58% compared to 49% for the remainder.

Converter slag is disposed of in many different ways. The most common method, irrespective of smelting technology, is to return the slag to the primary smelting vessel. This is the case for 59% of the smelters surveyed. Several smelters have techniques adapted to their particular situation such as Inco, who return slag to the nickel circuit for nickel recovery, Rönnskär who return slag to a zinc fuming furnace and Tsumeb who return slag for lead recovery. The remaining smelters treat converter slag for metal recovery either by cooling and milling, twelve smelters, or by treating the slag in an electric furnace, five smelters.

SLAG CLEANING

As can be seen by reference to Table I the majority of smelters using separate slag cleaning processes tend to be those operating at higher matte grades. With the exception of those operations where the primary smelting unit is a blast furnace, e.g. Kembla, Legnica, Glocow I, Luenen, and/or the CaO content of the slag is 7% or greater, smelters with a matte grade greater than 50% operate some form of separate slag cleaning process. This trend has also been noted by Floyd and Mackey (14).

A comparative review of the two slag cleaning methods, flotation and pyrometallurgical, either comparing one method with the other or by comparing different smelters using the same slag cleaning method, is difficult because of the varied nature of slag fed to the cleaning stage. For example some smelters treat converter and primary furnace slag concurrently in the same electric furnace e.g. Hidalgo, whereas others treat the primary furnace slag in an electric furnace, and the converter slag by cooling and milling, e.g. Ashio, Sagonoseki, Kosaka, Toyo.

Table II shows, for the two different slag cleaning processes, a comparison of the copper lost in slag or tailings compared to the iron-to-silica ratio of the slag. This table includes data for both converting and primary furnace slag. For each group the data have been arranged in order of discard slag or tailings copper analysis. This of course neglects many of the important factors such as power input and retention time but, for the electric furnaces, there is a definite trend of increasing copper losses with increasing iron-to-silica ratio. This trend was also noted in the previous study, (1).

The data for the slag flotation processes show no similar trend, the copper losses clearly being dependent on slag cooling regimes, fineness of grind and other milling parameters. There is nevertheless an interesting comparison that can be made. The electric furnace iron-to-silica ratio averages 1.27 and individual smelters all fall within 25% of this average. The corresponding average iron-to-silica ratio for the flotation case is 2.28 but with variations of 43% of this average for individual smelters. Flotation thus offers a clear advantage in terms of the reduced slag volume required to flux the iron associated with the copper bearing feed, an apparent insensitivity to slag analysis and a lower copper analysis of the discard product.

The number of operations practicing slag flotation appears to have changed little since the period of the previous study, (1). One must therefore conclude that the advantages of slag flotation are insufficient to overcome the differences in operating costs, the initial capital costs and other local factors such as the availability of metal bearing fluxes and the problem of slag or tailings disposal or sale etc. etc.

TABLE 11

SLAG CLEANING PROCESS DATA

FLOTATION			ELECTRIC FURNACE		
Smelter	Cu in Tailings	Slag Fe/SiO2	Smelter	Cu in final slag	Slag Fe/SiO2
Naoshima	0.30	2.44	Ashio	0.53	0.94
Utah	0.35	1.50	Kosaka	0.60	1.19
Horne	0.40	1.80	Sagonoseki	0.60	1.21
Harjavalta	0.40	1.90	Toyo	0.60	1.21
Toyo	0.40	2.19	Hamburg	0.60	1.25
Ashio	0.40	2.85	Onsan	0.60	1.46
Tamano	0.40	3.01	La Caridad	0.70	1.36
Isabel	0.44	3.25	Huelva	0.70	1.41
Kosaka	0.48	2.06	Hidalgo	1.00	1.43
Sagonoseki	0.48	2.19			
Isa	0.77	2.40			
Cannanea	0.80	2.29			
Ray	1.00	1.80			
Average:	0.51	2.28	Average:	0.66	1.27

COPPER CASTING AND REFINING

Rotary anode furnaces have become the standard as far as copper smelter fire refining equipment is concerned. Of the smelters producing anodes, only six retain reverberatory anode furnaces,the three Polish smelters as well as Brixlegg , Port Kembla and Inco. Unlike the furnace type, the reagent used for deoxidizing copper is far from standardized, fourteen smelters report the continuing use of green poles, a further thirteen use either natural or propane gas and five use ammonia. Bunker C, naptha or LPG are used by the remainder.

The majority of anodes continue to be cast on wheel type casting machines. The survey provides data for forty-two of these operations of which a few are double mould wheels e.g. Huelva, Tamamo and Harjavalta. Casting rates fall within a fairly narrow range, 80% within 30 to 50 MTPH, with an average rate of 40.4 MTPH for wheel cast anodes compared to 50 to 55 MTPH for Hazelett anodes of 135 to 155 kg. One might expect the weight of the anode to influence the casting rate but, for wheel cast anodes, this does not seem to be the case. In fact anode weights fall within a relatively narrow weight range from one smelter to another, i.e. 80% of smelters cast an anode within 60kg of the 307 kg anode average.

REFERENCES

1) Johnson R.E., Themelis N.J., Eltringham G.A. "A Survey of Worldwide Copper Converter Practices" Copper and Nickel Converters edited by R.E. Johnson, A.I.M.E. 1979.

2) Biswas A.K., Davenport W.G., "Extractive Metallurgy of Copper", 2nd edition, 1980, Pergamon Press, pp. 26-29.

3) Burger J.R., Sassos P.S. "Chile - Peru", Engineering and Mining Journal, Vol. 185, No. 11, November 1984, pp. 34-78.

4) Kohno H., Asao H., Amano T., "Use of Alternative Fuel at Onahama", Journal of Metals, Vol. 34, No. 7, (1982), pp. 36-40.

5) Verney L.R., "The Economics of Sulphide Smelting Processes" Sulfide Smelting Volume 2, Technology and Practice, edited by H.Y. John, D.B. George and A.D. Zunkel, The Metallurgical Society of AIME, Warrendale, Pennsylania, 1983. pp. 1065-1108.

6) Andersson B. et al. "Use of Oxygen in the Outokumpu Flash Smelting Processes" paper presented at the 20th Annual CIM Conference of Metallurgists, Hamilton, Ontario, August 1981.

7) Traulsen H.R., Taylor J.C., George D.B., "Copper Smelting - An Overview", Journal of Metals, Vol. 34, No. 8 (1982), pp. 35-40.

8) Eacott J.G., "The Role of Oxygen Potential and use of Tonnage Oxygen in Copper Smelting", Sulfide Smelting, Volume 2, Technology and Practice, op. cit., pp. 583-634.

9) Davenport W.G., Partelpoeg E.H., "Single-Step Smelting of Copper", Frontier Technology in Mineral Processing edited by Spisak J.F. and Jergensen G.V., 1985, the Society of Mining Engineers, A.I.M.E., pp. 75-84.

10) Opie W.R., "Pyrometallurgical Processes that Produce Blister Grade Copper Without Matte Smelting", Extraction Metallurgy '81, proceedings of London, U.K., Symposium, September 1981, The Institution of Mining and Metallurgy, 44 Portland Place, London, W.1.

11) Rodolff D.W., Murden K.B., "Suitability of the Outokumpu Direct Blister Process for Various Types of Copper Concentrates", paper presented at 112th A.I.M.E. Annual Meeting, Atlanta, Georgia, March 1983.

12) Partelpoeg E.H., "Energy Optimization in Flash Smelting", Ph.D. Thesis, submitted 1985 Department of Metallurgy, University of Arizona.

13) Persson H., Iwanic M., El-Barnachawy S., Mackey P.J., "The Noranda Process at Different Matte Grades" Journal of Metals, Vol. 38, No. 9, (1986), pp. 34-37

14) Floyd J.M., Mackey P.J., "Developments in the Pryometallurgical Treatment of Slag: A Review of Current Technology and Physical Chemistry". Extraction Metallurgy '81, op.cit. pp. 345-371.

COPPER SMELTERS
of
AFRICA

COMPANY NAME	Zambia Consolidated Copper Mines Limited			
PLANT	Nkana		Luanshya	Mufulira
Annual Production - Copper MTPY	264,000 (Gross)			
Flowsheet Page No.	89		90	91
FEED ANALYSIS				
Concentrate	Primary	Secondary		
% Cu	27	35	39	35-37
% Fe	20.6	14.5	15	14-17
% S	23	20	22	22-25
% SiO_2	11.6	14.0	11	15-20
Mesh	75% -200		70% -200	60% -325
% H_2O	-		-	-
Furnace Flux				
% SiO_2	2.6		1.4	1.0
% CaO	50.5		54	54
% Fe	0.9		0.5	-
Mesh	100% -10 mm		60% -200	-
Converter Flux				
% SiO_2	68		70-78	63-66
% CaO	0.7		-	3-5
% Fe	2.5		3	2
Mesh	20 mm		6-38 mm	70% -1.05
FEED PREPARATION				
Blending System	Front End Loader to Hoppers		Bin-Conveyors	Bins-Conveyors

PLANT - continued	Nkana	Luanshya	Mufulira
Dryer - Type	Rotary	Rotary	Rotary
- Number	1 Primary + 2 Secondary	1	1
- Dimensions - m	2.4 Ø x 23.3 & 2.6 Ø x 27.5	3 Ø x 30.5	3 Ø x 27.4
Average Feed Rate - WMTPH	Prim. 24; Sec. 110	75-80	60-70
Feed Moisture - % H_2O	16-18 11-12	10-12.5	8-12
Retention Time - Min.	20 20	45	5
Fuel Used - Type	Coal (90% -200M)	Coal	Heavy Oil
Calorific Value - kCal/kg	6,500	6,000	10,400
Inlet Gas Temp. - °C	850	560	800-850
Outlet Gas Temp. - °C	175	45-50	120
Product Moisture - % H_2O	Prim. 8-10; Sec. 7-8	8-8.5	0
Exhaust Gas Volume - Nm^3/hr	-	3,717	25,000
Cleaned by	Cyclone; Cyclone + Scrubber	Cyclones + ESP	Cyclones + Scrubber
SMELTING			
Furnace Type	Reverb	Reverb	36 MVA Elkem Electric
Number of Units	5	2	1
Nominal Capacity - MTPH Concentrate	33	23	80-95
Dimensions - m	9.75 x 29.6 x 3.7	8.33 x 33.3 x 4.1	8.5 x 29.6
Oxygen Enrichment - % O_2	30 by wt.	None	None
Auxiliary Fuel - Type	Coal & Oil	Coal	-
Calorific Value - kCal/kg	6,500 & 10,300	6,000-6,500	-
Operating Temperature - °C	1,250 (1,610 max)	1,250-1,570	800 (roof)
Flux as % of Concentrate	9-10	7-9	10
Campaign Life - Years	2.5	2-3	8.5+
Furnace Matte - MTPD	340/furnace	350-432	1,300
Analysis - % Cu	58.6	45-55	55
- % Fe	16	18-24	18
- % S	23	24-26	23
Temperature - °C	1,120	1,140-1,170	1,160

PLANT - continued	Nkana	Luanshya	Mufulira
Furnace Slag - MTPD	340/furnace	300	1,000
Analysis - % Cu	1.1	0.4-0.8	0.5
- % Fe	22.03	25-29	24-30
- % CaO	7.83	5-9	13
- % SiO_2	43.6	35-45	37
Temperature - °C	1,240	1,150-1,250	1,275
Disposition	Discard		
Offgas - Volume - Nm^3/hr	51,000	48,000	35,000
Temperature - °C	1,250	1,200-1,300	850
% SO_2	2	1-2	6-10
Cooled by	2 Boilers/Furnace	Evap. Spray	Air Infiltration
Inlet Temp. - °C	1,250	530	-
Outlet Temp. - °C	350	-	-
Disposition of Dust	-	Recycled	Recycled
CONVERTING			
Converter Type	P-S	P-S	P-S
Number - Hot	4	1	5
- Standby	2	3	1
Dimensions - m	4 Ø x 9.1	4 Ø x 9.1	4 Ø x 9.1
Number of Tuyeres	49-51	48	50
Size - mm	50	50	66.67
Average Blowing Rate - Nm^3/hr	-	-	-
Oxygen Enrichment - % O_2	26	-	-
Air Rate - Nm^3/min	516	544	530-550
Scrap added as % of matte charged	20	12-15	12-20
Flux as % of matte charged	20	15	15-16
Blister Production - MTPD	-	250	-
Analysis - % Cu	98.75	97	98.5
- % Fe	7 ppm	0.005	-
- % S	70 ppm	2	40 ppm

PLANT - continued	Nkana	Luanshya	Mufulira
Slag - Analysis - % Cu	4.1	3.5	2-3
- % Fe	41.6	40-45	40-45
- % SiO_2	23.4	23-28	24-25
- Disposition	To Reverb	To Reverb	To Smelting Furnace
FIRE REFINING			
Furnace - Type	Rotary	Rotary	Rotary
Number of Units	4	2	4
Dimensions - m	4 Ø x 9.1	4 Ø x 9.1	4 Ø x 9.1
Time per Charge for			
Oxidation - min.	150	60	90-108
- % O_2	21	21	21
Reduction - min.	150	240	180-240
Reductant - Used	Green Poles	Green Poles	Green Poles
- Consumption	3.5 MT/charge	12/charge	15-17/charge
PRODUCT CASTING			
Casting Machine - Type	4 Walker Wheels	Walker Wheel	Treadwell & Walker Wheels
Number of Molds	22	22	22
Average Casting Rate - MTPH	30	25	30-35
Normal Casting Temp. - °C	1,120-1,130	1,150	1,145-1,160
Cast Product - Type	Anodes	Anodes	Anodes
Dimensions - m	0.96 x 1.15	0.94 x 0.94 x 0.035	0.91 x 0.918
Average Weight - kg	272-300	300-330	285-320
Rejects as % of new material cast	2	-	3
SULPHUR FIXATION		None	None
Source of Gas	Converters		
Plant - Type	3 Single Contact Acid Plants		
- Rated Capacity - MTPD	850		
- Volume - Nm^3/hr	150,000		
- Average % SO_2	6.7		
- Average Conversion Efficiency %	96		

PLANT - continued	Nkana	Luanshya	Mufulira
Product Grade - % H_2SO_4	98		
KWH/MT Sulphur Fixed in Product	472		
Auxiliary Fuel	Kerosene 5.2 L/MT acid		
Tailgas Scrubbing	None		

COPPER SMELTERS of AFRICA

COMPANY NAME	Palabora Mining Company Ltd.	Tsumeb Corp. Ltd.
PLANT	Palabora	Tsumeb
Annual Production - Copper MTPY	-	46,000
Flowsheet Page No.	226	92
FEED ANALYSIS		
Concentrate		See Description Pages 229-233
% Cu	37.8	
% Fe	19.9	
% S	23.0	
% SiO_2	1-3	
Mesh	90% -325	
% H_2O	12-18	
Furnace Flux		
% SiO_2	90-95	
Mesh	-19 mm	
Converter Flux		
% SiO_2	60	
Mesh	-38 mm	
FEED PREPARATION		
Blending System	None	
Dryer - Type	Rotary	Rotary
Number	1	1
Dimensions - m	2.4 Ø x 14.7	1.5 Ø x 10.7
Average Feed Rate - WMTPH	-	25
Feed Moisture - % H_2O	13.5	15

PLANT - continued	Palabora	Tsumeb
Retention Time - Min.	-	5
Fuel Used - Type	Coal	Producer Gas
Calorific Value - kCal/kg	6,700	-
Inlet Gas Temp. - °C	800	-
Outlet Gas Temp. - °C	150	-
Product Moisture - % H_2O	6.9	5
Exhaust Gas Volume - Nm^3/hr	21,500	9,000
Cleaned by	Wet Scrubber	Baghouse
SMELTING		
Furnace Type	Reverb	Reverb
Number of Units	1	1
Nominal Capacity - MTPH Concentrate	45	-
Dimensions - m	10.7 x 36	9.7 x 27
Oxygen Enrichment - O_2	26 MTPD	None
Auxiliary Fuel - Type	Coal	Coal
Calorific Value - kCal/kg	6,700	-
Operating Temp. - °C	-	1150-1200
Flux as % of Concentrate	7	-
Campaign Life - Years	-	5
Furnace Matte - MTPD	805	325
Analysis - % Cu	47-51	-
- % Fe	23.7	-
- % S	23.9	-
Temperature - °C	-	1,100
Furnace Slag - MTPD	675	150
Analysis - % Cu	0.62	0.9
- % Fe	32.5	-
- % SiO_2	33.0	-
Temperature - °C	-	1100-1150
Disposition	Discard	-

PLANT - continued	Palabora	Tsumeb
Offgas Volume - Nm^3/hr	79,000	45,000
Temperature - °C	-	1150-1200
% SO_2	1.0	1.0
Cooled by	Boiler	-
Inlet Temp. - °C	-	-
Outlet Temp. - °C	350	-
Disposition of Dust	Recycled	-
CONVERTING		
Converter Type	P-S	-
Number - Hot	1	2
- Standby	1	-
Dimensions - m	4 Ø x 10.1	-
Number of Tuyeres	52	-
Size - mm	43	-
Oxygen Enrichment - % O_2	22.5 (occasionally)	None
Air Rate - $Nm^3/min.$	590-670	316
Scrap added as % of matte charged	15	-
Flux added as % of matte charged	11	-
Blister Production - MTPD	163	-
Analysis - % Cu	-	98.5
- % Fe	-	-
Slag - Analysis - % Cu	4.2	8
- % Fe	48.6	-
- % SiO_2	25.0	-
Disposition	To Reverb	To Lead Smelter
FIRE REFINING		
Furnace - Type	Rotary	None
Number of Units	2	
Dimensions - m	4 Ø x 9.1	

PLANT - continued	Palabora	Tsumeb
Time per Charge for		
Oxidation - min.	240	
- % O_2	21	
Reduction - min.	300	
Reductant - Used	Green Poles	
- Consumption	5-8/charge	
PRODUCT CASTING		
Casting Machine - Type	Wheel	-
Number of Molds	22	-
Average Casting Rate - MTPH	40	-
Normal Casting Temp. - °C	1,200	-
Cast Product - Type	-	Blister Cake
Dimensions - m	0.94 x 0.925 x 0.044	-
Average Weight - kg	320	1,650
Rejects as % of new material cast	5	15
SULPHUR FIXATION		None
Source of Gas	Converters	-
Plant - Type	Single Contact Acid Plant	
Rated Capacity - MTPD	400	
Volume - Nm^3/hr	80,000	
Average % SO_2	4.8	
Average Conversion Efficiency %	95.0	
Product Grade - % H_2SO_4	98.7	
Auxiliary Fuel	Diesel 80,000 L/year	
Tailgas Scrubbing	-	

COPPER SMELTERS
of
AUSTRALIA AND SOUTH EAST ASIA

COMPANY NAME	Mount Isa Mines Ltd.	Electrolytic Refining & Smelting Co. of Australia	Korea Mining & Smelting Co.Ltd.	Philippine Associated Smelting & Refining Corp.
PLANT	Mount Isa	Port Kembla	Onsan	Isabel
Annual Production - Copper MTPY	154,000	26,000	80,000	138,000
Flowsheet Page No.	-	93 & 94	-	95
FEED ANALYSIS				
Concentrate				
% Cu	26.4	20-25	27	28.6
% Fe	27.8	32-35	28	26.9
% S	32.2	28-32	30	30.1
% SiO_2	7.5	5-10	8	7.4
Mesh	96.25%-200	-	80%-200	40-80% -325
% H_2O	-	9-10	8	8-10
Furnace Flux				
% SiO_2	80.9	85	85	95
% CaO	4.65	1.5	-	2 max.
% Fe	2.05	-	-	0.5
Mesh	75%-100	+10-40mm	90%-28	75%-60
Converter Flux				
% SiO_2	70-95	97	90	95
% CaO	1-4	-	-	1.5 max.
% Fe	6	-	-	-
Mesh	+6mm-50mm	30	-	+9.5 mm -26mm
FEED PREPARATION				
Blending System	-	Gantry Storage	Day Bins	Bedding Plant
Number of Beds	-	-	-	2
Bed Size - MT	-	-	-	4,200
Method of Reclaim	-	Crane	-	Front End Loader

PLANT - continued	Mount Isa	Port Kembla	Onsan	Isabel
Equipment	Fluo-Solids Roaster	Sinter Machine	Rotary Dryer	Rotary-Flash
- Number	1	1	1	1
- Dimensions - m	6.7 Ø	1.5 x 20	2.5 Ø x 25	2.15 Ø x 16
Average Feed Rate - WMTPH	86.1 (60 min-110 max)	8-10	42	85
Feed Moisture - % H_2O	Slurry Feed	-	8	8-10
Air Rate Nm^3/hr	62,460	-	-	-
Fuel Used - Type	-	None	Heavy Oil	Heavy Oil
Calorific Value - kCal/kg	-	-	9,900	10,300
Inlet Gas Temp. - °C	-	-	600	350
Outlet Gas Temp. - °C	-	-	110	80-100
Product Moisture - % H_2O	0	-	0.2	0.3
Analysis - % Cu, % S	26.0, 19.6	27.0, 10.0	-	-
Exhaust Gas Volume - Nm^3/hr (wet)	100,000 (76,500 dry)	54,000	27,000	72,000
Cleaned by	-	Multiclone	-	Cyclone ESP
% SO_2	6	1-3	-	-
SMELTING				
Furnace Type	Reverb	Blast Furnace	Flash (OKO)	Flash with Electrodes
Number of Units	2	1	1	1
Nominal Capacity - MTPH	60 and 30 calcine	20-25 charge	-	-
Concentrate	-	-	38.2	75
Dimensions - m	11.6 x 35.1 6.1 x 27.4	1.4 x 7.8	4.88x20.07x2.15	20.25 x 7.5 x 2.93
Oxygen Enrichment - % O_2	Nil	Nil	50	27
Auxiliary Fuel - Type	Distillate	Coke	Heavy Oil	Heavy Oil
Calorific Value - kCal/kg	10,700	-	9,900	10,300
Operating Temperature - °C	1,500		1,350	1,250
Flux as % of Concentrate	4 max.	15	8.8	7-8
Campaign Life - Years	16+ and 12+	1-2	4	2

PLANT - continued	Mount Isa	Port Kembla	Onsan	Isabel
Furnace Matte - MTPD	1,220	140-160	403	900
Analysis - % Cu	44.0	50	60	50.0
- % Fe	26.8	26	14.2	22.3
- % S	25.8	21	22.2	23.8
Temperature - °C	1,080	1,140	1,200	1,200
Furnace Slag - MTPD	700	180-210	487	540
Analysis - % Cu	0.35	0.5	1.6	0.58
- % Fe	35.8	35	41.7	36.0
- % CaO	8.72	1.5	-	1.7
- % SiO_2	36.6	30	30	34.3
Temperature - °C	1,130-1,150	1,160	1,260	1,250
Disposition	Underground Fill			To Discard
Offgas - Volume - Nm^3/hr	93,600 and 50,400	85,000	27,000	44,000
Temperature - °C	1,150-1,250	100-300	1,350	-
% SO_2	1.5	0-1	20	10-12
Cooled by	Boilers(4)&Heat Exchangers(2)	Water Sprays	Boiler	Boiler
Inlet Temp. - °C	1,260	150-300	1,350	800
Outlet Temp. - °C	170	115	360	350
Disposition of Dust	Recycled	Recycled to Sinter Feed	Recycled	Recycled
CONVERTING				
Converter Type	P.S.	P.S.	P.S.	P.S.
Number - Hot	2	1	1	3
- Standby	1	2	1	1
Dimensions - m	3 Ø x 10.7	3 Øx7(2), 3 Øx6(1)	3.9 Ø x 9.1	3.96 Ø x 9.15
Number of Tuyeres	54	26	48	50
Size - mm	50.4	50	-	48
Average Blowing Rate - Nm^3/hr	-	-	-	-
Oxygen Enrichment	None	None	None	None
Slag - % O_2	-	-	-	-
Copper - % O_2	-	-	-	-

PLANT - continued	Mount Isa	Port Kembla	Onsan	Isabel
Air Rate - Nm^3/min.	-	430	600	610
Slag Blow - Nm^3/min.	750	-	-	-
Copper Blow - Nm^3/min.	840	-	-	-
Scrap added as % of matte charged	31	-	-	40
Flux as % of matte charged	13	-	9.6	10
Blister Production - MTPD	-	70-80	310	550
Analysis - % Cu	99.02	95-98	99	99.0
- % Fe	0.0006	-	-	-
- % S	0.0013	0.2	0.02	0.05
Slag-Analysis - % Cu	2.6	5-10	5	7-8
- % Fe	52.8	30-40	46	52.0
- % SiO_2	21.9	16	25	16.0
- Disposition	Returned to Concentrator	Returned to Blast Furnace	To Slag Cleaning	To Flotation
SLAG CLEANING				
Electric Furnace:				
Smelting Furnace Slag - MTPD	-	-	487	-
Converter Slag - MTPD	-	-	100	-
Power - KWH/MT Slag	-	-	90	-
Retention Time - Hours	-	-	6	-
Matte Produced - % Cu	-	-	67.6	-
Discard Slag - % Cu	-	-	0.6	-
- % Fe	-	-	44.3	-
- % SiO_2	-	-	30.4	-
Converter Slag - MTPD	510	-	-	450-500
Cooled in	Pits	-	-	Sand Bed
Duration - Hours	24	-	-	16
Flotation Concentrate - MTPD	41	-	-	90
Analysis - % Cu	42.48	-	-	32
- % Fe	23.2	-	-	37
- % S	16.5	-	-	8

PLANT - continued	Mount Isa	Port Kembla	Onsan	Isabel
Flotation Tailings - MTPD	600	-	-	520
Analysis - % Cu	0.77	-	-	0.44
- % Fe	-	-	-	57.0
Copper Recovery - %	79.2	-	-	95
FIRE REFINING				
Furnaces - Type	Rotary	Reverb	Rotary	Rotary
Number of Units	2	2 (1 standby)	2	2
Dimensions - m	3 Ø x 10.7	3.96 x 9.14	4.9 Ø x 9.2	4.4 Ø x 10
Number of Tuyeres	2	-	2	2
Size - mm	16 I.D.	-	-	25
Time per Charge for				
Oxidation - min.	167	240-300	-	90
- % O_2	21	21	21	21
Reduction - min.	132	120	-	150
Reductant - Used	Naphtha	Green Poles	Propane	LPG
- Consumption/MT anode	4.36 kg	3-5 poles/charge	-	5 kg
PRODUCT CASTING				
Casting Machine - Type	Mitsui	Walker	Wheel	Wheel
Number of Moulds	27	20	16	24
Average Casting Rate - MTPH	43	50	32	45
Normal Casting Temp. - °C	1,160	1,140	1,250	1,250
Cast Product - Type	Anodes	Anodes	Anodes	Anodes
Dimensions - m	-	0.87 x 0.91	-	1.08 x 1.33 x 0.04
Average Weight - kg	310	270	285	345
Rejects as % of new material cast	2.6	5	6.5	2

PLANT - continued	Mount Isa	Port Kembla	Onsan	Isabel
SULPHUR FIXATION	None	None		
Source of Gas	-	Sinter & Furnace	Flash + Converter	Flash + Converter
Plant - Type	-	Gases to 200 m stack after filtration	Double Contact	Double Contact
- Rated Capacity - MTPD	-		850	2,500
- Volume - Nm^3/hr	-	335,000	112,000	240,000
- Average % SO_2	-	0.5 to 1.0	10	10
- Average Conversion Efficiency %	-	-	99.6	99.9+
Product Grade - % H_2SO_4	-	-	98.5	98
KWH/MT sulphur fixed in product	-	-	-	85
Auxiliary Fuel	-	-	Light Oil	Bunker C
Tailgas Scrubbing	-	-	-	None

COPPER SMELTERS of EUROPE

COMPANY NAME	Norddeutsche Affinerie AG	Outokumpu Oy	Rio Tinto Minera SA	Quimigal
PLANT	Hamburg	Harjavalta	Huelva	Barreiro
Flowsheet Page No	96	97 & 299	-	-
Annual Production MTPY Copper	120 - 160,000	100,000	160,000	6,600
FEED ANALYSIS				
Concentrate				
% Cu	25 - 30	23 - 25	25	
% Fe	20 - 30	18 - 25	28	
% S	25 - 35	20 - 25	33	
% SiO2	4 - 8	13 - 15	3	
Mesh	- 200	- 270		
%	50 - 90	60 - 90		
% H2O	7 - 10	10 - 12	10	
Furnace Flux				
% SiO2	95	86 - 89	90	1.5
% CaO	0.3	0.5 - 1	0.06	$CaCO_3$ 92.1
% Fe		1 - 1.5	1	1.5
Mesh	- 16	- 50		
%	90	25 - 90		0 - 20 mm
Converter Flux				
% SiO2	95	86 - 89	94	92
% CaO	0.3	0.5 - 1	0.04	
% Fe		1 - 1.5	0.8	
Mesh	- 16			
%	90			0 - 20 mm
FEED PREPARATION				
Blending System	7 Bins	Bins	Bins	
Number of Beds			18	
Bed Size - MT			80 to 127	

PLANT - continued	Hamburg	Harjavalta	Huelva	Barreiro
Dryer - Type	Kiln	Kiln	Kiln	No Dryer
- Number	1	1	1	
- Dimensions, m	3 dia x 30	3.5 dia x 24	3 dia x 32	
Average Feed Rate - WMTPH	70 - 80	50 - 60	60	
Feed Moisture % H2O	7 - 10	10 - 12	10	
Fuel Used - Type	Natural Gas	Fuel Oil/Anode F.Gas	Fuel O./Gas Superh.	
Cal.Value - kcal/kg or Nm_3	8,800 kcal/Unit	9,500/ 96 MJ/Unit	9,600 / 60 kcal/Unit	
Inlet Gas Temperature oC	600 - 700	800 - 900	310 - 320	
Outlet Gas Temperature oC		100 - 115	95 - 100	
Product Moisture % H2O	0.2 - 0.3	0.1 - 0.3	0.2 - 0.3	
Exhaust Gas: - Nm3/hr	50 - 60,000	40 - 50,000		
- % SO2				
Cleaned by	EP	EP	EP	
SMELTING				
Furnace Type	Outokumpu Flash	Outokumpu Flash	Outokumpu Flash	Blast Furnace
Number of Units	1	1	1	1
Capacity, MTPH Conc.	70 - 80	60	53	0.76
Dimensions, Hearth or Settler, m	6 x 20 Shaft: 6.5x20	4.9 x 18.3 x 1.8 Shaft: 4.2 x 6.4	7.6 x 22.1 x 3.5 Shaft: 6.5 x 6.8	1.4 x 7 Tuyere Zone
Oxygen Enrichment % O2	30 - 50	75 - 95	36	-
Auxiliary Fuel -Type	Fuel Oil/nat.Gas	Fuel Oil	Fuel Oil	Coke
Cal.Value - kcal/kg or Nm_3	9,600 / 8,800 kcal/unit	9,500	9,600 kcal/kg	7,500
Operating Temperature oC	1,200 - 1,300	1,350 - 1,450	1,350	1,200
Flux as % of Concentrate	5 - 12	7 - 10	12 - 15	6
Campaign Life - Years		4 - 6	6	2

PLANT - continued	Hamburg	Harjavalta	Huelva	Barreiro
Furnace Matte - MTPD	550 - 750	400 - 600	510	12
Analysis - % Cu	60 - 63	65 - 72	57	40 - 60
- % Fe	13 - 16	7 - 8.5	17	31 - 27
- % S	21 - 23	20 - 28	21	19 - 22
Temperature - oC	1,180	1,289 +/-20	1,150	1,200 - 1,300
Furnace Slag - MTPD	650 - 950	580 - 650	744	10
Analysis - % Cu	1 - 1.5	2 - 3	2	0.5
- % Fe	37 - 42	36 - 44	41	38
- % CaO	1 - 8	1.5 - 3	0.1	10
- % SiO2	30 - 32	27 - 28	29	32
Temperature - oC	1,210 - 1,230	1,360 +/-20	1,250	1,200
Disposition	Electric Furnace	Flotation	Electric Furnace	Discarded
Off-Gas Volume - Nm3/hr	40,000 - 60,000	14,000 - 20,000	37,000	12,000
Temperature - oC	1,200 - 1,300	1,350	1,300	500
% SO2	15 - 20	25 - 50	21	0.5 - 2
Cooled by	Boiler	Boiler	Boiler	-
Inlet Temp. - oC		1,350	1,300	
Outlet Temp. - oC		350	390	
Disposition of Dust	Recycled to Feed	Recycled to Furnace	Recycled	Recycled
CONVERTING				
Converter Type	P.S.	P.S.	P.S	P.S.
Number - Hot	1	Blowing: 1 / hot:1	Blowing:1/hot: 1	1
- Standby	1	1	1	1
Dimensions - m	4.2 dia x 10.3	One: 3.6 x 7.5 Two: 3.7 x 7.9	3.9 dia x 9.1	2.226 dia x 2.87
Number of Tuyeres	50	32 / 2 x 40	44	14
Size - mm	44.5		51	68
Oxygen Enrichment - % O2		24 - 28		-
Slag Blow - % O2	23 - 24		23	
Copper Blow - % O2	-		-	

PLANT - continued	Hamburg	Harjavalta	Huelva	Barreiro
Air Rate - Nm3/min		330 - 350	567	250
Slag Blow - Nm3/min	600 - 650			
Copper Blow - Nm3/min	700 - 750			
Stacktime - %			70	
Scrap Added as % of Matte		17 - 20	40	10 - 15
Flux Added as % of Matte	8 - 10	9 - 11	10	14 - 16
MT of Matte per Blow	200	100 - 110	175	
MT of Matte per Day	600 - 650	400 - 450	510	12
Blister Production - MTPD	350 - 400	140 - 160	435	5
- Mt per Blow				
Analysis - % Cu	98	98 - 98.5	99.2	98
- % Fe	-	< 0.05	-	0.0002
- % S	0.03	0.06 - 0.03	0.01	0.008
Slag Analysis - % Cu	3.5 - 4.5	6 - 10	3.8	2 - 4
- % Fe	45	32 - 36	44	45.38
- % SiO2	23 - 25	26 - 28	23	25
Disposition	Flash Furnace	Flotation	El. Furnace	Blast Furnace

SLAG CLEANING	Hamburg	Harjavalta	Huelva	Barreiro
Electric Furnace:	1		1	No special slag cleaning
Furnace Slag - MTPD	) 650 - 950		744	
Converter Slag - MTPD	)		282 (36% Solid)	
Power - kWh/MT Slag			90	
Retention Time - Hours			1.6	
Matte Produced - % Cu	63 - 65		70	
Discard Slag - % Cu	0.5 - 0.7		0.65 - 0.75	
- % Fe	37 - 42		38	
- % SiO2	31 - 32		27	
Copper Recovery - %				
Flotation:				
Converter Slag - MTPD		130 - 150		
Cooled in		Pits		
Duration - Hours		20 - 24		
Furnace Slag - MTPD		580 - 650		

PLANT - continued	Hamburg	Harjavalta	Huelva	Barreiro
Flotation Concentrate - MTPD		50 - 60		
Analysis - % Cu		35 - 42		
- % Fe		18 - 24		
- % S		6 - 8		
Flotation Tailings - MTPD		650 - 720		
Analysis - % Cu		0.3 - 0.5		
- % Fe		36 - 43		
Copper Recovery - %		96		
FIRE REFINING				
Furnaces - Type	Rotary	Rotary	Rotary	Rotary
Number of Units	2	2	2 (1 hot)	1
Dimensions, m	4.2 dia x 10.2	4 dia x 9.2	3.9 dia x 9.2	3.1 dia x 6.8
Number of Tuyeres per Furnace	2	2	1	-
Size, mm			20 ID	
Time per Charge for				
Oxidation - min	30 - 60	60 - 120	57	180
- % O2	21	21	21	
Reduction - min	180 - 240	30 - 90	100	60
Reductant Used	Natural Gas	Propane	Propane	Poles
Consumption / MT Anode		4 - 6 kg	8 kg	647 MTPY
PRODUCT CASTING				
Casting Machine - Type	Wheel	Wheel	Wheel	Wheel
Number of Moulds	24	28	20	12
Weight Contolled Casting	Single Spoon	Double Mould	Double Mould	
Average Casting Rate - MTPH	50 - 70	60 - 80	36	6
Casting Temperature - oC	1,150 - 1,200	1,265 - 1,275	1,160	1,200
Cast Product - Type	Anode	Anode	Anode	Anode
Dimensions - m	0.9x0.65/1.05x0.9	0.89 x 0.925	0.905 x 0.915	0.9 x 0.66
Average Weight - kg	175 / 275	285 +/- 3 kg	300	165
Rejects as % of Material Cast	4 -6	< 0.5	1.5	3

PLANT - continued	Hamburg	Harjavalta	Huelva	Barreiro
SULPHUR FIXATION				
Source of Gas	Smelter	Smelter	Smelter	
Plant:	2 Line	See Description Page 291	2 Plants:	
Type	Double Contact		Normal / Double Contact	
Rated Capacity - MTPD	1,800		1,070 / 600	
Volume - Nm3/hr	204,000		150,000/60,000	
Average % SO2	6.5		4 - 6.8 /6 -8.5	
Conversion Eff. - %	99.5		98 / 99.6	
Product Grade - % H2SO4	96		98.5 / 98.8	
kwh/MT Sulphur in Product			44.3 / 23.9	
Auxiliary Fuel			-	
Tailgas Scrubbing				

COPPER SMELTERS of EUROPE

COMPANY NAME	Boliden Metall AB	Boliden Metall AB	RTB Bor	Metallurgie Hoboken - Overpelt
PLANT	Roennskaer	Roennskaer	Bor	Hoboken
	Copper Plant	Kaldo Plant		
Flowsheet Page No	410	410	98	396 & 398-399
Annual Production -MTPY	105,000		150,000 - 160,000	40 - 50,000
FEED ANALYSIS				The smelter operates on a complex flowsheet: Cu and Pb act as collectors. Feed: Low-grade and complex concentrates, matte, scrap, residues, dust, cement copper.
Concentrate				
% Cu	20	24.5	21.3	
% Fe	14.5	24	28.7	
% S	28	28.5	34	
% SiO2	12	5	8,7	
% H2O	12	6.5	10 - 13	
Furnace Flux				
% SiO2	99.1	99.1	90	
% CaO			-	
% Fe			4	
Mesh			- 0.15:	
Converter Flux				
% SiO2	99.1	-	90	
% CaO			-	
% Fe			4	
Mesh			6 - 50 mm	
FEED PREPARATION				
Blending System	Beds	Beds	Beds	
Number of Beds	4	2	2	
Bed Size - MT	1x1,500 / 3x2,000	3,000	12,000 / 10,500	
	Wheel Carrier	Wheel Carrier	Reclaimer	

PLANT - continued	Roennskaer	Roennskaer	Bor	Hoboken
Dryer - Type	Tubular, electrically heated circul. air	Multicoil Steam D.	No Dryer	
- Number	2 parallel	1		
- Dimensions - m	2.75 dia. x 6	2 dia. x 6.5		
Average Feed Rate - WMTPH	12 - 15	25		
Feed Moisture -% H2O	12	6.5		
Fuel Used - Type	1.8 MW / Dryer	18 bar Steam		
Calorific Value				
Inlet Gas - Temp. - oC	425			
Outlet Gas Temp. - oC	90 (vented air)	110		
Product Moisture -% H2O	< 1	< 0.5		
Exhaust Gas - Nm3/hr	5,700	6,500		
- % SO2				
Cleaned by	1 Baghouse / Dryer	Baghouse		
Roasters -Number	1	No Roaster	2	No Roaster
- Type	Fluo-Solid		Fluo-Solid	
- Dimensions	4.25 dia. x 9.25		Dia: 4.8/5.5 ,H:7.15	
- Capacity, MTPH	38		35.42 / 40	
- Fuel Used - Type	-		-	
Calorific Value				
- Calcine Composition				
% Cu	26.1		21.3	
% Fe	27.7		29.2	
% S	19		16.7	
- Discharge Temp.	700		550	
- Offgas - Nm3/hr	37,000		32,500 / 37,200	
- % SO2	7		12.95	
- Air Rate, Nm3/hr	20,000		26,000 / 29,000	

PLANT - continued	Roennskaer	Roennskaer	Bor	Hoboken
SMELTING				
Process Comments	Calcine and dryed Cu bearing materials are fed.		Only calcine feed.	Blast Furnace: Average charge contains 35 % Pb-Cu-matte from Pb-blast furnace, 20% Cu-slags and varying addtions of cement copper, slimes, sludges, scrap and concentrates.
Furnace - Type	Electric	Kaldo	Reverb	Blast Furnace
Number of Units	1	1	2	1 (36 Tuyeres)
Capacity, MTPH Conc.	50 (Roaster+Dryer)	27	35,42 / 40	21
Dimensions, hearth or settler - m	25 x 6.5	3.6 dia x 6	(.26x29.62 / 8.68x29.1	At Tuyere Zone: 1.13 x 5.5
Oxygen Enrichment - % O2	-	100		
Auxiliary Fuel - Type	21 MVA	-	No1: Coal /No2: Fuel Oil	Natural Gas
Cal. Value - kcal/kg			6,500 / 9500	8,400 /Nm3
Operating Temperature - oC	1,300	1,300	1,300	400 (Blast Air)
Flux as % of Concentrate	8.3		15 (Silica+Limestone)	
Campaign Life - Years	Furnace: 8 End Walls: 2	8 weeks	2	3

PLANT - continued	Roennskaer	Roennskaer	Bor	Hoboken
Furnace Matte - MTPD	505	315	463 + 518	300
Analysis - % Cu	45	45	40	Cu: 45 - 50 ,Pb: 20
- % Fe	18.8	19	32.5	8 - 10
- % S	22.5	19	25	15
Temperature - oC	1,175	1,300	1,150	1,000
Furnace Slag - MTPD	800	140	666 + 745	175
Analysis - % Cu	0.95	1.3	0.5	Cu: 0.6, Pb: 1.5
- % Fe	33	32.4	37	22
- % CaO	Zn: 8	Pb+Zn: 11	5.7	13
- % SiO2	< 33	31.3	37	29
Temperature - oC	1,300	1,200	1,200	1,100
Disposition	Slag Fuming	Slag Fuming	Discarded	
Offgas Volume - Nm3/hr	65,000	35,000	39,000 + 44,000	18,000
Temperature - oC	1,300	1,300	1,300	200
% SO2	'2 - 3	15	1.5	0
Cooled by	Water Spray Tower	Water Spray Tower	Boiler	Tubular Heat Exchanger
Inlet Temp. - oC		600	1,300	100 - 400
Outlet Temp. - oC	170	225	350	80 - 100
Disposition of Dust	Recycled	Recycled	Recycled	Recycled as filter Cake
CONVERTING		Matte further treated in Copper Plant.		
Converter Type	P.S		P.S.	Hoboken
Number - Hot	2		3	3
- Standby	1		1	1
Dimensions, m	3.9 dia x 9.1		3.96 dia x 9.14	3 dia x 6
Number of Tuyeres	52		48	18
Size, mm	50		50	38
Oxygen Enrichment - % O2	21 - 25		23 - 27	-
Slag Blow - % O2				
Copper Blow - % O2				

PLANT - continued	Roennskaer	Roennskaer	Bor	Hoboken
Air Rate -Nm3/hr	550		500 - 600	180
Slag Blow - Nm3/hr				
Copper Blow - Nm3/hr				
Matte per Converter - MTPD	326		180-190 MT: 5-5.5 Cycles	100 - 130
Scrap Added as % of Matte	7		40	10/50: Matte+Scrap
Flux Added as % of Matte	14.2		20	
Blister Production - MTPD	234 / Converter		180 - 190	45-50 per Converter
Analysis - % Cu	98.5		98 - 99.3	97.5 Cu / 0.3 Pb
- % Fe	-			
- % S	60 - 70 g/t			
Slag Analysis - % Cu	6		3 - 5	13 Cu / 30 Pb
- % Fe	35.4		50	10 -14
- % SiO2	24		24 - 26	13 - 17
	Zn+Pb: 13			
Slag Disposition	Recycled to Elec.Furnace		Recycled to Reverbs	Recycled to Blast F.
SLAG FUMING	From Elect.Furnace Kaldo, Pb-Smelter:		No Slag Fuming	No Slag Fuming
Slags Treated, MTPD	1,010			
Analysis:				
- % Cu	0.9			
- % Pb	1.6			
- % Zn	9.2			
- % SiO2	32.2			
Furnace Dimensions - m	8.2 x 2.4			
Reductant Used - MTPD	193			
- Type	Coal			
Final Slag - % Cu	0.5			
- % Pb	0			
- % Zn	1.7			
- SiO2	36.8			
Fume Analysis - % Pb	12.2			
- % Zn	61.6			
Fume Produced - MTPD	116			

PLANT - continued	Roennskaer	Roennskaer	Bor	Hoboken
FIRE REFINING		See Copper Plant		
Furnace - Type	Rotary		Rotary	
Number of Units	2		3	
Dimensions, m	2.5 dia x 6.3		2: 3.9x7.6 + 1: 3.9x9.1	
Tuyeres / Furnace - No	-		2 Lances at Mouth	
Size, mm			32	
Time per Charge for				
Oxidation - min.	-		90	
- % O_2			21	
Reduction - min.	180		240 - 300	
Reductant Used	Poles		Poles	
Comsumption/ Mt Anode	26 kg		0.055 m^3	
PRODUCT CASTING				
Casting Machine - Type	Wheel		2 Wheels	
Number of Moulds	22		26	
Casting Rate - MTPH	35		40 / 30	
Cast Product Type	Anode		Anode	
Dimensions, m	0.875 x 0.95		0.785 x 0.94 x 0.041	
Casting Temp. - oC	1,170		1,130 - 1,160	
Average Weight - kg	315		260	
Rejects in % of Anodes			1	
SULPHUR FIXATION				
Source of Gas	SO_2 Containing	Smelter Gases	Roaster / Converter Gas	Smelter Gas
Plant Type	Normal/ Double Cont. Acid Plant +	Liquid SO_2 Plant	Normal Contact	Normal Contact
Capacity - MTPD	120 / 1,000	SO_2: 330	1,650	200
Volume - Nm^3/hr	25,000 / 145,000	60,000 / 100,000	300,000	38,000
Average % SO_2	4.5 / 6.5	3.2 / 4.8	4 - 5	5
Product Grade - % H_2SO_4	95.5	100	98	66 o Be
kWh/MT Sulphur in Product	200	665	550 - 600	160
Auxiliary Fuel	-	-	3 1/MT MH Fuel Oil	100 MTPY Gasoil
Conversion Efficiency - %	97 / 99.7	99.5	97 - 98	98 - 98.5
Tailgas Scrubbing				Demister

COPPER SMELTERS
of
EUROPE

COMPANY NAME	Huettenwerke Kaiser AG	Austria Metall AG Montanwerke Brixlegg	La Metalli Industriale SpA
PLANT	Luenen	Brixlegg	Firenze
Flowsheet Page No	-	-	-
Annual Production - MTPY	105,000 Copper 3,000 Tin Alloy 6,000 Zinc Oxide 2,500 Nickel Sulfate	40,000	
Copper Products	Anodes	Anodes	Anodes Cakes (1000x200x7100 mm)
FEED MATRIALS			
Type of Materials Treated	Residues, Ashes, Drosses, Slimes, Dusts, Alloy-Scrap, Cu-Fe-Scrap, Refinery Scrap.	Scrap, Alloy Scrap, Oxide Residues etc.	Scrap
Composition			
% Cu		60 - 70	
% Zn		5 - 10	
% Pb		1 - 5	
% Sn			
% Fe		10 - 15	
% S			
FEED PREPARATION			
Type	Stockyard	Stockyard	

PLANT - continued	Luenen	Brixlegg	Firenze
SMELTING			
Smelting Furnace			
Type	Blast Furnace	Blast Furnace	
Number	3	1	
Capacity -MTPH	6	6.7	
Dimensions - m	1,3 x 2.8 at Tuyere Zone	1.25 x 2.5 at Tuyere Zone	
Oxygen Enrichment - % O_2	23	24	
Auxiliary Fuel - Type	Coke	Coke	
Calorific Value	7,000 kcal/kg		
Operating Temperature - °C	1,250		
Campaign Life - Years	2 - 3		
Flux as % of Feed	10	16	
Furnace Metal - MTPH	90	52	
% Cu	70 - 80	80	
% Fe	5		
% S	1 - 2		
Temperature - °C	1,250		
Furnace Slag - MTPH	90	83	
% Cu	1		
% Fe	45		
% CaO	15		
% SiO_2	20		
Temperature - °C	1,250		
Return Dust - MTPH		5.7	
Bleed-Off Dust	Is practiced.	5.4	

PLANT - continued	Luenen	Brixlegg	Firenze
Furnace Offgas - Temp. oC	950	800	
Volume - Nm3/hr	20,000	20,000	
% SO2	0.02		
Cooling Method	Boiler	Boiler+ Gas Cooler	
Inlet Temp. oC		650 - 950	
Outlet Temp. oC		100	
Cleaned by	Baghouse	Baghouse	
Dust - MTPD		6	
Analysis - % Cu		3 - 4	
- % Zn		50	
- % Pb		12	
- % Sn		0.8	
- % Cl		3	
Settler - Type		Rotary	
Dimensions - m		1.75 dia x 3.9	
Capacity - MT		Metal: 25 Slag: 15	
Retention Time - Hours		4	
CONVERTING			
Converter - Type	P.S.	P.S.	
Number - Hot	2	1	
- Standby			
Dimensions - m		2.1 dia x 3.8	
Tuyeres - Number	16 / 9	10	
Size - mm	38	25	
Oxigen Enrichment - % O2	-		
Air Rate - Nm3/min	140 / 100	67	
Cycles per Day		4	
Flux Added - MTPD		1.2	
Scrap Added - MTPD		40	
Metal Charged - MTPD	120 - 168	52	

PLANT - continued	Luenen	Brixlegg	Firenze
Auxiliary Fuel - Type		Fuel Oil by Burner	
Quantity - MTPH		2	
Copper Produced - MTPD			
Analysis - % Cu	95	97	
- % Sn		0.3	
- % Pb		0.3	
- % Sb		0.1	
- % Ni		1.3	
Converter Slag - MTPD			
Analysis - % Cu	40	5 - 6	
- % Fe	20	25	
- % Ni		20	
- % Sn		5	
- % Zn		5	
- % Pb		3	
- % SiO2	5	5	
Offgas - Nm3/hr		10 to 35,000	
Temp. after Coolig - oC		180	
Cooled by		Air-to- Gas Cooler 500 to 180 oC	
Cleaned by		Baghouse	
Dust - MTPD			
Analysis - % Cu		0.9	
- % Zn		56	
- % Sn		10	
- % Pb		14	
SO2-Scrubbing by		Mg(OH)2 - Scrubbing	
Residual SO2 - mg/m3		200	

PLANTE - continued	Luenen	Brixlegg	Firenze
FIRE REFINING			
Furnace - Type		Reverb	Rotary
Number of Units		2	2
Dimensions		120 / 160 MT	1.9 dia x 6.23 m
Tuyeres - Number		1	1
Size - mm		75	20
Size of Carge - MT		120 / 160	12 -22
Time per Charge for			
Oxidation - min.		120	15 - 20
- % O2			
Reduction - min.		240	
Reductant Used		Poles	Poles
Consumption/MT Copper		0.05 m3	40 kg
PRODUCT CASTING			
Casting Machine - Type	Wheel	Wheel	Stationary Moulds for Anodes. Semi-continous Casting for Cakes.
Number or Moulds	18	14	
Casting Rate - MTPH	30	25	
Cast Product - Type	Anode	Anode	Anode
Dimensions - m	o.95 x 0.95	o.81 x o.76	0.88 x 0.71 x 0.055
Casting Temperature - oC	1,200	1,200	
Average Weight - kg	330	220	
Rejects in % of Cast Product	15	4	

COPPER SMELTERS of EUROPE

COMPANY NAME	Huta Miedzi Legnica	Huta Miedzi Glogow I	Huta Miedzi Glogow II	Black Sea Copper Works	Ergani Copper Plant
PLANT	Legnica	Glogow	Glogow	Murgul-Goektas	Maden - Elazig
Flowsheet Page No	-	99	100	-	-
Annual Production - MTPY	85,000	190,000 + Lead Bullion	102,000		
FEED ANALYSIS					
Concentrate					
% Cu	22 - 24	22 - 24	28	17.79	
% Fe	5	5	3	39.9	
% S	10	10	10	35.77	
% SiO2	19	18	18	3.98	
Mesh	-	-	-0.075 mm: 70 % -0.15 mm: 13 %	- 325	
% H2O	10 - 12	10 - 12	8	13	
Furnace Flux					
% SiO2	Converter slag is used as shaft furnace flux.		Not used.	1.21	70
% CaO				52.97	22
% Fe				1.21	3
Mesh				- 20 mm	- 15 mm
Converter Flux					
% SiO2				80.62	70
% CaO				0.46	2
% Fe				6.4	3
Mesh				- 20 mm	5 - 7 mm
FEED PREPARATION	Furnace feed is briquetted.				

PLANT - continued	Legnica	Glogow I	Glogow II	Murgul	Maden
Dryer - Type/No Inst.	Kiln/3	Kiln/5	Kiln/1	Kiln/2	-
Number Operating	1.5	3	1	2	
Dimensions - m	3 x 24	3 x 24	4,3 x 49	2.4 dia x 16	
Feed Rate - Wet MTPH	60	60	100	50	
Moisture - % H2O	10 - 12	10 - 12	8	13	
Retention Time - min	25	25	45	6	
Fuel Used - Type	Natural Gas	Natural Gas	Natural Gas	Fuel Oil	
Cal.Value-kcal/ kg or Nm_3	5,900	5,900	5,900	10,200-10,800	
Gas Temp. - oC Inlet	750 - 800	750 - 800	550 - 650	870	
- oC Outlet	110	110	130	127	
Product - % H2O	4 - 6	4 - 6	0.3	6	
Exhaust Gas - Nm3/hr	35,000		60 - 80,000	14,950	
Cleaned by	3 Cyclones	2 Wet Systems	E.P.		
Roasters - Number	-	-	-	-	1
Type					Fluo-Solid
Capacity - MTPH					10
Air Rate - Nm3/hr					11,000
Calcine - % Cu					18
- % Fe					42
- % S					19
Discharge Temp. - oC					500
Offgas - % SO2					5 - 15
SMELTING					
Furnace Type	Shaft Furnace	Shaft Furnace	Outokumpu Flash	Reverb	Reverb
Number of Units	3	3	1	1	1
Capacity - MTPH Conc	25	50	87.5	6.7	12.5
Dimensions, Hearth or	10 m2	20 m2	214 m2	5.79x25.6x2.65	8 x 28 x 3.5
Settler - m	Tuyere Zone	Tuyere Zone	Shaft: 7 dia x 7		
Oxygen Enrich. - % O2	-	-	60 - 70	-	-
Aux. Fuel - Type	Coke	Coke	Fuel Oil	Fuel Oil	Fuel Oil
Cal.Value-kcal/kg	9,650				10,000
Operating Temp. - oC	1,320 - 1,350	1,320 - 1,350	1,300	1,180	1,270
Campaign Life - Years	8	8	10 -11	1	1

PLANT - continued	Legnica	Glogow I	Glogow II	Murgul	Maden
Furnace Matte -MTPD	210	480		184	125 - 150
Black Copper - MTPD	per Furnace	per Furnace	300		
Analysis - % Cu	60 - 63	60 - 65	98.6	19.26	
- % Fe	10 - 12	8 - 12	-	41.38	
- % S	2 - 4			25.42	
Temperature - oC	1,230	1,230	1,260 - 1,280	1,250	1,270
Furnace Slag - MTPD	Per Furnace:320	Per Furnace:720	800 - 1,000	120	270
Analysis - % Cu	0.5	0.5	14.5	0.55	0.2
- % Fe	n/a	n/a	5.5	61.78	40 - 45
- % CaO	15	17	12	-	1 - 2
- % SiO2	38	39	30	27.7	
Temperature - oC	1,250	1,250 - 1,280	1,280 - 1,300	1,150	1,100 - 1,150
Disposition	Settled	and discarded.	Electric Furnace	Discarded	Discarded
Off-Gas Volume - Nm3/hr	35,000	70,000	60,000 - 80,000	20,000	38,000
Temperature - oC	500	500	1,300	350	1,250 - 1,300
% SO2	-	-	9 - 14	1.87	1 - 1.5
% CO	10	12	-		
Cooled by	Cyclones and	Scrubbing	Boiler	Economizer	Boiler
Number of Units	2	3	1		1
Inlet Temp. - oC	350 at Cyclone	Outlet	1,300		1,250 - 1,300
Outlet Temp. - oC	60	60	330 at EP Outlet		345 - 375
Disposition of Dust	Smelting for	lead recovery.	Recycled	Recycled	Recycled
CONVERTING			Metal from Slag Cleaning Furnace:		
Converter - Type		Hoboken	Hoboken	P.S.	P.S.
- Hot/Blowing	3 / 3	3 / 2	1	1	2
- Standby	1	2	1	1	1
Dimensions - m	2.4 dia x 5	3 dia x 7.6	3 dia x 7.6	3.06 x 6.1	3.06 x 6.1
Number of Tuyeres	20	32 Used	32 Used	32	32
Size - mm	42	38	38	38	38

PLANT - continued	Legnica	Glogow I	Glogow II	Murgul	Maden
Air Rate - Nm3/min				230	100
Slag Blow - Nm3/min	160	400	300		
Copper Blow - Nm3/min	160	320	300		
Scrap Added: % of Matte				62.5	6 - 8
Flux Added: % of Matte	15	15	5	83	30
Matte/Converter - MTPD	48 + 36	150	120	70	60 - 65
Blister Produced - MTPD		180	140		12 - 13
- % Cu					99
Slag Analysis - % Cu	4	4 - 6	6 - 10	2.02	2 - 3
- % Fe					35 - 40
- % SiO2	26	26	25	20.37	20 - 25
Disposition	Recycled to shaft furnace.		Electric Furnace		
SLAG CLEANING	Only settling in shaft furnace settler.			-	-
Electric Furnace:			1		
Furnace Slag - MTPH			1,000		
Converter Slag - MTPD					
Power - kWh/MT Slag			230 - 250		
Retention Time - Hours			8 - 10		
Alloy Produced - % Cu			80		
Discard Slag - % Cu			0.5 - 0.6		
- % Fe			-		
- % SiO2			40		
FIRE REFINING					
Furnaces - Type	Reverb	Reverb	Reverb/Rotary	-	-
Number of Units	3	3	2 / 2		
Capacity - MT	2x200 + 1x150	2x250 + 1x300	2x250 + 2x250		
Number of Tuyeres	Lances	Lances	2		
Diameter, mm	38	38	38		
Time for Oxidation - min.	30	60	60		
Time for Reduction - min.	240	240	120 / 240		
Reductant Used	Poles	Poles	Poles / Oil		
Poles - m3/MT Anode	0.025	0.025	0.025		
Oil - kg/MT Anode	-	-	7		

PLANT - continued	Legnica	Glogow I	Glogow II	Murgul	Maden
PRODUCT CASTING					
Casting Machine - Type	Wheel	Wheel	Wheel	-	-
Number of Moulds	20	28	28		
Casting Rate -MTPH	30	35	35		
Casting Temp. - oC	1,180	1,150	1,180		
Cast Product - Type	Anode	Anode	Anode		
Dimesions, m	0.755 x 0.9	0.88 x 0.825	0.9 x 0.86		
Average Weight - kg	260 +/-5	245 / 250	370 +/- 5		
Rejects as % of Material Cast	4	4	3.5		
SULPHUR FIXATION				-	
Source of Gas	Converter	Converter	Smelter		
Plant - Type	Normal Contact	Normal Contact	Double Contact		3 - Stage Normal Contact
- Capacity, MTPD	200	500	600		400
- Volume, Nm3/hr	60,000	90,000	40 + 90,000		56,000
- Conversion Efficiency %	90	91	98 + 99		97
- % SO2 in Gas					7
- Product: % H2SO4					93.19
- Auxiliary Fuel MTPY/Type					50 / Diesel Oil
Tailgas Scrubbing	Demister	Demister	Demister		

COPPER SMELTERS
of
JAPAN

COMPANY NAME	Mitsubishi Metal Corporation		Hibi Kyodo Smelting Co. Ltd.	Furukawa Mining Co. Ltd.
PLANT	Naoshima	Naoshima	Tamano	Ashio
Annual Production - Copper MTPY	132,000	90,000	144,200	36,000
Flowsheet Page No.	101	101	102	-
FEED ANALYSIS				
Concentrate				
% Cu	27.7	27.7	27.6	25
% Fe	23.6	23.6	22	22
% S	28.0	28.0	27	21.4
% SiO_2	9.3	9.3	11.5	8.6
Mesh	80%-200	80%-200	88%-200	80%-325
% H_2O	8	8	8	10
Furnace Flux				
% SiO_2	85	90	80	90
% CaO	0.5	-	5	-
% Fe	4.0	2	-	-
Mesh	-3mm	-5mm	70%-200	80%-25
Converter Flux				
% SiO_2	87	-	90	92
% CaO	0.3	54	-	-
% Fe	1.5	-	-	-
Mesh	-50mm	-5mm	-	80%-40mm
FEED PREPARATION				
Blending System	yard	yard	-	bins
Number of Beds	2	2	2	24
Bed Size - MT	1,300	700	4,500	1,000
Method of Reclaim	Automatic Rake Reclaimer		-	Bucket Crane

PLANT - continued	Naoshima	Naoshima	Tamano	Ashio
Dryer - Type	Fluo-solid Roaster	Rotary-Flash	Rotary-Flash	Rotary-Flash
- Number	1	1	1	1
- Dimensions - m	5.3 Ø x 8	1.6 Ø x 10	2 Ø x 8	1 Ø x 25
Average Feed Rate - WMTPH	55	55	68	24
Feed Moisture - % H_2O	8	8	8	10
Fuel Used - Type	Air Rate	Bunker Oil	Bunker Oil	Oil & Shredded Tires
Calorific Value	19,200 Nm^3/hr	9,500 kCal/ℓ	10,300 kCal/ℓ	9,800 kCal/ℓ 5,000 kCal/kg
Inlet Gas - Temp. - °C	-	400	400	600-700
Outlet Gas - Temp - °C	630	80	60	85
Product Moisture - % H_2O	Calcine 28% Cu 20.3% S	0.5	0.3	0.3
Exhaust Gas Volume-Nm^3/hr	33,000	100,000	80,000	33,000
-% SO_2	10	-	-	-
Cleaned by	-	Cyclone, Baghouse	Cyclone Precipitator	Cyclone, Precipitator Scrubber
SMELTING				
Furnace Type	Reverb	MMC Continuous	Flash with Electrodes	Flash (OKO)
Number of Units	1	1	1	1
Nominal Capacity-MTPH Conc.	Calcine 44	40	80	27
Dimensions, hearth or settler - m	33 x 9	8.25 Ø x 3.3	7x19.751x2.6	4.26x2.9x12.7
Oxygen Enrichment - % O_2	21	43	27.5	40
Auxiliary Fuel - Type	Coal	Coal	Oil, Coke, Coal	Oil
Calorific Value	7,000 kCal/kg	7,000 kCal/kg	10,300 kCal/ℓ 6,650 kCal/kg	9,800 kCal/kg
Operating Temperature - °C	1,300	1,220	1,180	1,250
Flux as % of Concentrate	-	10-15	7	9.5
Campaign Life - Years	10	10	1	1

PLANT - continued	Naoshima	Naoshima	Tamano	Ashio
Furnace Matte - MTPD	650	420	563	216
Analysis - % Cu	42.8	65.7	60.6	50
- % Fe	26.4	10.1	14.6	22
- % S	25.5	21.3	21.6	23
Temperature - °C	1,100	1,220	1,150	1,180
Furnace Slag - MTPD	330	-	505	270
Analysis - % Cu	0.5	-	0.53	1.5
- % Fe	30.9	-	37.8	33
- % CaO	9.6	-	4.2	-
- % SiO_2	34.5	-	32.0	28.2
Temperature - °C	1,230	-	1,190	1,240
Disposition	Discard	Electric Furnace	Discard	Electric Furnace
Offgas - Volume - Nm^3/hr	60,000	22,800	53,000	13,000
Temperature - °C	1,300	1,220	-	1,250
% SO_2	1.6	23	14	20-35
Cooled by	Boiler (2)	Boiler	Boiler	Boiler
Inlet Temp. - °C	1,300	1,200	1,000	1,250
Outlet Temp. - °C	340	350	300	350
Disposition of Dust	Recycled	Recycled	Recycled	Recycled
CONVERTING				
Converter Type	P.S.	MMC Continuous	P.S.	P.S.
Number - Hot	2	1	2	2
- Standby	1	0	1	1
Dimensions - m	4 Ø x 9	6.65 Ø x 2.9	3.96 Ø x 10.69 (2) 3.96 Ø x 10.39 (1)	2.64 Ø x 8.44
Number of Tuyeres	48	Lances 5	52	39
Size - mm	44.5	76.2	48	52.9
Oxygen Enrichment - % O_2	-	30	-	-
Slag - % O_2	21	-	25	-
Copper - % O_2	21	-	21	-

PLANT - continued	Naoshima	Naoshima	Tamano	Ashio
Air Rate - Nm^3/min.	-	180	-	-
Slag Blow - Nm^3/min.	650	-	675	260
Copper Blow - Nm^3/min.	650	-	633	260
Scrap added as % of matte charged	34	-	28	17
Flux as % of matte charged	14	0.5	7	5
Blister Production - MTPD	450	250	483	109
Analysis - % Cu	99.2	98.5	98.9	99
- % Fe	-	-	0.03	.007
- % S	0.3	0.8	0.025	.003
Slag - Analysis - % Cu	3.6	13.2	3.4	4-6
- % Fe	50.0	43.9	52.7	46.1
- % SiO_2	20.5	-	17.5	16.2
- Disposition	Slag Cleaning	Recycle to Smelting Furnace	Slag Cleaning	24% to Flash Furnace
SLAG CLEANING				
Electric Furnace:				
Smelting Furnace Slag - MTPD	-	680	-	246
Power - kWh/MT slag	-	30	FSFE Electrodes 44.2	103
Retention Time - Hours	-	2.5	8	6
Matte Produced - % Cu	-	65.7	-	54.5
Discard Slag - % Cu	-	0.6	See Furnace	0.53
- % Fe	-	38.1	Slag	34
- % SiO_2	-	33.4	-	36
Copper Recovery - % (overall)	-	98.7	-	-
Flotation:				
Converter Slag - MTPD	370	-	326	114
Cooled in	Mould Conveyor	-	Sand Bed	Pits
Duration - Hours	0.5	-	16	12

PLANT - continued	Naoshima	Naoshima	Tamano	Ashio
Flotation Concentrate - MTPD	40	-	53	27
Analysis - % Cu	31	-	28.4	25.5
- % Fe	33	-	34.6	-
- % S	6.5	-	6.9	-
Flotation Tailings - MTPD	320	-	273	87
Analysis - % Cu	0.3	-	0.4	0.4
- % Fe	53	-	55.6	51
Copper Recovery - %	93	-	-	95
FIRE REFINING				
Furnaces - Type	Rotary	Rotary	Rotary	-
Number of Units	2	2	2	
Dimensions - m	4 Ø x 10	2.4 Ø x 7	4.4 Ø x 10	-
Number of Tuyeres per Furnace	2	2	2	
Size - mm	70	51	25.4	-
Time per Charge for				
Oxidation - min.	70	120	80	-
- % O_2	21	23	21	-
Reduction - min.	70	30	150	-
Reductant - Used	Ammonia	Ammonia	LPG	-
- Consumption/MT Anode	7 kg	6 kg	390 kg/hr.	-
PRODUCT CASTING				
Casting Machine - Type	Wheel	Wheel	Double Mould Wheel	Wheel
Number of Moulds	24	16	24 Outside 12 Inside	15
Average Casting Rate - MTPH	50	50	45	30
Normal Casting Temp. - °C	1,180	1,200	1,200	1,100
Cast Product - Type	Anode		Anode	Blister Cake
Dimensions - m	0.98 x 0.96 x 0.044		.97x.99x.043 (outside) .69x.91x.040 (inside)	
Average Weight - kg	360	380	350 outside-250 inside	980
Rejects as % of new material cast	2.3	2	2	-

PLANT - continued	Naoshima	Naoshima	Tamano	Ashio
SULPHUR FIXATION				
Source of Gas	All Conventional and Continuous		Flash + Converter	Flash + Converter
Plant - Type	Lurgi Single(2) and Double Contact(1)		Double Contact	Single Contact
- Rated Capacity - MTPD	240, 590 and 720		1,201	320
- Volume - Nm^3/hr	42,000, 144,000 and 108,000		132,000	60,000
- Average % SO_2	6.0, 4.7 and 8.2		8	7
- Avg. Conversion Efficiency %	96.5, 96.5 and 99.8		99.7	96.4
Product Grade - % H_2SO_4	98.5		98	98
KWH/MT sulphur fixed in product	72, 97 and 82		233	135
Auxiliary Fuel	None		Oil	-
Tailgas Scrubbing	Slaked lime and basic aluminum sulphate on single contact plants.		-	TCA Lime Scrubbing

COPPER SMELTERS
of
JAPAN

COMPANY NAME	Nippon Mining Co. Ltd.	Dowa Mining Co. Ltd.	Sumitomo Metal Mining Co. Ltd.	Onahama Smelting & Refining Co. Ltd.
PLANT	Saganoseki	Kosaka	Toyo	Onahama
Annual Production - Copper MTPY	300,000	50,000	172,000	230,000
- Lead MTPY	-	25,000	-	-
Flowsheet Page No.	103	-	-	-
FEED ANALYSIS				
Concentrate				
% Cu	27.5	21	30	30
% Fe	21.4	27	21.5	23
% S	24.2	33	27	30
% SiO_2	12.0	13	-	6
Mesh	80%-200	-	-	90%-200
% H_2O	8-10	14	7-10	8
Furnace Flux				
% SiO_2	92	85	88	80
% CaO	-	-	0.2	-
% Fe	2	-	3.7	2
Mesh	55%-100	-	95%-28	100%-80mm
Converter Flux				
% SiO_2	88	93	88	80
% CaO	-	-	0.2	-
% Fe	7	-	37	2
Mesh	52% 20-50mm	-25mm	15%-50mm	100%-80mm
FEED PREPARATION				
Blending System	Conveyor	Crane	Bins	Beds
Number of Beds	-	-	14	2-3
Bed Size - MT	-	-	150 MT/bin	10,000
Method of Reclaim.	-	-	Belt feeders	Shovel loader

PLANT - continued	Saganoseki	Kosaka	Toyo	Onahama
Dryer - Type	Rotary+Flash	Rotary+Flash	Rotary+Flash	Rotary
- Number	2	1	1	1
- Dimensions - m	1.74Ø x 36.0 1.63Ø x 36.9	2.8Ø x 12	2 Ø x 7	3.28 Ø x 13
Average Feed Rate - WMTPH	100 and 110	45	-	60
Feed Moisture - % H_2O	8-10	14	7-10	8
Fuel Used - Type	Bunker Oil	Bunker Oil	Bunker Oil	Bunker Oil
Calorific Value - kCal/ℓ	9,500-10,000	9,800	9,800	10,500
Inlet Gas - Temp - °C	420	700	430	500
Outlet Gas - Temp - °C	73	90	80	100
Retention Time - min.	6-10	15	-	40
Product Moisture - % H_2O	0.4	0.2	<0.3	6
Exhaust Gas Volume - Nm^3/hr	90,000	60,000	86,000	21,000
Cleaned by	Cyclone-Baghouse Cyclone-Precipit.	Cyclone, Precipit., Wash Tower	Precipitator	Precipitator
SMELTING				
Furnace Type	Flash (OKO)	Flash (OKO)	Flash (OKO)	Reverberatory
Number of Units	2	1	1	2
Nominal Capacity - MTPH Concentrate	70 and 75	50	56	70
Dimensions - m (settler)	8.3x22.17x5.3 8.8x22.17x4.6	6 x 15	7 x 20	33.55x9.73x3.69 33.27x 11 x 4
Oxygen Enrichment - % O_2	25-30	32	35	21
Auxiliary Fuel - Type	Coke & Bunker Oil	Bunker Oil	Coal & Oil	Coal
Calorific Value	6,500-7,000 kCal/kg & 9,500-10,000 kCal/ℓ	9,800 kCal/ℓ	6,700 kCal/ℓ 9,800 kCal/ℓ	7,000 kCal/kg
Operating Temperature - °C	1,300	1,250	1,300	1,300
Flux as % of Concentrate	6	13	6	20
Campaign Life - Years	1 to 1.5	1.0	1-1.3	1.0

PLANT - continued	Saganoseki	Kosaka	Toyo	Onahama
Furnace Matte - MTPD	1,070	250	700	1,400
Analysis - % Cu	60	55-60	58	38
- % Fe	14	12	14	28
- % S	22	23	23.4	26
Temperature - °C	1,200	1,150	1,200	1,050-1,100
Furnace Slag - MTPD	910	300	460	1,100
Analysis - % Cu	1.0	-	0.9	0.5
- % Fe	40.0	-	37.4	36
- % CaO	1-3	-	1.5	3
- % SiO_2	33	-	33	31
Temperature- °C	1200-1250	1,190	1,240	1,250
Disposition				
Offgas - Volume - Nm^3/hr	54,000	22,000	59,000	180,000
Temperature - °C	1,300	1,250	-	1,280
% SO_2	15	12.4	16.8	1.5-2.2
Cooled by	Boiler	Boiler	Boiler	Boiler (4)
Inlet Temp. - °C	1,300	1,250	1,300	1,280
Outlet Temp - °C	350	300	350	350
Disposition of Dust	Recycle to Furnace	Recycle to Furnace & Bleed	Recycle to Furnace	Recycle to Reverb
CONVERTING				
Converter Type	P-S	P-S	P-S	P-S
Number - Hot	4	1	2	4
- Standby	2	1	1	1
Dimensions - m	4.2Ø x 11.51	3.3Ø x 10	4.2Ø x 11.9	3.69Ø x 9.15(4) 3.96Ø x 11.0(1)
Number of Tuyeres	46	44	54	51/60
Size - mm	50.8 I.D.	39 I.D.	50.8 I.D.	50 I.D.
Average Blowing Rate - Nm^3/hr	33,000 to 36,000	25,500	36,000	-
Oxygen Enrichment				
Slag - % O_2	25	25	21	21
Copper - % O_2	21	21	26	21

PLANT - continued	Saganoseki	Kosaka	Toyo	Onahama
Air Rate - Nm^3/MT matte	873	-	-	-
Slag Blow - Nm^3/min.		400	600	550
Copper Blow - Nm^3/min.		400	600	600
Scrap added as % of matte charged	31	5	30	24
Flux as % of matte charged	9	12	8	20
Blister Production - MTPD	883	220	580	820
Analysis - % Cu	98.8	98.5	98.6	98-99
- % Fe	0.04	0.01	0.02	0.01
- % S	0.02	0.007	0.03	0.03
Slag - Analysis - % Cu	4.9	7.0	4.5	2.5
- % Fe	46	37	46	45
- % SiO_2	21	18	21	26
- Disposition	To slag cleaning circuit			Return to Reverb
SLAG CLEANING				
Electric Furnace:				
Smelting Furnace Slag - MTPD	910	350	460	-
Power - KWH/MT Slag	37.6	65	83	-
Retention Time - hours	6	8	8	-
Matte Produced - % Cu	60	55-60	58	-
Discard Slag - % Cu	0.6	0.6	0.6	-
- % Fe	40	38	38	-
- % SiO_2	33	32	31.5	-
Copper Recovery - %	70	-	35	-
Flotation:				
Converter Slag - MTPD	515	280	230	-
Cooled in	Mould Conveyor	Pits	Mould Conveyor	-
Duration - hours	0.7	48	1.5	-

PLANT - continued	Saganoseki	Kosaka	Toyo	Onahama
Flotation Concentrate - MTPD	81	50	28	-
Analysis - % Cu	30	40	36	-
- % Fe	30	17	26	-
- % S	8	10	6	-
Flotation Tailings - MTPD	434	230	202	-
Analysis - % Cu	0.48	0.48	0.4	-
- % Fe	47	40	50	-
Copper Recovery - %	-	95.2	92.5	-
FIRE REFINING				
Furnaces - type	Rotary	Rotary	Rotary	Rotary
Number of Units	2	2	2	3
Dimensions	4.5 Ø x 11.2	3.3 Ø x 6.0	4.2 Ø x 11.7	4.1 Ø x 9.1 (2) 4.4 Ø x 10 (1)
Number of Tuyeres	2/furnace	2/furnace	2/furnace	2/furnace
Size - mm	28 I.D.	39 I.D.	25 I.D.	45 I.D.
Time per Charge for				
Oxidation - min.	15	10	10	30
- % O_2	-	-	21	21
Reduction - min.	80	105	120	135
Reductant - Used	Ammonia	Bunker Oil	Ammonia	Bunker Oil
- Consumption/MT anode	4.0 kg	9ℓ	5 kg	5.7ℓ

PLANT - continued	Saganoseki	Kosaka	Toyo	Onahama
PRODUCT CASTING				
Casting Machine - Type	Walker	Walker	Wheel	Walker Wheel & Hazelett Caster
Number of Moulds	60	18	30	24
Average Casting Rate - MTPH	80	30	50	50 & 50
Normal Casting Temp - °C	1150	1180	1160	1150
Cast Product - Type	Anodes	Anodes	Anodes	Anodes
Dimensions - m	1.01x0.914x0.38	1.27 x 0.67	1.04x1.05x.035	.96 x .98 x .037 .96 x 1.07 x .015
Average Weight - kg	325	330	350	330 & 135
Rejects as % of new material cast	6.5	2	2.5	4 & 8
SULPHUR FIXATION				
Source of Gas	Flash+Converter	Flash+Converter	Flash+Converter	Reverb+Converter
Plant - Type	Lurgi Single&Double Contact	Single Contact	Chemico	Lurgi S.& D. Contact
- Rated Capacity - MTPD	3,150	750	1,300	500/800
- Volume - Nm^3/hr	381,000	110,000	165,000	78,000/120,000
- Average % SO_2	8	5	8.5	5-6
- Average Conversion Efficiency %	99.7	97	99.8	97.0/99.8
Product Grade - % H_2SO_4	98	98	98	98.5
KWH/MT sulphur fixed in product	90	90	100	-
Auxiliary Fuel	None	None	None	19730 Nm^3/day tire distillate
Tailgas Scrubbing	On single contact plant only	Basic aluminum sulphate	NaOH	Lime/caustic soda

COPPER SMELTERS
of
NORTH AND CENTRAL AMERICA (CANADA & MEXICO)

COMPANY NAME	Noranda Inc.	Hudson Bay Mining and Smelting	Compania Minera de Cananea S.A.
PLANT	Horne	Flin Flon	Cananea
Flowsheet Page No.	104	105	106
Annual Production - Copper, MTPY	187,000	67,000	50,000
FEED ANALYSIS			
Concentrate		12.5% Zn	
% Cu	-	9.0	22.5
% Fe	-	12.6	28.8
% S	-	16	36.2
% SiO_2	-	25	6.6
Mesh	-	70%-325	88%-150
% H_2O	-	20	12-15
Furnace Flux			
% SiO_2	60	91.5	65-75
% CaO	0.2	-	2-5
% Fe	8-10	0.8	4
Mesh	75%-10 mm	15%-200	82%+10
Converter Flux			
% SiO_2	68	76	65-75
% CaO	1.5	-	2-5
% Fe	7-8	2.5	4
Mesh	75%-13 mm	5%-200	82%+10

PLANT - continued	Horne	Flin Flon	Cananea
FEED PREPARATION			
Blending System	-	Bedding Plant	Bedding Plant
Number of Beds	-	5	3
Bed Size - MT	-	4 @ 200 + 1 @ 1,250	2,500
Method of Reclaim	-	Clam to Belt Conveyor	Front End Loader
Dryer - Type	None	Taylor Rotary	Rotary
- Number	-	2	1
- Dimensions - m	-	2.74 Ø x 24.4	3.05 Ø x 18.3
Average Feed Rate - WMTPH	-	57	50
Feed Moisture - % H_2O	-	20	12-15
Fuel Used - Type	-	Bunker C	Bunker C or Natural Gas
Calorific Value - kCal/kg	-	10,200	8900 or 9362 kCal/m^3
Retention Time - min.	-	-	20
Inlet Gas Temp. - °C	-	320	950
Outlet Gas Temp. - °C	-	200	110
Product Moisture - % H_2O	-	10	8
Exhaust Gas Volume - Nm^3/hr	-	-	41,240
Cleaned by	-	Cyclones	Scrubber
Roasters - Type	-	Nichols Herreshoff	-
- Number	-	5 (4 operating)	-
- Dimensions	-	6.55 Ø x 9.14	-
Average Feed Rate - WMTPH	-	13	-
Fuel Used - Type	-	Coal	-
Calorific Value - kCal/kg	-	7,000	-
Calcine Analysis - % Cu	-	17.7	-
- % Fe	-	28.3	-
- S	-	15.0	-
Temperature - °C	-	650	-
Offgas Volume - Nm^3/hr	-	160,000	-
- % SO_2	-	1.5	-

PLANT - continued	Horne		Flin Flon	Cananea
SMELTING				
Furnace Type	Reverb	Reactor	Reverb	Reverb
Number of Units	1	1	1	1
Nominal Capacity - MTPH			50 DSC	
Concentrate	42	75	-	30
Dimensions - m	3.5 x 11 x 35	4.2 Ø x 21	9.14 x 31.7	9.14 x 33.5
Oxygen Enrichment - % O_2	-	35	None	None
Auxiliary Fuel - Type	Oxy-Fuel (Gas)	Coal	Bunker C	Bunker C or Natural Gas
Calorific Value - kCal/kg	-	7,800	10,200	8900 or 9362 kCal/m^3
Operating Temperature - °C	1,100	1,175	1,200	1,350
Flux as % of Concentrate	10	10-14	23	5-7.5
Campaign Life - Years	2	0.8-1.1	1.5	3
Furnace Matte - MTPD	900	680	560	440
Analysis - % Cu	32	65	36	32-35
- % Fe	31	6.5	28	30-33
- % S	27	20	24	28-30
Temperature - °C	1,200	1,200	1,150	1,100
Furnace Slag - MTPD	1,150	1,300	960	225
Analysis - % Cu	0.57	4.6	0.68 (8.3% Zn)	0.6
- % Fe	33	40	33.2	18-20
- % CaO	0.5	2	-	10-12
- % SiO_2	35	22	37.8	36-38
Temperature - °C	1,150	1,120	1,200	1,150
Disposition	Discarded	To Flotation	To Slag Fuming	Discarded
Offgas - Volume - Nm^3/hr	401,000	467,000	110,400	86,800
Temperature - °C	1,100	1,100	1,370	1,250-1,350
% SO_2	1.5	6-8	1.5	1.5
Cooled by	Boiler	Water Spray	2 Boilers	Spray Chamber
Inlet Temp. - °C	1,100	1,100	1,370	1,250-1,350
Outlet Temp. - °C	700	270	300	450-500
Disposition of Dust	Recycled	Recycled & Bled	Recycled	Recycled

PLANT - continued	Horne	Flin Flon	Cananea
CONVERTING			
Converter Type	P-S	P-S	P-S
Number - Hot	3	2	2
- Standby	2	1	2
Dimensions - m	4 Ø x 9.1	3.96 Ø x 9.14	1 @ 3.96 x 9.14 3 @ 3.66 Ø x 7.62
Number of Tuyeres	48	44	48 and 38
Size - mm	49 (ID)	50	50
Average Blowing Rate - Nm^3/hr	42,600	30,600	-
Oxygen Enrichment - % O_2	21	None	None
Air Rate	-	510	455
Scrap added as % of matte charged	10	31	15
Flux as % of matte charged	23	32.5	27.5
Blister Production - MTPD	225	160	-
Analysis - % Cu	98	98.5	99.5
- % Fe	low	-	0.02
- % S	1.5	0.2	0.02
Slag - Analysis - % Cu	4.7	3.6 (7.2% Zn)	7.8
- % Fe	40	35	45.8
- % SiO_2	25	29	20
- Disposition	Returned to Smelting	Returned to Smelting	To Slag Cleaning
SLAG CLEANING			
Smelting Furnace Slag - MTPD	1,000 (Reactor)	Reverb Slag held in two 2000 KVA holding furnaces prior to slag fuming. Fume 72% Zn	-
Converter Slag - MTPD	0		335
Cooled in	Ladles		Pits
Duration - hours	26-30		60
Flotation Concentrate - MTPD	275		80
Analysis - % Cu	38		30
- % Fe	20		35
- % S	13		12.5

PLANT - continued	Horne	Flin Flon	Cananea
Flotation Tailings - MTPD	865		275
Analysis - % Cu	0.4	Slag: 0.6% Cu	0.8
- % Fe	42	1.0% Zn	45
Copper Recovery - %	-	43.0% SiO_2	92
FIRE REFINING			
Furnace - Type	Rotary	Rotary	None
Number of Units	3	2	-
Dimensions - m	4 Ø x 9.1; 4 Ø x 10.7 4 Ø x 11	3.96 Ø x 9.14	-
Number of Tuyeres	1	1	-
Size - mm	25 (ID)	19	-
Time per Charge for			
Oxidation - min.	30	20	-
- % O_2	21	21	-
Reduction - min.	180	96	-
Reductant - Used	Natural Gas	Propane	-
- Consumption	50 Nm^3/hr	12 L/MT Cu	-
PRODUCT CASTING			
Casting Machine - Type	2 Wheels	Treadwell Wheel	Straight Line
Number of Molds	25	22	20
Average Casting Rate - MTPH	32	40	30
Normal Casting Temp. - °C	1,150	1,200	1,150
Cast Product - Type	Anodes	Anodes	Blister Cake
Dimensions - m	1.07x0.76x0.04	0.90 x 0.86	.635 x .889 x .152
Average Weight - kg	290	290	420-450
Rejects as % of new material cast	3	4.5	2-5
SULPHUR FIXATION	None	None	None

COPPER SMELTERS
of
NORTH AND CENTRAL AMERICA (CANADA & MEXICO)

COMPANY NAME	INCO Limited	Falconbridge Ltd.		McLaren Forest Products	Mexicana de Cobra S.A.
PLANT	Copper Cliff	Kidd Creek		Gaspé	La Caridad
Annual Production - Copper MTPY	136,000	90,000		-	180,000
Flowsheet Page No.	107 & 334	-		-	-
FEED ANALYSIS					
Concentrate	0.9% Ni				
% Cu	29	26		22	32
% Fe	32	28		-	21
% S	34	31		-	31
% SiO_2	1.9	4		-	6.5
Mesh	60% -325	100% -270		-	-200
% H_2O	8-9	6.5		5-10	6-8
Furnace Flux		#1	#2		
% SiO_2	80	73	85	70	85
% CaO	1-2	-	-	14	-
% Fe	1.9	6	1	3	2.5
Mesh	70% -48	100% -170	-3 mm	100% -7 mm	-1.68 mm
Converter Flux					
% SiO_2	94	-		70	85
% CaO	1	54		14	-
% Fe	2	-		3	2.5
Mesh	100% -32 mm	-3 mm		100% -19 mm	-12.7+6.35 mm
FEED PREPARATION					
Blending System	Belt Mixing	None		-	Bedding Plant
Number of Beds	-	-		-	2
Bed Size - MT	-	-		-	20,000
Method of Reclaim	From Bins	-		-	Front End Loader

PLANT - continued	Copper Cliff	Kidd Creek	Gaspé	La Caridad
Dryer - Type	Fluid Bed	Rotary-Flash	None	Rotary-Flash
- Number	2	1	-	1
- Dimensions - m	-	2.4 Ø x 12.0	-	2.4 Ø x 13.5 (Rotary) 2.7 Ø x 33.0 (Flash)
Average Feed Rate - WMTPH	37	50-70	-	110
Feed Moisture - % H_2O	8-9	6.5	-	6-8
Fuel Used - Type	Natural Gas	Natural Gas	-	Bunker C
Calorific Value	8900 kCal/m³	8900 kCal/m³	-	9700 kCal/kg
Retention Time - min.	0.02	-	-	
Inlet Gas Temp. - °C	Ambient	300	-	Ambient
Outlet Gas Temp. - °C	100	100	-	90
Product Moisture - % H_2O	0.8	<0.5	-	0.2
Exhaust Gas Volume - Nm³/hr	50,000	62,000	-	132,000
Cleaned by	Baghouse	Cyclones & Baghouse	-	Cyclones & Precipitator
SMELTING				
Furnace Type	Flash (Inco)	Mitsubishi	Reverb	Flash (OKO)
Number of Units	1	1	1	1
Nominal Capacity - MTPH Concentrate	75	55-60	30	80
Dimensions - m	22.2 x 5.5 x 5.2	10.16 Ø x 3.6	9.14 x 30.5	7.92 x 25.4
Oxygen Enrichment - % O_2	95-97	40-45	None	43
Auxiliary Fuel - Type	None	Bunker C	Bunker C	Bunker C
Calorific Value - kCal/kg	-	10,800	10,000	9,700
Operating Temperature - °C	1200-1220	1,200	1,350	1,350
Flux as % of Concentrate	12	30	5	5.6
Campaign Life - Years	3	-	-	-
Furnace Matte - MTPD	1,150	-	576	820
Analysis - % Cu	45 (1.5% Ni)	68	35	60
- % Fe	26	7	25	16.2
- % S	24.5	22	26	22.5
Temperature - °C	1,170	1,200	1,150	1,180

PLANT - continued	Copper Cliff	Kidd Creek	Gaspé	La Caridad
Furnace Slag - MTPD	700	700	600	730
Analysis - % Cu	0.6 (0.06% Ni)	0.7	0.45	1.3
- % Fe	40	40	32-33	35.5
- % CaO	-	2-3	4-6	-
- % SiO_2	33	32	38-40	28.35
Temperature - °C	1,220	1,200	1,200	1,280
Offgas - Volume - Nm^3/hr	13,000	84,000 max.	-	50,700
- Temperature - °C	1,260	1,200	1,170	1,350
- % SO_2	70-80	-	1.0-1.5	25.1
Cooled by	Venturi Scrubbing	Boiler	2 Boilers	Boiler
Inlet Temp. - °C	800	1,200	950	1,350
Outlet Temp. - °C	37	350	360	350
Disposition of Dust	Recycled or Bled	Recycled	Recycled	Recycled
CONVERTING				
Converter Type	P-S	Mitsubishi	P-S	P-S
Number - Hot	4	1	1	2
- Standby	1	0	1	1
Dimensions - m	3.97 Ø x 9.15	8.16 Ø x 3.3	3.96 Ø x 9.14	4.57 Ø x 10.67
Number of Tuyeres	48	lances	42	60
Size - mm	60.3 O.D.	-	63.5	44
Average Blowing Rate - Nm^3/hr	31,800	15,000	37,400	40,000
Oxygen Enrichment - % O_2	up to 30%	30	None	24
Air Rate - Nm^3/min.	520	215	628	634
Slag Blow - Nm^3/min.	-	N/A	-	-
Copper Blow - Nm^3/min.	-	N/A	-	-
Scrap added as % of matte charged	12	-	None	0.7
Flux as % of matte charged	18	6.5	40	11.2
Blister Production - MTPD	-	280-300	180	596
Analysis - % Cu	98	98.5	98.4	98.5
- % Fe	0	-	0.2-0.3	0.5
- % S	<0.01	0.8	0.3-0.5	-

PLANT - continued	Copper Cliff	Kidd Creek	Gaspé	La Caridad
Slag - Analysis - % Cu	4 (1% Ni)	13-15	3	6
- % Fe	48	40	35	47
- % SiO_2	24	15-18% CaO	30	25
- Disposition	Returned to Ni Reverb	Returned to Smelting Fce.	Returned to Reverb	Returned to Flash or Slag Cleaning
SLAG CLEANING	None		None	
Electric Furnace:				
Smelting Furnace Slag MTPD	-	700	-	726
Converter Slag MTPD	-	-	-	245
Power - KWH/MT Slag	-	55-62	-	86
Retention Time - Hours	-	2	-	2
Matte Produced - % Cu	-	68	-	60
Discard Slag - % Cu	-	0.7	-	0.7
- % Fe	-	40	-	43.5
- % SiO_2	-	32	-	32.0
FIRE REFINING				
Furnace - Type	Reverb	Rotary	Rotary	Rotary
Number of Units	2 Optg.+ 1 Standby	2	1	2
Dimensions - m	5.19 x 15.25	4.57 Ø x 10.67	3.96 Ø x 9.14	4.57 Ø x 10.67
Number of Tuyeres	N/A	2	1	2
Size - mm	-	38	19	19
Time per Charge for				
Oxidation - min.	-	180	20	120
- % O_2	23	21	-	21
Reduction - min.	360	60	30	120
Reductant - Used	Green Poles	Ammonia	Propane	Propane
- Consumption/MT Cu	15/charge (1 MT/pole)	7 kg	10.9 L	4 kg

PLANT - continued	Copper Cliff	Kidd Creek	Gaspé	La Caridad
PRODUCT CASTING				
Casting Machine - Type	Clark Wheel (cable drive)	Hazelett	Wheel	OKO Wheel
Number of Molds	22	N/A	22	28
Average Casting Rate - MTPH	55 (two wheel casting)	55	36	70
Normal Casting Temp. - °C	1170-1180	1,150	1,160	1,200
Cast Product - Type	Anode	Anode	Anode	Anode
Dimensions - m	0.914 x 0.914 x 0.038	1.105 x .760 x .017	0.91 x 0.91	0.8 x 0.04 x 0.93
Average Weight - kg	272	155	290	285
Rejects as % of new material cast	3	4	4	1
SULPHUR FIXATION				
Source of Gas	Flash Furnace	Smelting & Converter	Converter	None at present
Plant - Type	SO_2 Liquefaction	Double Contact	Single Contact	-
- Rated Capacity - MTPD	360 SO_2	1,280 acid	1,100	-
- Volume - Nm^3/hr	9,000	-	155,000	-
- Average % SO_2	75	12	4.8-8.0	-
- Average Conversion Efficiency %	97.5	99.7	96	-
Product Grade - %	99 SO_2	93 H_2SO_4	93 & 96	-
KWH/MT Sulphur Fixed in Product	54	-	150-165	-
Auxiliary Fuel	None Required	Natural Gas	No. 2 Oil	-
Tailgas Scrubbing	None	None	None	-

COPPER SMELTERS
of
NORTH AMERICA (UNITED STATES)

COMPANY NAME	Phelps Dodge Corporation		Inspiration Consolidated Copper Co.
PLANT	Hidalgo	Douglas	Inspiration
Annual Production - Copper MTPY	150,000	106,000	110,000
Flowsheet Page No.	108	109	-
FEED ANALYSIS			
Concentrate			
% Cu	22	-	25
% Fe	28	-	21
% S	36	-	22
% SiO_2	8	-	16
Mesh	-200	-	-
% H_2O	6	-	9
Furnace Flux			
% SiO_2	75	60-90	65 / 15
% CaO	1.6	0.2-6	10 / 42
% Fe	2	0.2-6	3 / <1
Mesh	-200	100% -7 mm	100% -19 mm
Converter Flux			
% SiO_2	80	63-75	85
% CaO	0.2	0.2-3	2
% Fe	2.5	2-7	2
Mesh	-19 +6 mm	95% -26 mm	-51 +9 mm

PLANT - continued	Hidalgo	Douglas	Inspiration
FEED PREPARATION			
Blending System	Bedding Plant	Bedding Plant	Bedding Plant
Number of Beds	4	8	2
Bed Size - MT	5,500	2 @ 4,100 & 6 @ 6,600	5,900
Method of Reclaim	Bucket Wheel Reclaimers	Reclaimers	Front End Loader
Equipment - Type	Rotary Dryer	Herreshoff	Rotary Dryer
- Number	1	24	1
- Dimensions - m	3.35 Ø x 34.1	6.58 Ø	4.9 Ø x 24.4
Average Feed Rate - WMTPH	100	5-4	91
Feed Moisture - % H_2O	6	-	9
Retention Time - min.	30	-	20
Fuel Used - Type	No. 6 Oil	Natural Gas & Oil	Natural Gas
Calorific Value	7,200 kCal/kg	-	9,400 kCal/m^3
Inlet Gas Temp. - °C	650	Calcine: 20-25% Cu	925
Outlet Gas Temp. - °C	93	22-28% Fe	82
		13-19% S	
		Temp. 600°	
Product Moisture - % H_2O	0.1		0.25
Exhaust Gas Volume - Nm^3/hr	45,000	300,000	71,250
Cleaned by	Precipitator	1.6% SO_2	Cyclone & Baghouse
SMELTING			
Furnace - Type	Flash (OKO)	Reverb	Submerged Arc Electric
Number of Units	1	3	1
Nominal Capacity - MTPH		34 calcine	
Concentrate	92	-	95
Dimensions - m	10.36 x 25.3 x 5.79	7.92 x 32.61	10.7 x 36
Oxygen Enrichment - %	21	None	None
Auxiliary Fuel - Type	No. 6 Oil	No. 6 Fuel Oil	None
Calorific Value	7200 kCal/kg	-	-

PLANT - continued	Hidalgo	Douglas	Inspiration
Operating Temperature - °C	1,275	1,450	680
Flux as % of Concentrate	8	7	12
Campaign Life - Years	5	10	5
Furnace Matte - MTPD	785	666	885
Analysis - % Cu	60	37-53	37
- % Fe	16	20-30	34
- % S	22.5	24-26	25
Temperature - °C	1,200	1,120	1,121
Furnace Slag - MTPD	1,300	611	800
Analysis - % Cu	1.5	0.5-0.8	0.75
- % Fe	42	32-36	33
- % CaO	1.3	2-4	8
- % SiO_2	30	38-40	38
Temperature - °C	1,250	1,175	1,230
Disposition	Slag Cleaning	Discard	Discard
Offgas - Volume - Nm^3/hr	190,000	42,000	39,580
Temperature - °C	1,275	295	680
% SO_2	∿10	1	5
Cooled by	Boiler	Boilers	Radiation + Spray
Inlet Temp. - °C	1,275	1,230	680
Outlet Temp. - °C	425	270	540
Disposition of Dust	Recycled	Recycled	Recycled
CONVERTING			
Converter Type	P-S	P-S	Hoboken-Syphon
Number - Hot	3	4	3
- Standby	0	1	1 (plus 1 on repair)
Dimensions - m	3.96 Ø x 9.14	3.96 Ø x 9.14	4.3 Ø x 14.0
Number of Tuyeres	52	4 @ 42 & 1 @ 52	48
Size - mm	48	44	49
Average Blowing Rate - Nm^3/hr	28,200	33,000	39,540
Oxygen Enrichment - % O_2	None	None	None

PLANT - continued	Hidalgo	Douglas	Inspiration
Air Rate - Nm^3/hr	470	550	659
Slag Blow - Nm^3/min	-	-	-
Copper Blow - Nm^3/min	-	-	-
Scrap added as % of matte charged	30	28	17.7
Flux added as % of matte charged	8	32	17.4
Blister Production - MTPD	480	-	-
Analysis - % Cu	98.5	98.6	98.5
- % Fe	-	Nil	-
- % S	0.6	Nil	-
Slag - Analysis - % Cu	5	3-4	2.2
- % Fe	45	47-49	50
- % SiO_2	27	24-26	23
- Disposition	Slag Cleaning	Returned to Reverbs	Returned to Electric Fce.
SLAG CLEANING			
Electric Furnace:			
Smelting Furnace Slag - MTPD	1,300	-	-
Converter Slag - MTPD	200	-	-
Power: KWH/MT Slag	50	-	-
Matte Produced - % Cu	65	-	-
Discard Slag - % Cu	1.0	-	-
- % Fe	43	-	-
- % SiO_2	30	-	-
Retention Time - Hours	3	-	-
FIRE REFINING			
Furnace - Type	Rotary	Rotary	Rotary
Number of Units	2	2	1
Dimensions - m	3.96 Ø x 9.14	3.96 Ø x 8.53	3.6 Ø x 6.4

PLANT - continued	Hidalgo	Douglas	Inspiration
Number of Tuyeres	4	2	2
Size - mm	38	38	19
Time per Charge for			
Oxidation - min.	60	60	120
- % O_2	21	21	21
Reduction - min.	120	200	150
Reductant - Used	Reformed Propane	Reformed Natural Gas	Natural Gas
- Consumption	185 Nm^3/hr	21 kg/min	-
PRODUCT CASTING			
Casting Machine - Type	Taylor Wheel	Wheel	Wheel
Number of Molds	26	22	10
Average Casting Rate - MTPH	32	30	23
Normal Casting Temp. - °C	1,140	1,140	1,140
Cast Product - Type	Anode	Anode	Anode
Dimensions - m	0.914 x 0.914 x 0.051	0.914 x 0.914	1.01 x 1.105 x 0.057
Average Weight - kg	340	340	545
Rejects as % of new material cast	1.1	1.3	3
SULPHUR FIXATION			
Source of Gas	Flash + Converter	None	Furnace + Converter
Plant - Type	2 Double Contact	-	Lurgi Double Contact
- Rated Capacity - MTPD	1,070 and 2,000	-	1,210
- Volume - Nm^3/hr	145,000 & 210,000	-	205,838
- Average % SO_2	8	-	4-6
- Avg. Conversion Efficiency %	99.5	-	99.3
Product Grade - % H_2SO_4	93 and 98	-	93
KWH/MT Sulphur Fixed in Product	-	-	-
Auxiliary Fuel	None	-	Natural Gas
Tailgas Scrubbing	None	-	None

COPPER SMELTERS
of
NORTH AMERICA (UNITED STATES)

COMPANY NAME	Kennecott Corporation		Chino Mines Company
PLANT	Utah	Ray Mines	Chino
Annual Production - Copper MTPY	200,000	80,000	115,000
Flowsheet Page No.	-	-	110
FEED ANALYSIS			
Concentrate			
% Cu	27.9	-	25
% Fe	26.2	-	30
% S	25.3	-	35
% SiO_2	11.7	-	5
Mesh	70% -270	-	60% -325
% H_2O	10-12	-	11
Furnace Flux			
% SiO_2	75.5	-	70
% CaO	-	-	<0.2
% Fe	4.6	-	9
Mesh	100% -9 mm	-	80% -10
Converter Flux	Washed Gravel		
% SiO_2	80.0	80	80
% CaO	-	1	<0.2
% Fe	5.0	4	9
Mesh	7 mm	-38 mm +6 mm	-19 mm +6

PLANT - continued	Utah	Ray Mines	Chino
FEED PREPARATION			
Blending System	Batch	-	Bedding Plant
Number of Beds	-	-	2
Bed Size - MT	-	-	18,000
Method of Reclaim	Clam or Front End Loader	-	Front End Loader
Equipment - Type	Rotary Dryer	Dorr Roaster	Fluid Bed Dryer
- Number	2	1	2
- Dimensions - m	3.35 Ø x 21.3	4.11 Ø	3.51 Ø x 8.53
Average Feed Rate - WMTPH	118	60	45
Feed Moisture - % H_2O	10-12	-	11
Fuel Used - Type	Natural Gas & Oil	Air Rate	Natural Gas
Calorific Value	8400 kCal/m^3	24,000 Nm^3/hr	9200 kCal/m^3
Inlet Gas Temp. - °C	1,090	Calcine: 30% Cu	290
Outlet Gas Temp. - °C	100	33.3% Fe	88
		15.0% S	
Product Moisture - % H_2O	7	Temp. 500-600°C	0.1
Exhaust Gas Volume - Nm^3/hr	18,400	23,000	73,000
Cleaned by	Cyclone & Wet Scrubbers	-	Baghouse
SMELTING			
Furnace Type	Noranda Reactors	Reverb	Flash (Inco)
Number of Units	3	1	1
Nominal Capacity - MTPH Concentrate	85	60	80
Dimensions - m	5.18 Ø x 21.34	10.67 x 36.58	7.32 x 24.38
Oxygen Enrichment - % O_2	34	None	98
Auxiliary Fuel - Type	Gas, Oil & Coal	Natural Gas	None
Calorific Value	-	9200 kCal/m^3	-
Operating Temperature - °C	1,190	1,380	1,260
Flux as % of Concentrate	4	-	12
Campaign Life - Years	0.74	5	2

PLANT - continued	Utah	Ray Mines	Chino
Furnace Matte - MTPD	900	800	800
Analysis - % Cu	73	45	48
- % Fe	4.2	28	25
- % S	20.8	24	23
Temperature - °C	1,190	1,200	1,180
Furnace Slag - MTPD	1,748	640	1,100
Analysis - % Cu	7.0	0.6	0.7
- % Fe	42.2	30	44
- % CaO	-	2	<1.0
- % SiO_2	22.8	40	33
Temperature - °C	1,190	1,220	1,235
Offgas - Volume - °C	118,900	51,000	14,000
Temperature - °C	1,320	1,380	1,260
% SO_2	17	0.25	80
Cooled by	Water Cooled Hood + Boiler	Boiler	Air-to-Gas
Inlet Temp. - °C	705	-	1,260
Outlet Temp. - °C	310	-	45
Disposition of Dust	Recycled	Recycled	Recycled
CONVERTING			
Converter Type	P-S	P-S	P-S
Number - Hot	3	2	3
- Standby	1	1	1
Dimensions - m	3.96 Ø x 9.14	3.96 Ø x 9.14	3.96 Ø x 9.14
Number of Tuyeres	50	42	44
Size - mm	51	51	51
Average Blowing Rate - Nm^3/hr	34,000	82,000	36,600
Oxygen Enrichment - % O_2	23	None	23
Slag - % O_2	-	-	-
Copper - % O_2	-	-	-

PLANT - continued	Utah	Ray Mines	Chino
Air Rate - Nm^3/min.	570	540	600
Scrap added as % of matte charged	10	0	25
Flux added as % of matte charged	2.6	35	18
Blister Production - MTPD	-	270	-
Analysis - % Cu	98.7	99	99.3
- % Fe	-	0.2	-
- % S	0.04	0.5	0.025
Slag - Analysis - % Cu	12.0	5	6
- % Fe	41.9	45	50
- % SiO_2	21.0	25	25
- Disposition	Slag Cleaning	Slag Cleaning	Returned to Flash

SLAG CLEANING	Utah	Ray Mines	Chino
Flotation:			
Smelting Furnace Slag - MTPD	1,748	0	-
Converter Slag - MTPD	94	500	-
Cooled in	Ladles & Piles	Pits	-
Duration - Hours	48	24	-
Flotation Concentrate - MTPD	297	100	-
Analysis - % Cu	40	25	-
- % Fe	23	30	-
- % S	10	32	-
Flotation Tailings - MTPD	1,451	400	-
Analysis - % Cu	0.35	1.0	-
- % Fe	46.1	45	-
Copper Recovery - %	95	80	-

PLANT - continued	Utah	Ray Mines	Chino
FIRE REFINING			
Furnace Type	Rotary	-	Rotary
Number of Units	4	-	-
Dimensions - m	3.96 Ø x 8.53	-	3.96 Ø x 12.8
Number of Tuyeres	2	-	2
Size - mm	50	-	-
Time per Charge for			
Oxidation - min.	45	-	-
- % O_2	-	-	-
Reduction - min.	60	-	-
Reductant - Used	Natural Gas	-	Natural Gas
- Consumption	-	-	-
PRODUCT CASTING			
Casting Machine - Type	Wheel	Walker Wheel	-
Number of Molds	26	22	22
Average Casting Rate - MTPH	30	40	45
Normal Casting Temp. - °C	1,110	1,200	1,170
Cast Product - Type	Anode	Anode	-
Dimensions - m	.802 x 1.0 x 0.046	0.889 x 1.1	-
Average Weight - kg	345	317	344
Rejects as % of new material cast	6	5	-
SULPHUR FIXATION			
Source of Gas	Reactor & Converters	-	Flash + Converter
Plant - Type	Acid Plants	Double Contact	Double Contact
- Rated Capacity - MTPD	2,083	1,250	2,200
- Volume - Nm^3/hr	378,000	175,000	187,000
- Average % SO_2	5	7.5	9.2
- Average Conversion Efficiency %	97.5	99.4	99.5
Product Grade - % H_2SO_4	94	95	95 or 98
KWH/MT Sulphur Fixed in Product	-	600	200
Auxiliary Fuel	Natural Gas	Natural Gas	Natural Gas

COPPER SMELTERS
of
CHILE

COMPANY NAME	Empresa Nacional de Mineria		Codelco Chile
PLANT	Ventanas	Paipote	El Teniente(1)
Annual Production - Copper MTPY	194,000	70,000	345,000
Flowsheet Page No.	-	-	-
FEED ANALYSIS			
Concentrate			
% Cu	32-36	27-36	36.5
% Fe	-	-	22.6
% S	-	-	31.2
% SiO_2	-	-	5.7
Mesh	-	-	70% -325
% H_2O	-	-	13.8
Furnace Flux			
% SiO_2	2-3	5	-
% CaO	50-55	49.2	-
% Fe	-	-	-
Mesh	-	100% -9.5 mm	-
Converter Flux			
% SiO_2	93-96	93	89.0
% CaO	0.2-0.4	-	-
% Fe	1-2	-	-
Mesh	-	100% -26 mm	3 to 19 mm TMC 19 to 38 mm P.S.
FEED PREPARATION			
Blending System	Beds by Truck	Mixing Bins	Belts
Number of Beds	2	3	-
Bed Size - MT	8,000	2,000	-
Method of Reclaim	Front End Loader	Conveyor	-
Dryer - Type	None	None	3 Rotary 2.57 Ø x 15.24 m

PLANT - continued	Ventanas	Paipote	El Teniente
SMELTING			
Furnace Type	Reverb	Reverb	Reverb
Number of Units	1	1	2
Nominal Capacity - MTPH conc.	32	27	33-40 (7.5% H_2O)
Dimensions - m	8 x 35	8 x 32 x 4.6	8.2 x 36.2 x 4.3 9.1 x 35.3 x 3.8
Oxygen Enrichment - % O_2	None	None	Oxy-Fuel
Auxiliary Fuel - Type	Coal & No. 6 Oil	Coal	Bunker C
Calorific Value - kCal/kg	6,900 & 9,800	7,200	10,350
Operating Temperature - °C	1100-1200	1,200	-
Flux as % of Concentrate	4	4	Reverts only
Campaign Life - Years	-	10	-
Furnace Matte - MTPD	600	450	880/Fce.
Analysis - % Cu	44-49	50	50.3
- % Fe	22-24	24	23.0
- % S	25-28	26	25.0
Temperature - °C	1100-1150	1,100	1,145
Furnace Slag - MTPD	300	400	920/Fce.
Analysis - % Cu	0.8-1.0	0.8	0.95
- % Fe	35-38	32	42
- % CaO	5-6	6.7	1.0
- % SiO_2	32-34	35	31.3
Temperature - °C	1150-1200	1,235	1,230
Disposition	Discard	Discard	Discard
Offgas - Volume - Nm^3/hr	80,000	66,000	33,000/Fce.
Temperature - °C	-	1,200	1,300
% SO_2	2-3	1.0	9.5-11.5
Cooled by	3 Boilers	3 Boilers	Spray Chamber
Inlet Temp. - °C	1,200	1,200	1,260
Outlet Temp. - °C	400	400	600

PLANT - continued	Ventanas	Paipote	El Teniente	
			Teniente Modified Conv. - 750 MTPD Copper Concentrate	
CONVERTING				
Converter Type	1 Modified + 2 P-S	2 Hoboken + 1 P-S	TMC	P-S
Number - Hot	3	3	2	3
- Standby	1 P-S	-	1	1
Dimensions - m	Mod. 4 Ø x15;P-S 3 Ø x 8	3 Ø x 6.1	4Øx16.8 - 4 Ø x 17.7	4 Ø x 10.6
Number of Tuyeres	36 (60 available) & 36	21 and 31	41 and 36	48-50
Size - mm	51 and 38	38 and 51	54 and 64	51
Average Blowing Rate - Nm^3/hr	-	-	-	-
Oxygen Enrichment - % O_2	None	None	30-32	None
Air Rate				
Modified - Nm^3/min.	530-580	Hoboken 180	505-535	-
P-S - Nm^3/min.	320-360	250	-	605-630
Stack Time - %	-	Hoboken 86 P-S 77	88-90 -	- 50-55
Scrap added as % of matte charged	Modified 5 P-S 20	28 -	- -	- 17
Flux added as % of matte charged	Modified 18	6	-	None
Blister Production - MTPD	150 per Conv.	-	875 (White Metal)	345-380
Analysis - % Cu	98.5	-	73-75	99.32
- % Fe	50-100 ppm	-	3.4-5.3	0.004
- % S	200-300 ppm	-	22-23	0.08
Slag - Analysis - % Cu	4-10	2.32	6.5-8.0	14.0
- % Fe	44-50	50.66	43-45	47
- % SiO_2	19-23	19.98	22-23	10
- Disposition	Recycled	Recycled	Recycled to Reverb	Recycled TMC
SLAG CLEANING	None	None	None	

PLANT - continued	Ventanas		Paipote		El Teniente	
FIRE REFINING						
Furnaces - Type	1 Rotary & 2 Reverb		Rotary		Reverb	Rotary
Number of Units	3		1		1	1
Dimensions - m	4.1 Ø x 9 & 3 x 5 x 15		3.96 Ø x 9.14		12.9x4.9x3.8	4 Ø x 10.7
Number of Tuyeres	Rotary 2; 4 lances/reverb		4 (2 operating)		-	6 (2 operating)
Size - mm	32		51		-	-
Time per Charge for	Rotary	Reverb				
Oxidation - min.	90	180	190		360-480	360
- % O_2	21	21	21		21	21
Reduction - min.	60	50-200	90		240-360	150-210
Reductant - Used	Kerosene	Poles	Kerosene		Poles	Kerosene + Steam
- Consumption	16 L/min.	14/charge	6 L/MT anode		0.5 MT/MT Cu	6 kg/MT Cu
PRODUCT CASTING						
Casting Machine - Type	2 Outokumpu Wheels		Wheel		Wheel	Wheel
Number of Moulds	20/wheel		16		15	15
Average Casting Rate - MTPH	35		40		32	50
Normal Casting Temp. - °C	-		1,180		1,130	1,100
Cast Product - Type	Anodes		Anodes		Cake	Ingot
			Enami	Codelco		
Dimensions - m	-		.89 x .895	1.21 x .85	1.115x.61x.05	0.69x.085x.065
Average Weight - kg	275-300		292	365	450	25
Rejects as % of new material cast	2		0.35-1.8		2.5	1.5
SULPHUR FIXATION	None		-		-	
Source of Gas	-		Hoboken Converters		TMC Gas	
Plant - Type	-		Single Contact Acid Plant		Single Contact	
- Rated Capacity - MTPD	-		140		85-90	
- Volume - Nm^3/hr.	-		23,000		21,500	
- Average % SO_2	-		6.1		4.2	
- Average Conversion Efficiency - %	-		96		92	

PLANT - continued	Ventanas	Paipote	El Teniente
Product Grade - % H_2SO_4	-	98	98
KWH/MT Sulphur Fixed in Product	-	116	135
Auxiliary Fuel	-	Diesel Oil 90,000 L/year	- -
Tailgas Scrubbing	-	None	-

(1) G. Munoz, G. Veva and H. Salazar; "Codelco Chile: A Realistic Way to Increase Copper Smelting Capacity", Copper Smelting - An Update, edited by David B. George and John C. Taylor, TMS-AIME, Warrendale, PA, February 1982, pp 151-155.

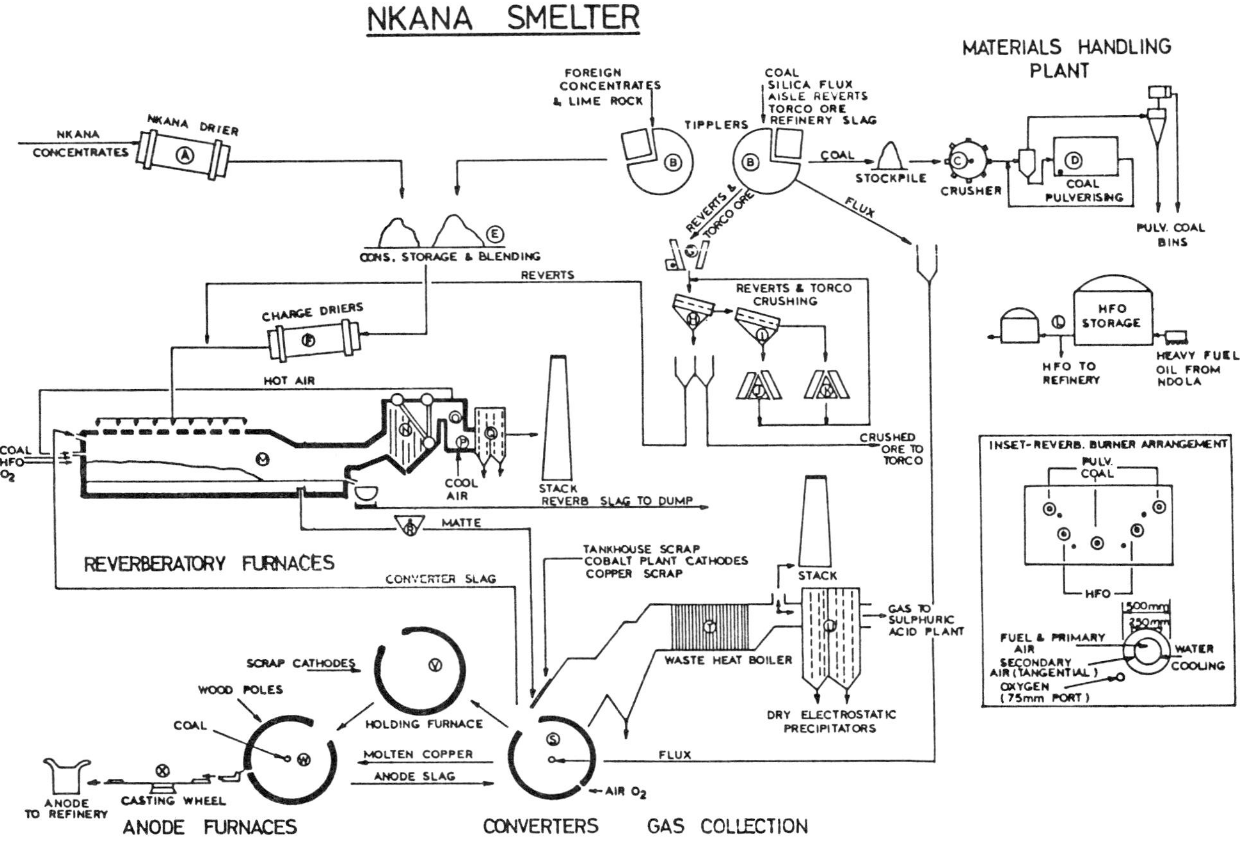
NKANA SMELTER
MATERIALS HANDLING PLANT
NKANA CONCENTRATES
NKANA DRIER
FOREIGN CONCENTRATES & LIME ROCK
TIPPLERS
COAL
SILICA FLUX
AISLE REVERTS
TORCO ORE
REFINERY SLAG
COAL
STOCKPILE
CRUSHER
COAL PULVERISING
PULV. COAL BINS
HFO STORAGE
HFO TO REFINERY
HEAVY FUEL OIL FROM NDOLA
FLUX
REVERTS & TORCO ORE
CONS. STORAGE & BLENDING
REVERTS
REVERTS & TORCO CRUSHING
CHARGE DRIERS
HOT AIR
CRUSHED ORE TO TORCO
COAL
HFO
O2
COOL AIR
STACK
REVERB SLAG TO DUMP
MATTE
INSET-REVERB. BURNER ARRANGEMENT
PULV. COAL
HFO
500mm
250mm
FUEL & PRIMARY AIR
SECONDARY AIR (TANGENTIAL)
OXYGEN (75mm PORT)
WATER COOLING
REVERBERATORY FURNACES
CONVERTER SLAG
TANKHOUSE SCRAP
COBALT PLANT CATHODES
COPPER SCRAP
STACK
GAS TO SULPHURIC ACID PLANT
WASTE HEAT BOILER
DRY ELECTROSTATIC PRECIPITATORS
SCRAP CATHODES
WOOD POLES
COAL
HOLDING FURNACE
MOLTEN COPPER
ANODE SLAG
FLUX
AIR O2
ANODE TO REFINERY
CASTING WHEEL
ANODE FURNACES
CONVERTERS
GAS COLLECTION

Zambia Consolidated Copper Mines Ltd.

Luanshya Smelter

Simplified Flow Sheet

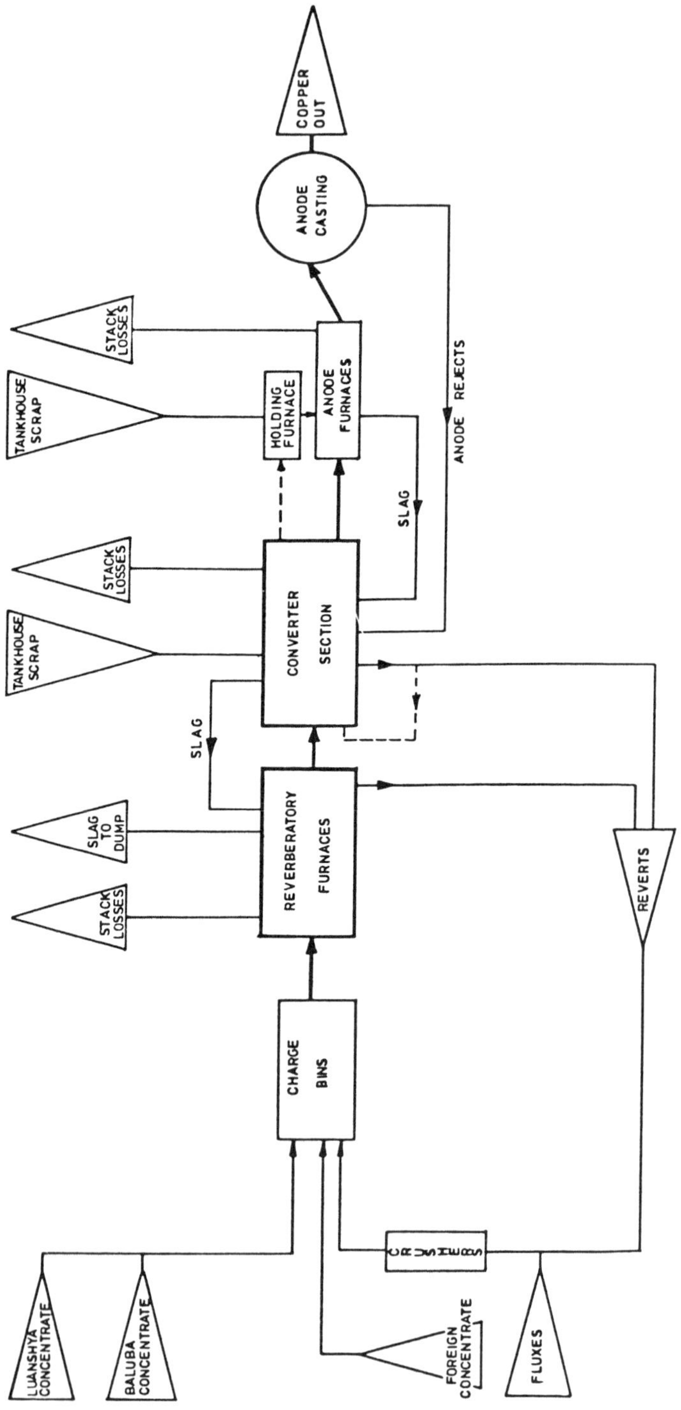

LUANSHYA CONCENTRATE
BALUBA CONCENTRATE
FOREIGN CONCENTRATE
FLUXES
CRUSHERS
CHARGE BINS
STACK LOSSES
SLAG TO DUMP
REVERBERATORY FURNACES
SLAG
TANKHOUSE SCRAP
STACK LOSSES
CONVERTER SECTION
TANKHOUSE SCRAP
HOLDING FURNACE
STACK LOSSES
ANODE FURNACES
SLAG
ANODE CASTING
ANODE REJECTS
COPPER OUT
REVERTS

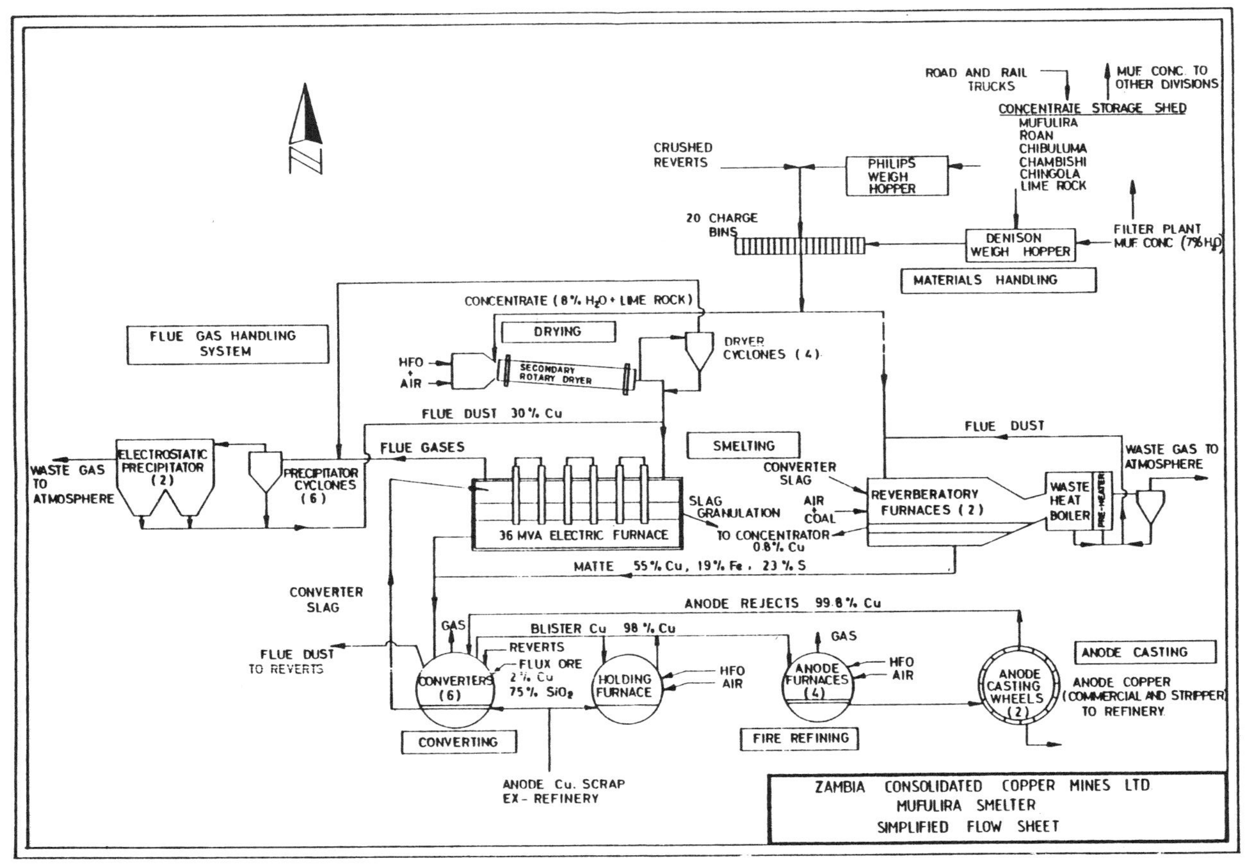
ROAD AND RAIL TRUCKS
MUF. CONC. TO OTHER DIVISIONS
CONCENTRATE STORAGE SHED
MUFULIRA
ROAN
CHIBULUMA
CHAMBISHI
CHINGOLA
LIME ROCK
CRUSHED REVERTS
PHILIPS WEIGH HOPPER
20 CHARGE BINS
DENISON WEIGH HOPPER
FILTER PLANT MUF. CONC (7% H_2O)
MATERIALS HANDLING
FLUE GAS HANDLING SYSTEM
CONCENTRATE (8% H_2O + LIME ROCK)
DRYING
HFO + AIR
SECONDARY ROTARY DRYER
DRYER CYCLONES (4)
FLUE DUST 30% Cu
SMELTING
FLUE DUST
WASTE GAS TO ATMOSPHERE
ELECTROSTATIC PRECIPITATOR (2)
PRECIPITATOR CYCLONES (6)
FLUE GASES
CONVERTER SLAG
REVERBERATORY FURNACES (2)
WASTE HEAT BOILER
PRE-HEATER
SLAG GRANULATION
AIR + COAL
TO CONCENTRATOR 0.8% Cu
36 MVA ELECTRIC FURNACE
MATTE 55% Cu, 19% Fe, 23% S
CONVERTER SLAG
ANODE REJECTS 99.8% Cu
GAS
BLISTER Cu 98% Cu
GAS
REVERTS
FLUX ORE 2% Cu 75% SiO_2
FLUE DUST TO REVERTS
CONVERTERS (6)
HOLDING FURNACE
HFO AIR
ANODE FURNACES (4)
HFO AIR
ANODE CASTING WHEELS (2)
ANODE CASTING
ANODE COPPER (COMMERCIAL AND STRIPPER) TO REFINERY
CONVERTING
FIRE REFINING
ANODE Cu. SCRAP EX-REFINERY
ZAMBIA CONSOLIDATED COPPER MINES LTD
MUFULIRA SMELTER
SIMPLIFIED FLOW SHEET

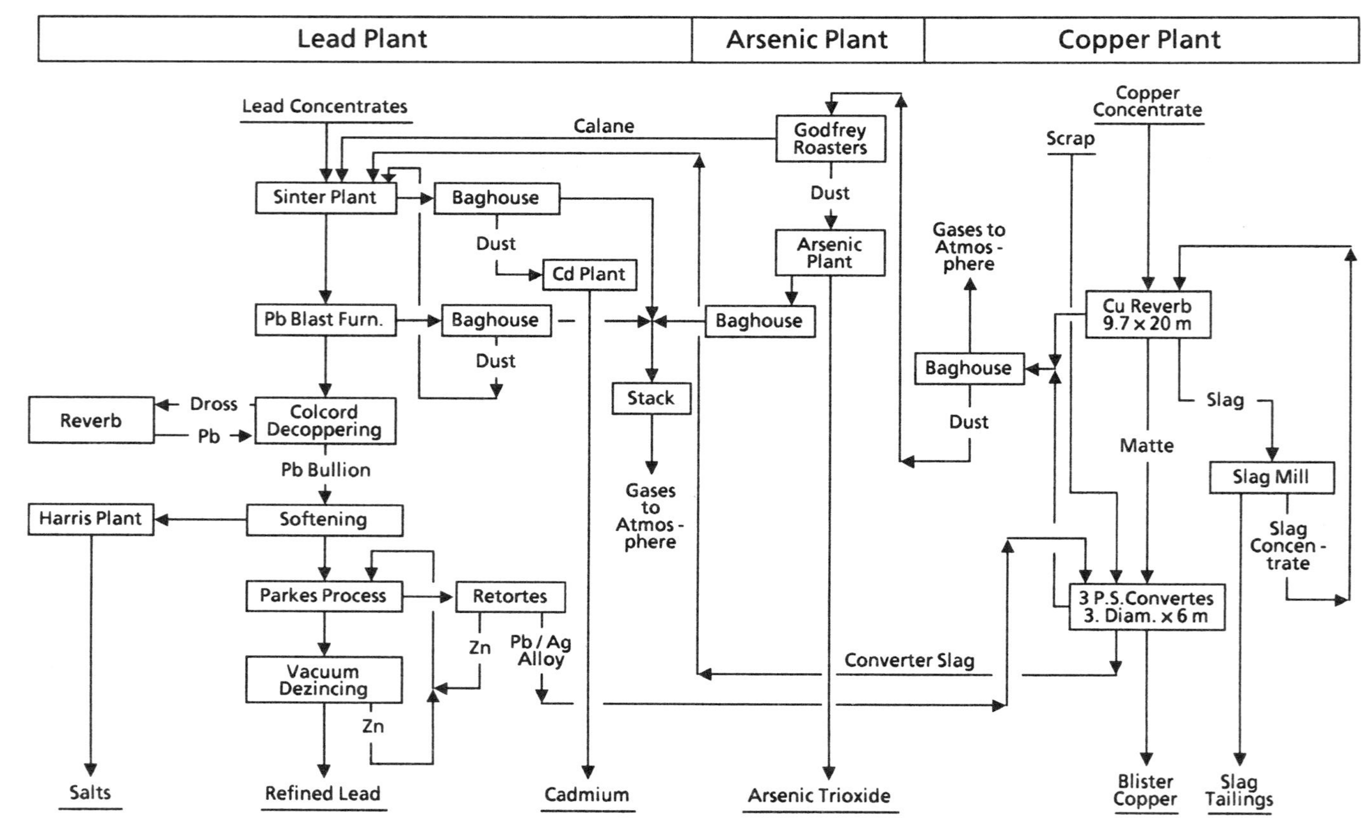

Tsumeb Smelter
Simplified Flowsheet
Lead Plant
Arsenic Plant
Copper Plant
Lead Concentrates
Calane
Godfrey Roasters
Copper Concentrate
Scrap
Sinter Plant
Baghouse
Dust
Cd Plant
Arsenic Plant
Gases to Atmos - phere
Cu Reverb 9.7 x 20 m
Pb Blast Furn.
Baghouse
Baghouse
Dust
Baghouse
Stack
Dust
Slag
Reverb
Dross
Pb
Colcord Decoppering
Matte
Pb Bullion
Slag Mill
Gases to Atmos - phere
Harris Plant
Softening
Slag Concen - trate
Parkes Process
Retortes
3 P.S. Convertes 3. Diam. x 6 m
Zn
Pb / Ag Alloy
Vacuum Dezincing
Converter Slag
Zn
Salts
Refined Lead
Cadmium
Arsenic Trioxide
Blister Copper
Slag Tailings

THE ELECTROLYTIC REFINING & SMELTING CO. OF AUST. LTD.

SMELTER DEPARTMENT

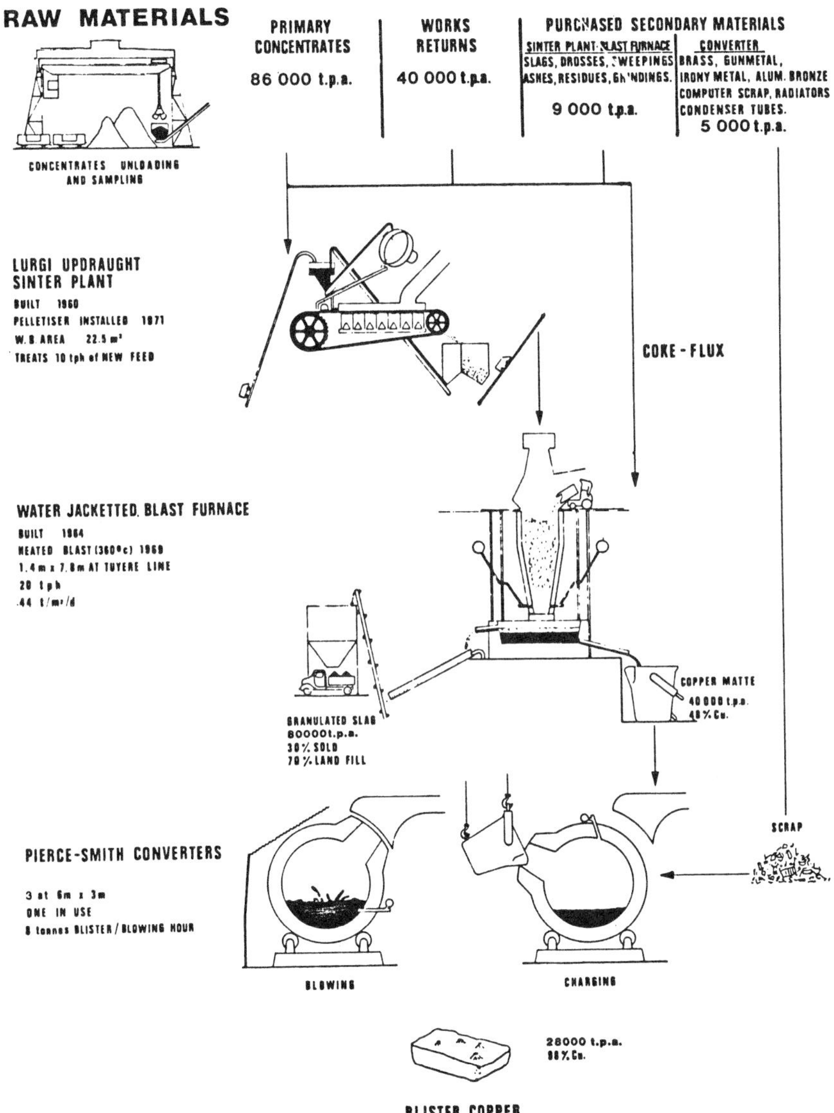

THE ELECTROLYTIC REFINING & SMELTING CO. OF AUST. LTD.

CASTING DEPARTMENT

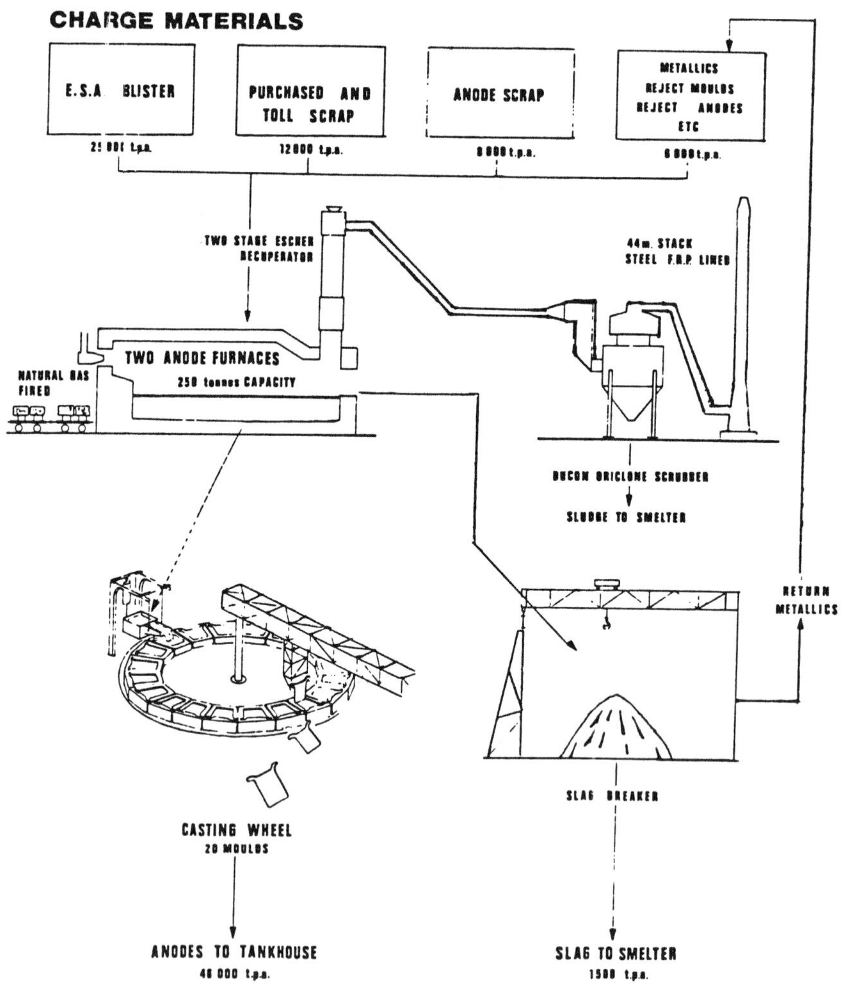

SMELTING

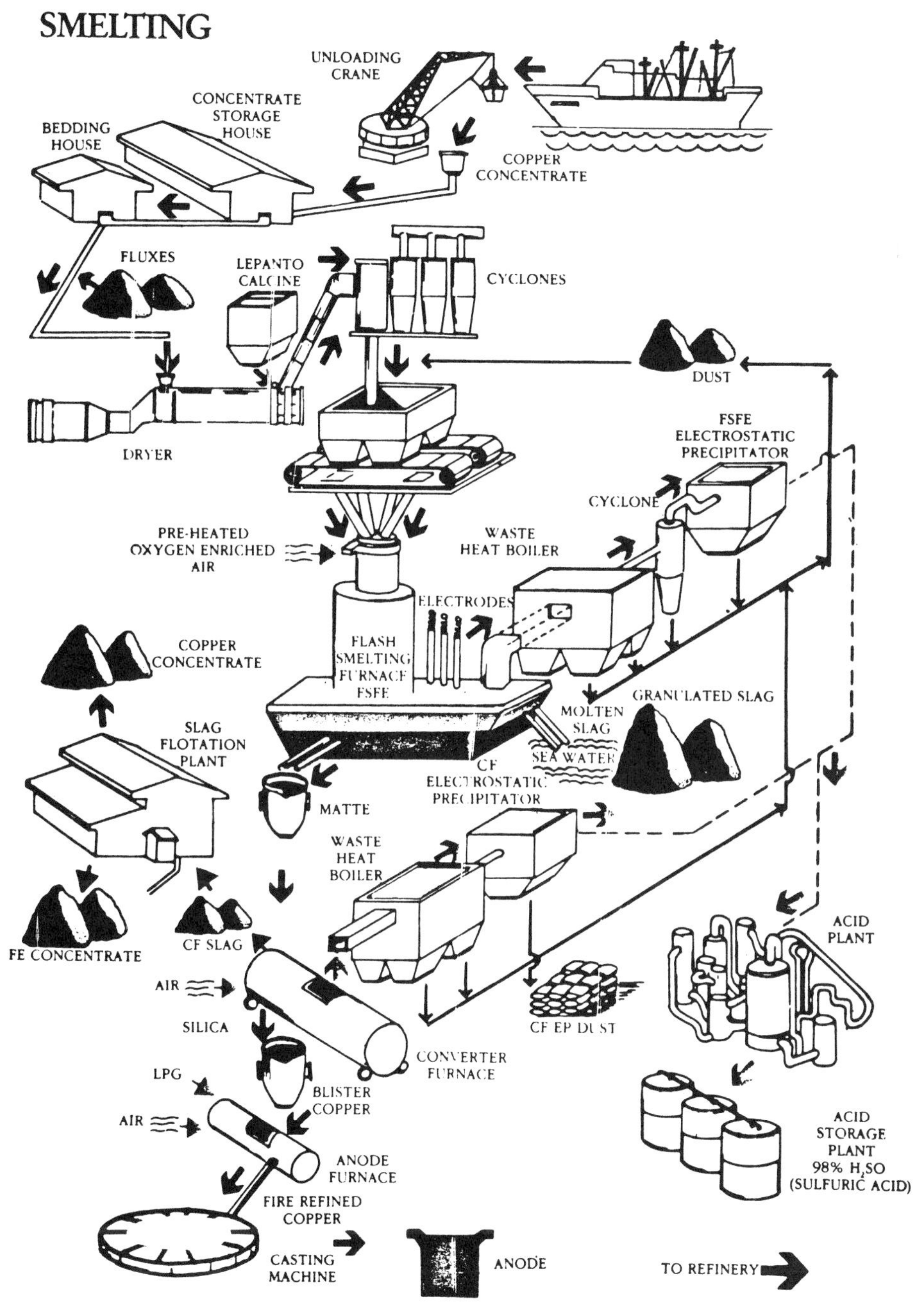

PHILIPPINE ASSOCIATED SMELTING & REFINING CORPORATION

Flowsheet
Norddeutsche Affinerie AG, Hamburg / F. R. Germany

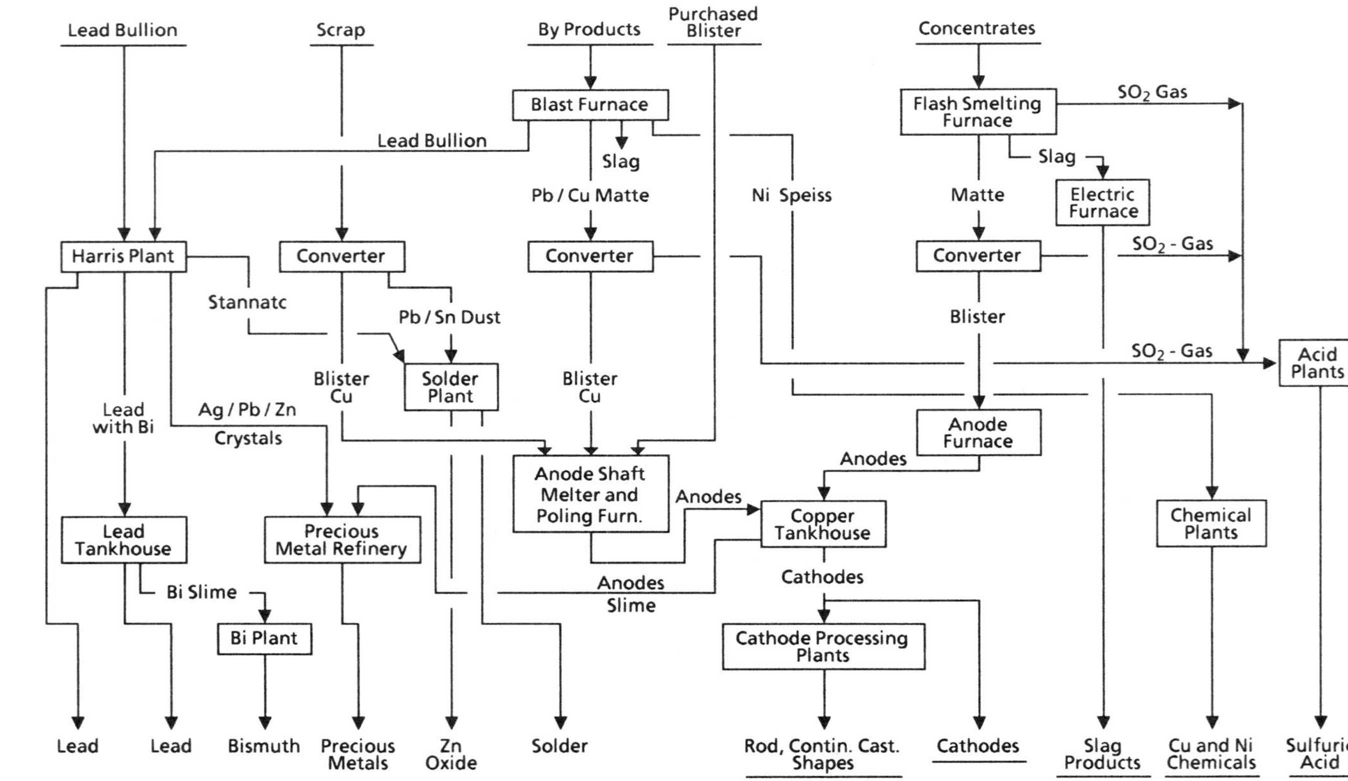

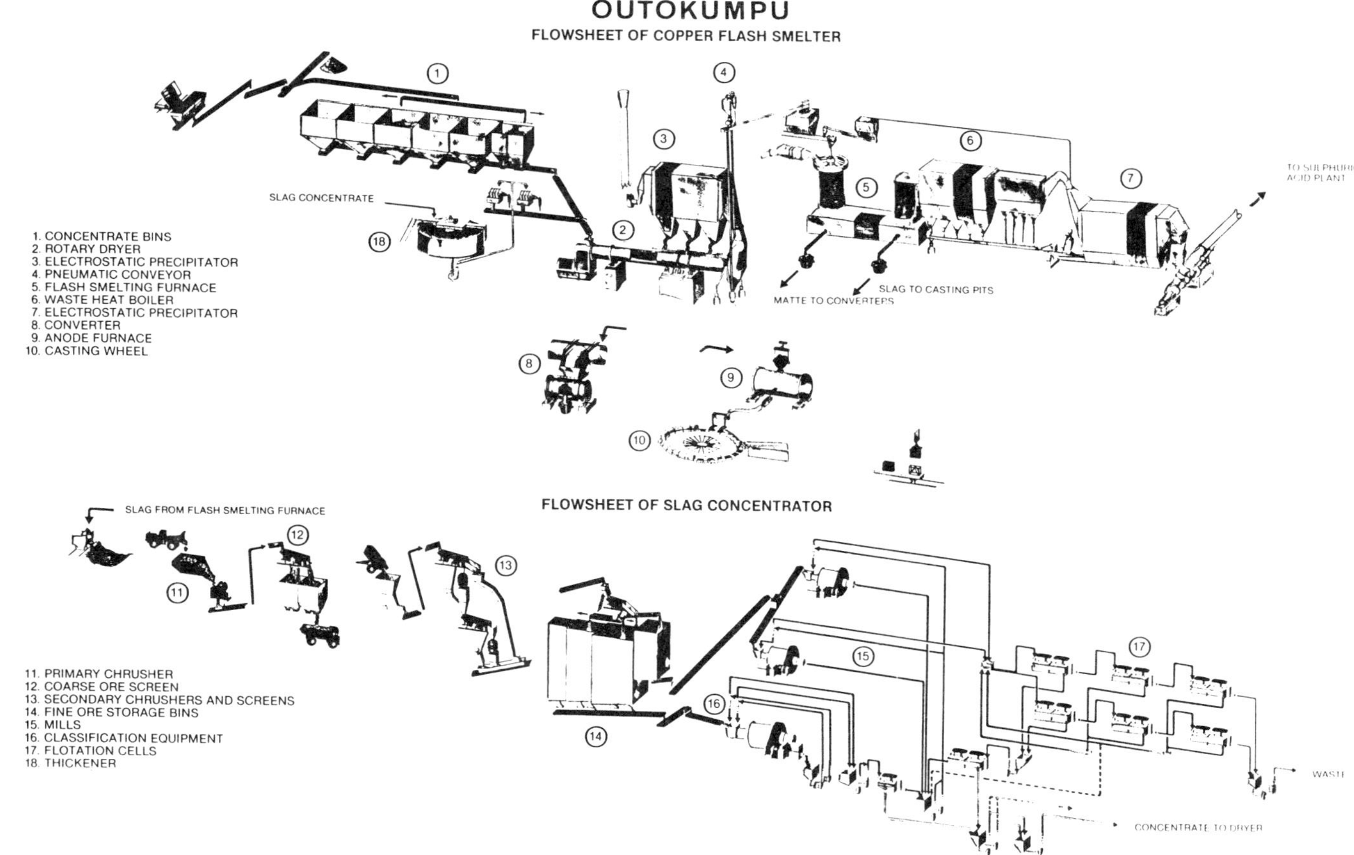
OUTOKUMPU
FLOWSHEET OF COPPER FLASH SMELTER
SLAG CONCENTRATE
TO SULPHURIC ACID PLANT
SLAG TO CASTING PITS
MATTE TO CONVERTERS
1. CONCENTRATE BINS
2. ROTARY DRYER
3. ELECTROSTATIC PRECIPITATOR
4. PNEUMATIC CONVEYOR
5. FLASH SMELTING FURNACE
6. WASTE HEAT BOILER
7. ELECTROSTATIC PRECIPITATOR
8. CONVERTER
9. ANODE FURNACE
10. CASTING WHEEL
FLOWSHEET OF SLAG CONCENTRATOR
SLAG FROM FLASH SMELTING FURNACE
CONCENTRATE TO DRYER
11. PRIMARY CHRUSHER
12. COARSE ORE SCREEN
13. SECONDARY CHRUSHERS AND SCREENS
14. FINE ORE STORAGE BINS
15. MILLS
16. CLASSIFICATION EQUIPMENT
17. FLOTATION CELLS
18. THICKENER

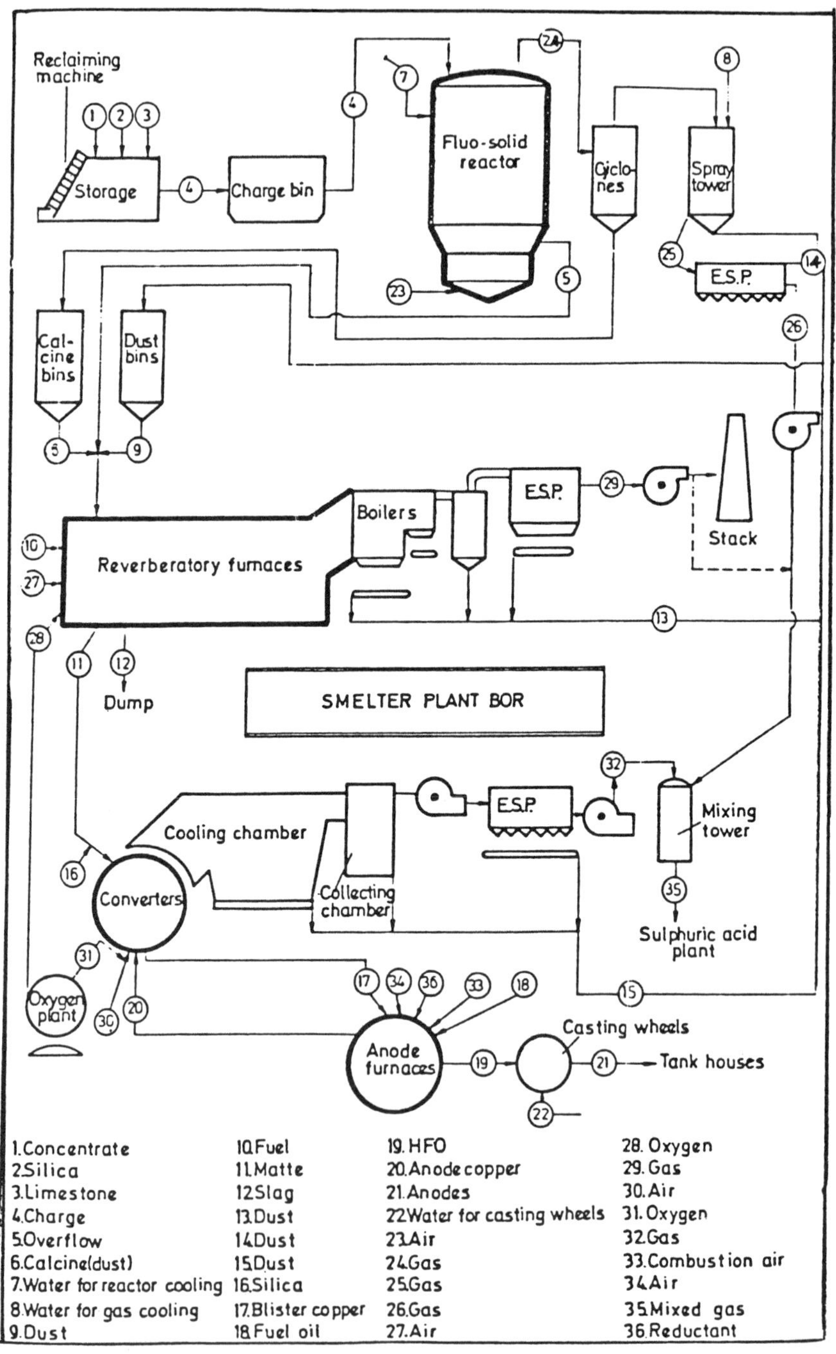
Reclaiming machine
Storage
Charge bin
Fluo-solid reactor
Ciclones
Spray tower
E.S.P.
Cal-cine bins
Dust bins
Reverberatory furnaces
Boilers
E.S.P.
Stack
Dump
SMELTER PLANT BOR
Cooling chamber
Collecting chamber
E.S.P.
Mixing tower
Sulphuric acid plant
Converters
Oxygen plant
Anode furnaces
Casting wheels
Tank houses
1.Concentrate
2.Silica
3.Limestone
4.Charge
5.Overflow
6.Calcine(dust)
7.Water for reactor cooling
8.Water for gas cooling
9.Dust
10.Fuel
11.Matte
12.Slag
13.Dust
14.Dust
15.Dust
16.Silica
17.Blister copper
18.Fuel oil
19.HFO
20.Anode copper
21.Anodes
22.Water for casting wheels
23.Air
24.Gas
25.Gas
26.Gas
27.Air
28.Oxygen
29.Gas
30.Air
31.Oxygen
32.Gas
33.Combustion air
34.Air
35.Mixed gas
36.Reductant

GENERAL FLOW DIAGRAM FOR GLOGOW I

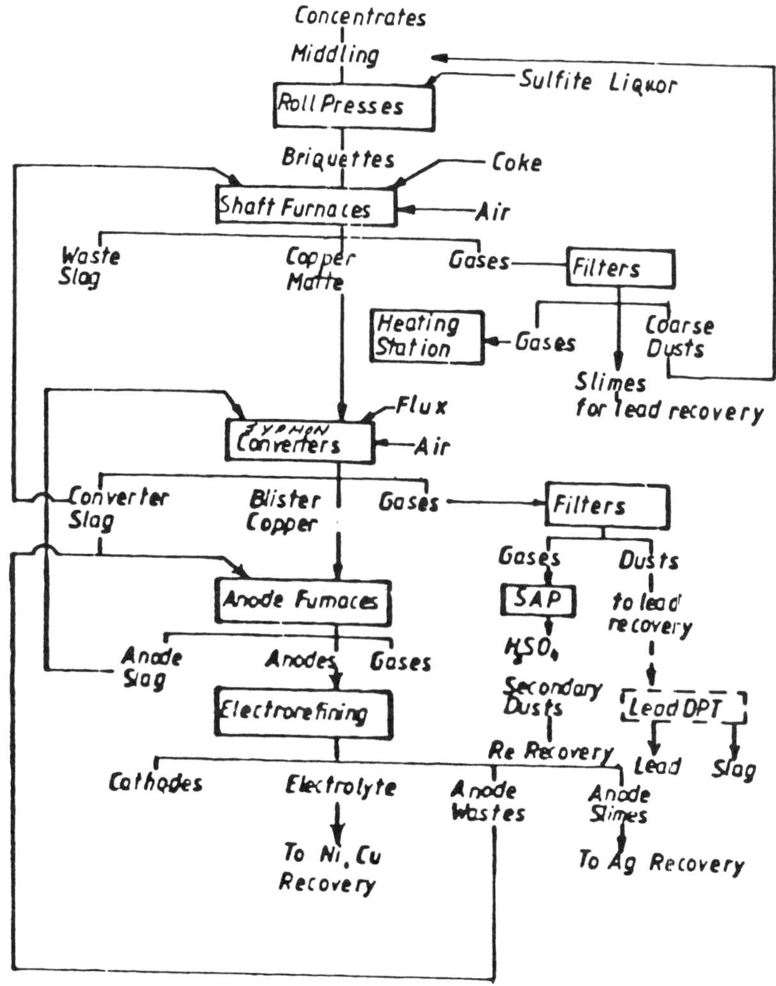

Reference:

"Modern Copper Smelting and Refining Practice in Poland", IMM-CSM Conference on Mineral Processing and Extractive Metallurgy, Kunming, People's Republic of China, October 1984.

GENERAL FLOW DIAGRAM FOR GLOGOW II

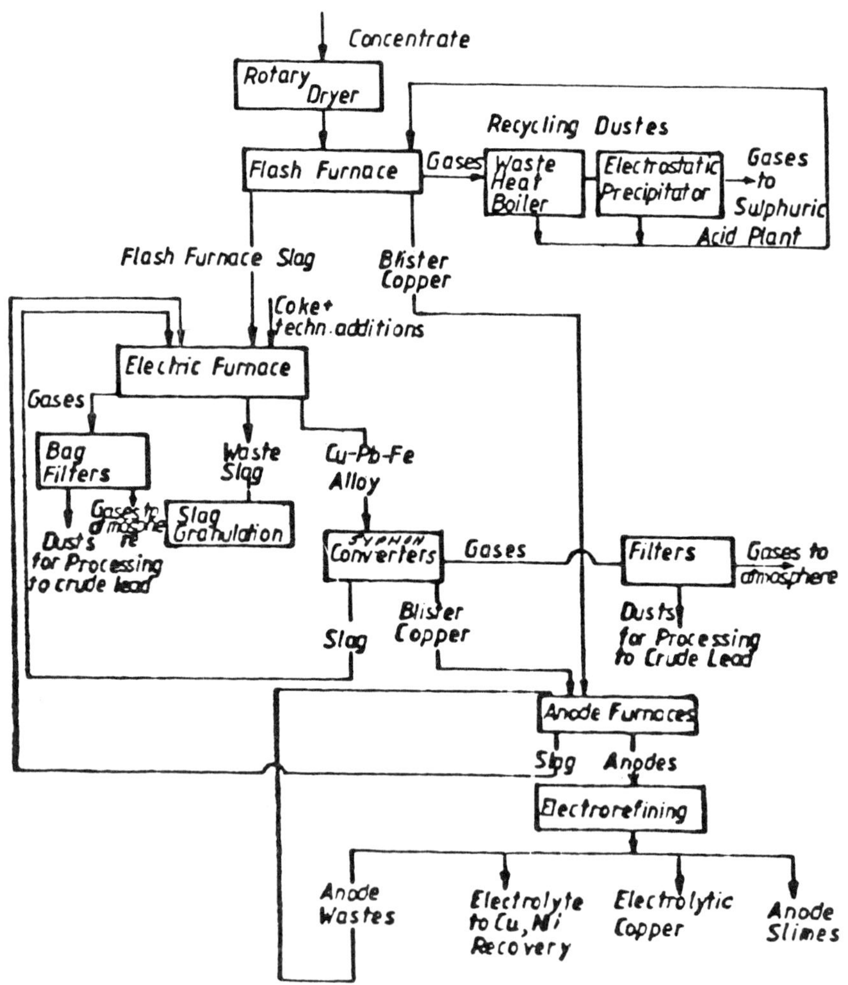

Reference:

"Modern Copper Smelting and Refining Practice in Poland", IMM-CSM Conference on Mineral Processing and Extractive Metallurgy, Kunming, People's Republic of China, October 1984.

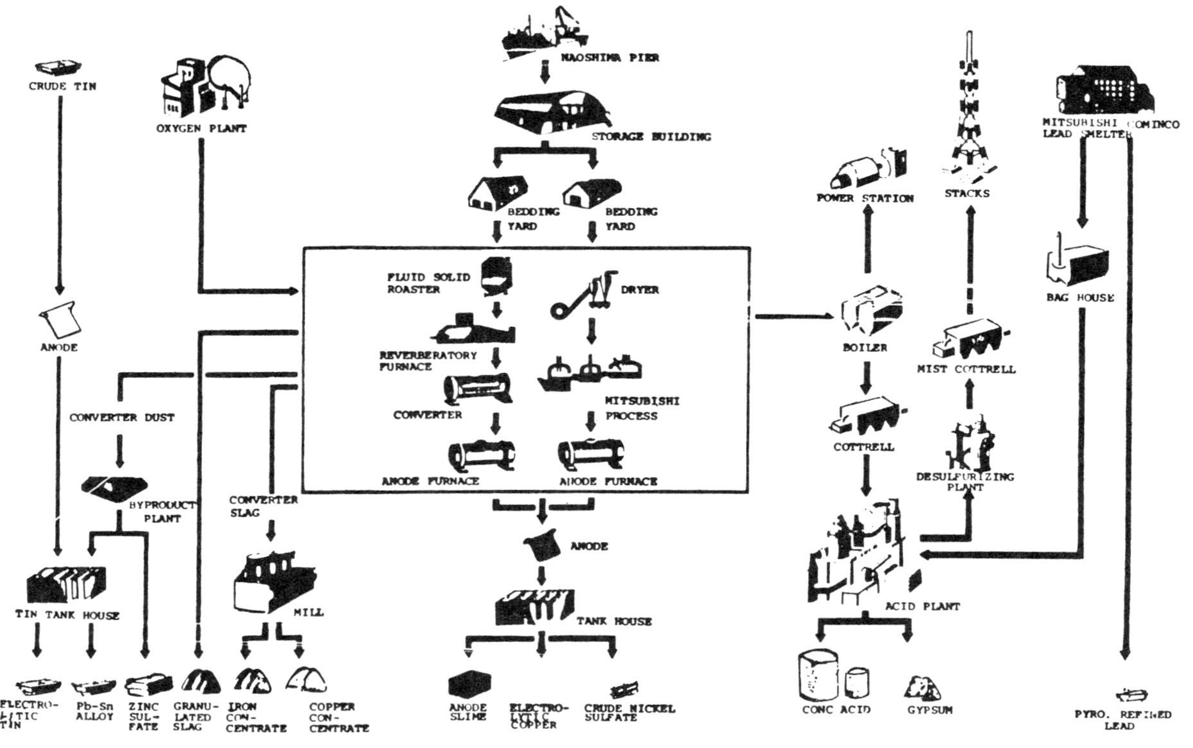

General Flow Diagram of Naoshima Smelter

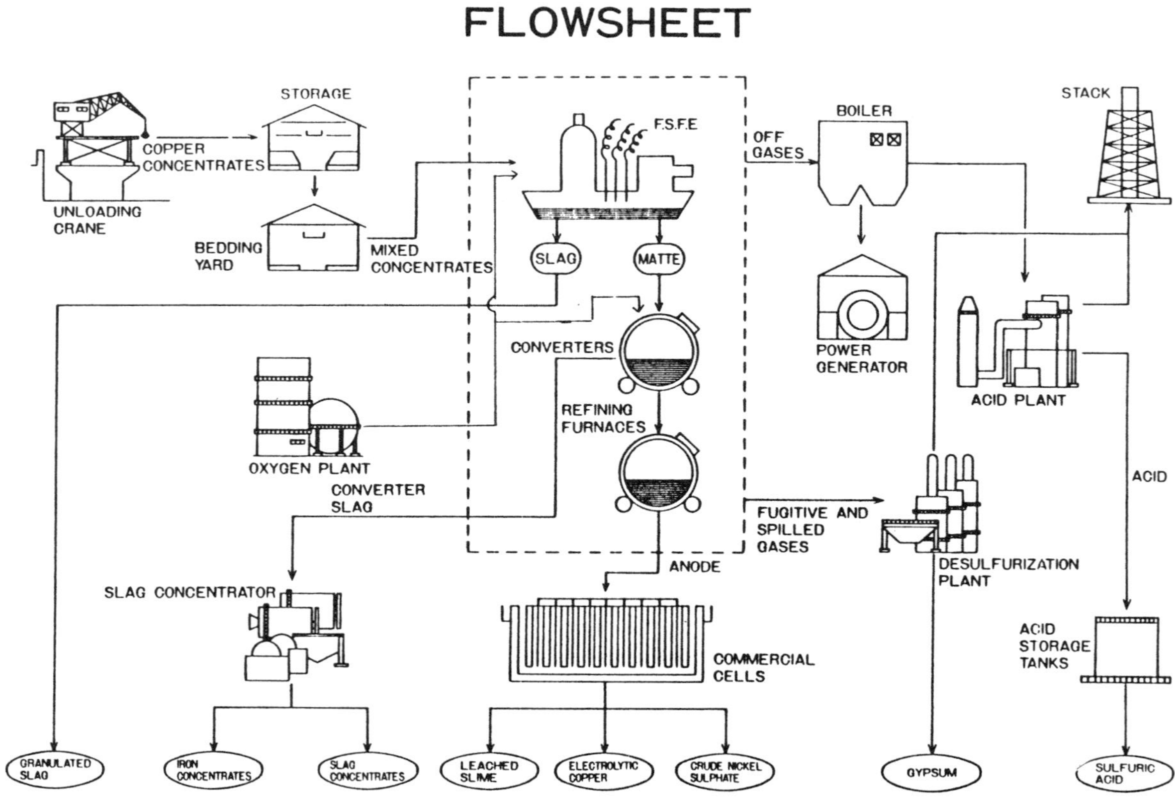

HIBI KYODO SMELTING CO. LTD. - TAMANO SMELTER

Flowsheet Nippon Mining Co. LTD. SAGANOSEKI

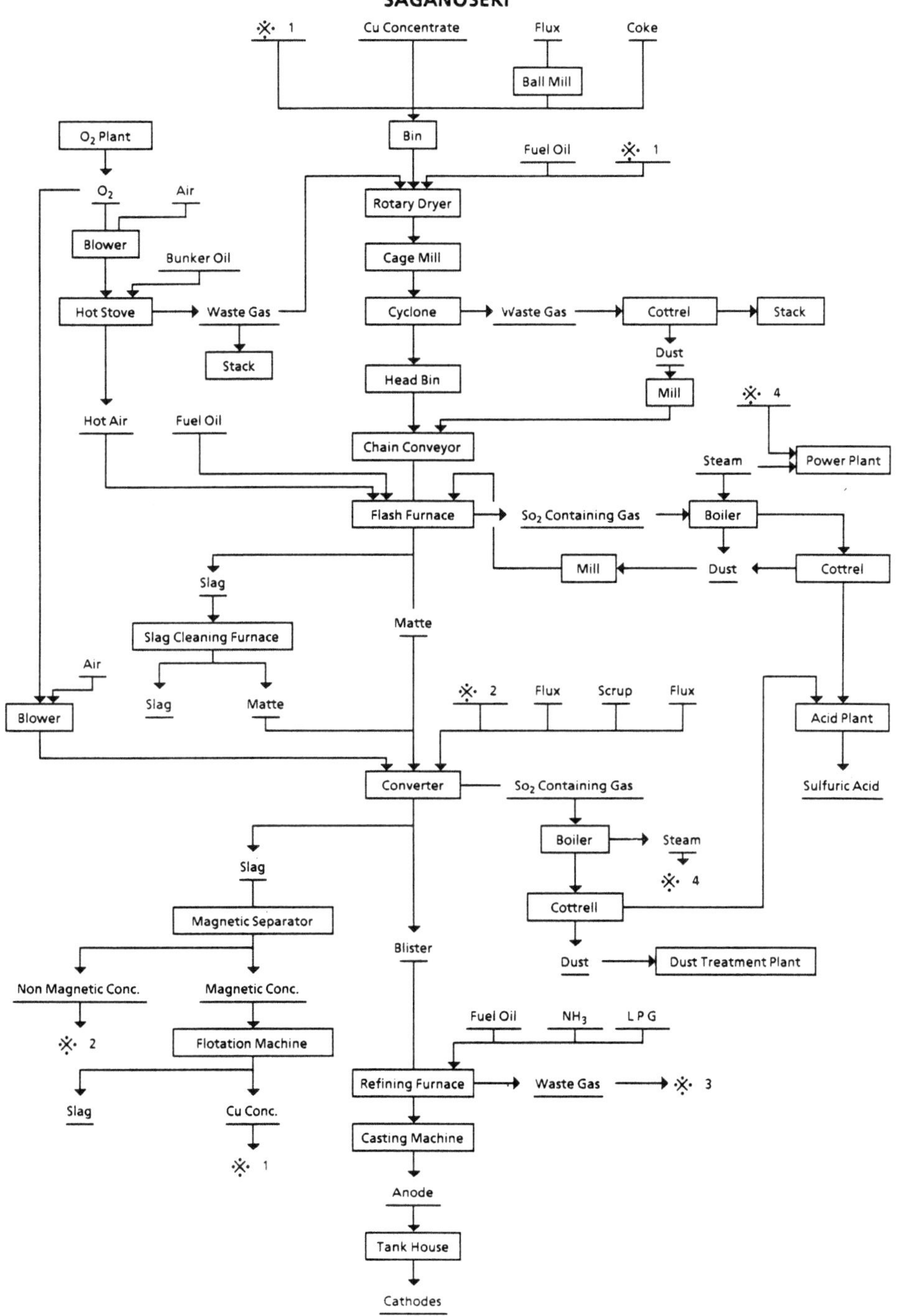

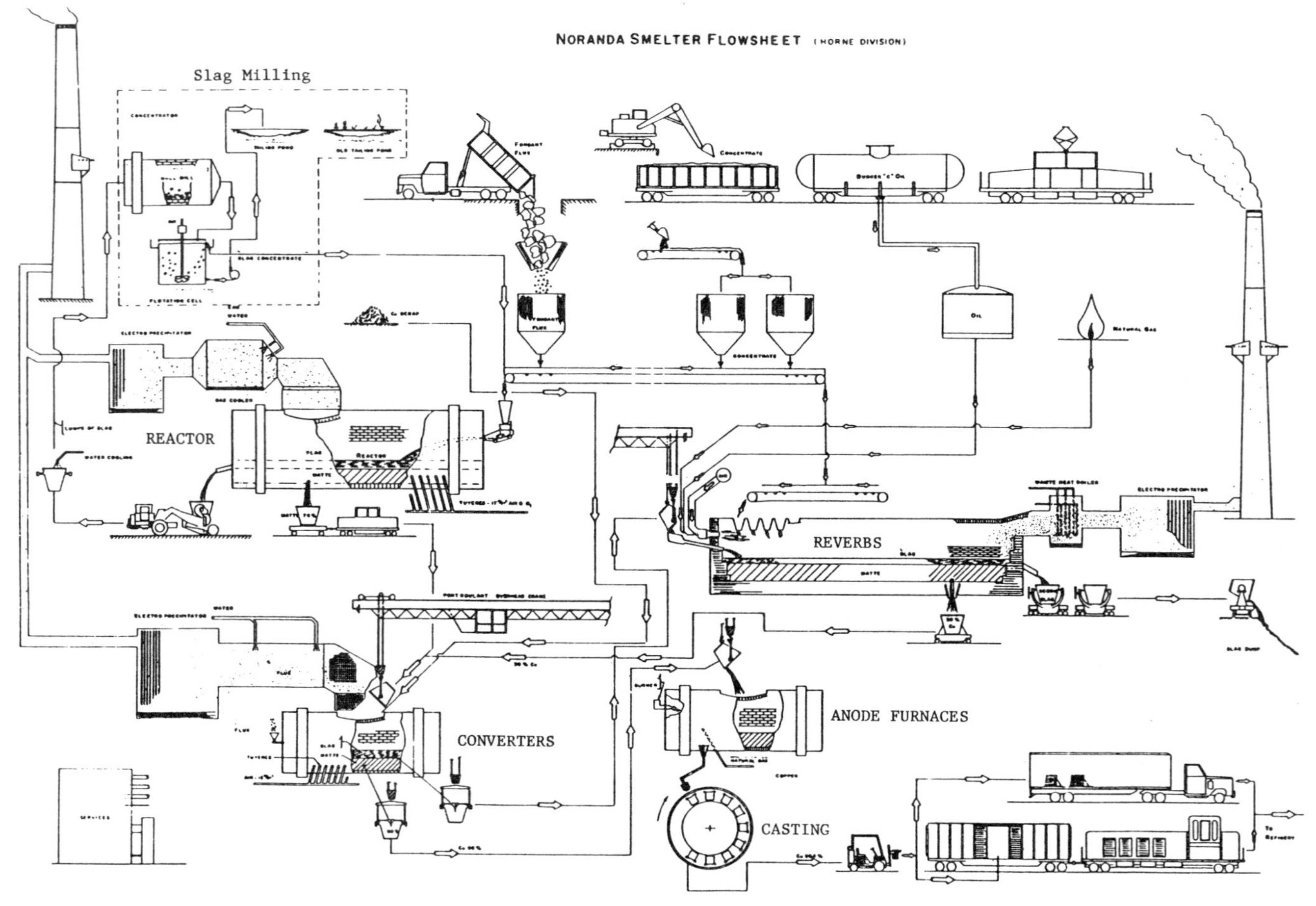
NORANDA SMELTER FLOWSHEET (HORNE DIVISION)
Slag Milling
REACTOR
REVERBS
CONVERTERS
ANODE FURNACES
CASTING

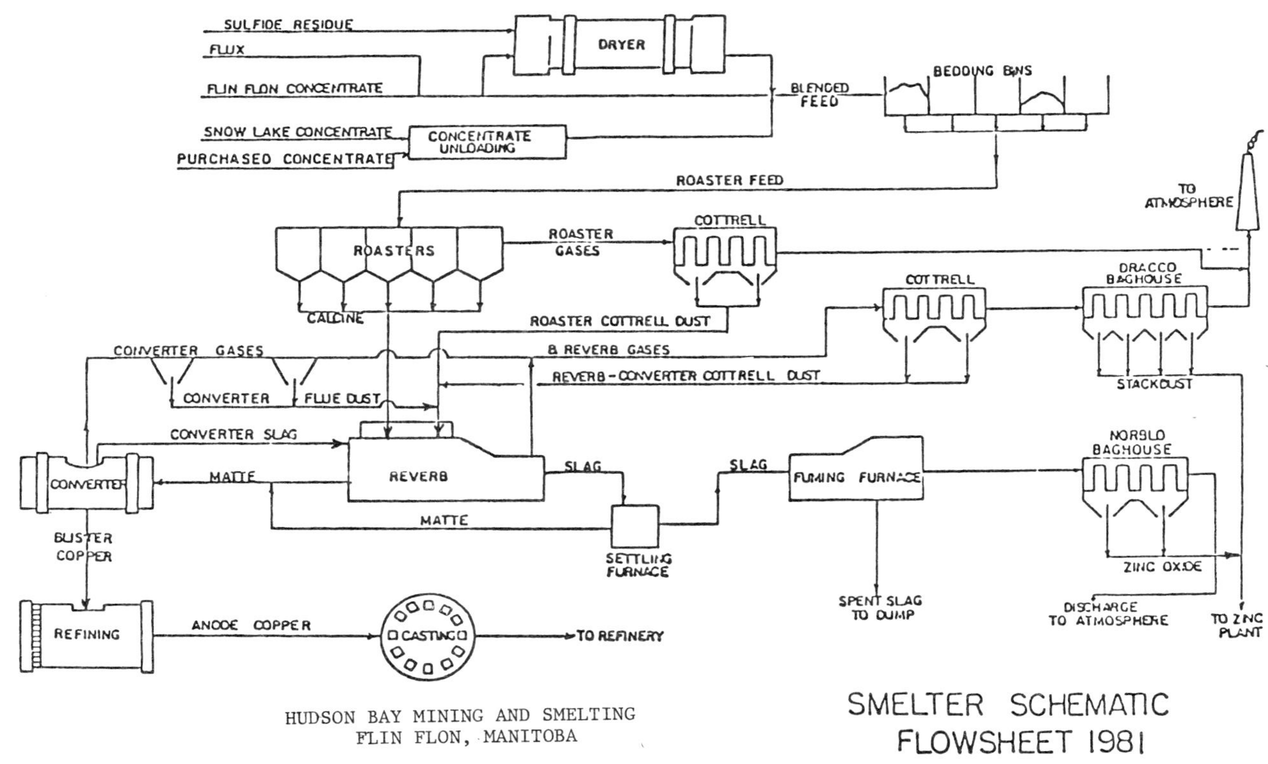

SULFIDE RESIDUE
FLUX
FLIN FLON CONCENTRATE
SNOW LAKE CONCENTRATE
PURCHASED CONCENTRATE
DRYER
CONCENTRATE UNLOADING
BLENDED FEED
BEDDING BINS
ROASTER FEED
TO ATMOSPHERE
ROASTERS
ROASTER GASES
COTTRELL
CALCINE
COTTRELL
DRACCO BAGHOUSE
ROASTER COTTRELL DUST
CONVERTER GASES
B REVERB GASES
REVERB-CONVERTER COTTRELL DUST
CONVERTER
FLUE DUST
STACKDUST
CONVERTER SLAG
NORBLO BAGHOUSE
CONVERTER
MATTE
REVERB
SLAG
SLAG
FUMING FURNACE
MATTE
SETTLING FURNACE
BLISTER COPPER
ZINC OXIDE
SPENT SLAG TO DUMP
DISCHARGE TO ATMOSPHERE
TO ZINC PLANT
REFINING
ANODE COPPER
CASTING
TO REFINERY
HUDSON BAY MINING AND SMELTING
FLIN FLON, MANITOBA
SMELTER SCHEMATIC
FLOWSHEET 1981

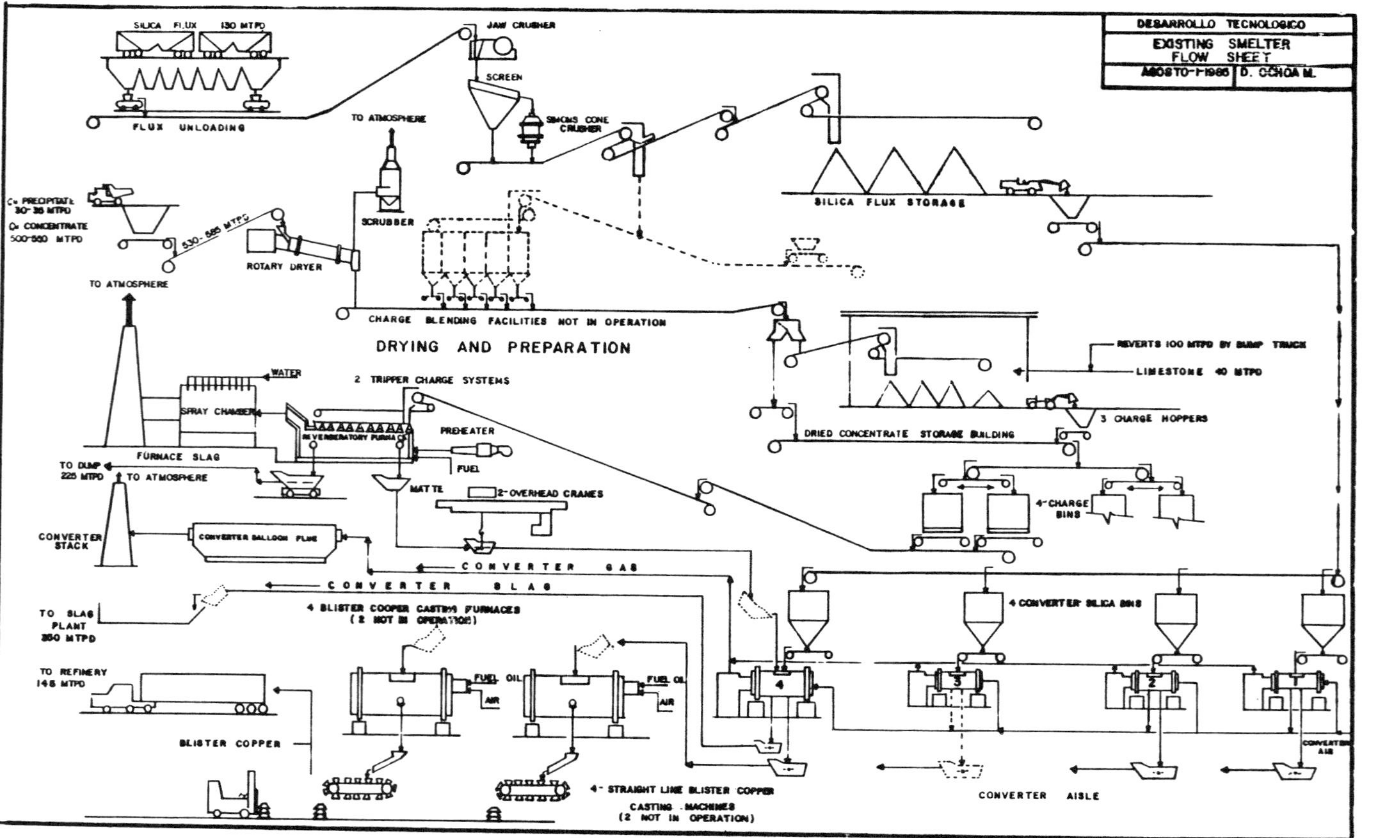

GENERAL FLOW DIAGRAM OF THE CANANEA SMELTER

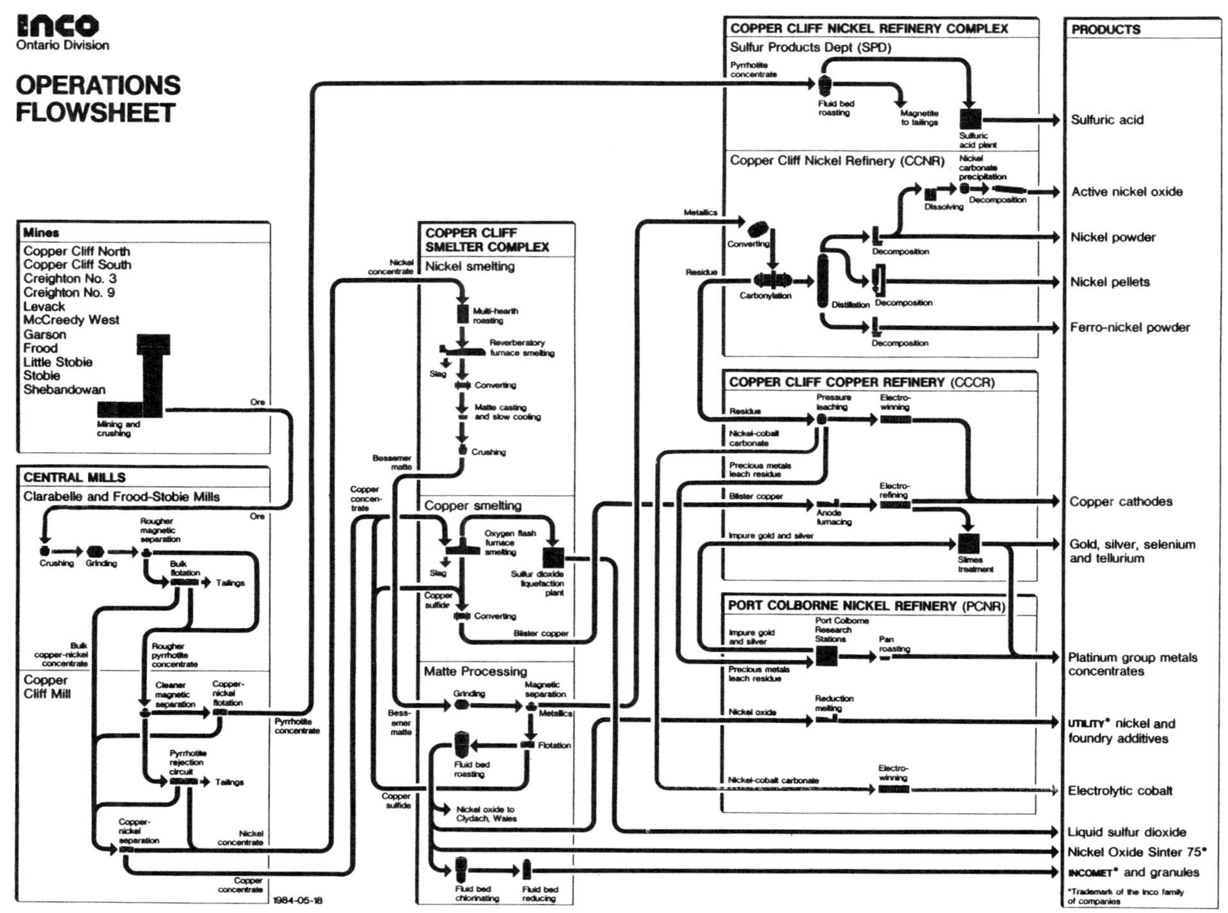

INCO
Ontario Division
OPERATIONS FLOWSHEET
Mines
Copper Cliff North
Copper Cliff South
Creighton No. 3
Creighton No. 9
Levack
McCreedy West
Garson
Frood
Little Stobie
Stobie
Shebandowan
Mining and crushing
Ore
CENTRAL MILLS
Clarabelle and Frood-Stobie Mills
Crushing
Grinding
Rougher magnetic separation
Bulk flotation
Tailings
Bulk copper-nickel concentrate
Copper Cliff Mill
Rougher pyrrhotite concentrate
Cleaner magnetic separation
Copper-nickel flotation
Pyrrhotite concentrate
Pyrrhotite rejection circuit
Copper-nickel separation
Nickel concentrate
Copper concentrate
1984-05-18
COPPER CLIFF SMELTER COMPLEX
Nickel smelting
Multi-hearth roasting
Reverberatory furnace smelting
Slag
Converting
Matte casting and slow cooling
Crushing
Bessemer matte
Copper concen-trate
Copper smelting
Oxygen flash furnace smelting
Sulfur dioxide liquefaction plant
Copper sulfide
Blister copper
Matte Processing
Grinding
Magnetic separation
Metallics
Flotation
Bess-emer matte
Fluid bed roasting
Nickel oxide to Clydach, Wales
Fluid bed chlorinating
Fluid bed reducing
COPPER CLIFF NICKEL REFINERY COMPLEX
Sulfur Products Dept (SPD)
Pyrrhotite concentrate
Fluid bed roasting
Magnetite to tailings
Sulfuric acid plant
Copper Cliff Nickel Refinery (CCNR)
Nickel carbonate precipitation
Dissolving
Decomposition
Metallics
Converting
Residue
Carbonylation
Distillation
COPPER CLIFF COPPER REFINERY (CCCR)
Residue
Pressure leaching
Electro-winning
Nickel-cobalt carbonate
Precious metals leach residue
Blister copper
Anode furnacing
Electro-refining
Impure gold and silver
Slimes treatment
PORT COLBORNE NICKEL REFINERY (PCNR)
Impure gold and silver
Port Colborne Research Stations
Pan roasting
Precious metals leach residue
Nickel oxide
Reduction melting
Nickel-cobalt carbonate
Electro-winning
PRODUCTS
Sulfuric acid
Active nickel oxide
Nickel powder
Nickel pellets
Ferro-nickel powder
Copper cathodes
Gold, silver, selenium and tellurium
Platinum group metals concentrates
UTILITY* nickel and foundry additives
Electrolytic cobalt
Liquid sulfur dioxide
Nickel Oxide Sinter 75*
INCOMET* and granules
*Trademark of the Inco family of companies

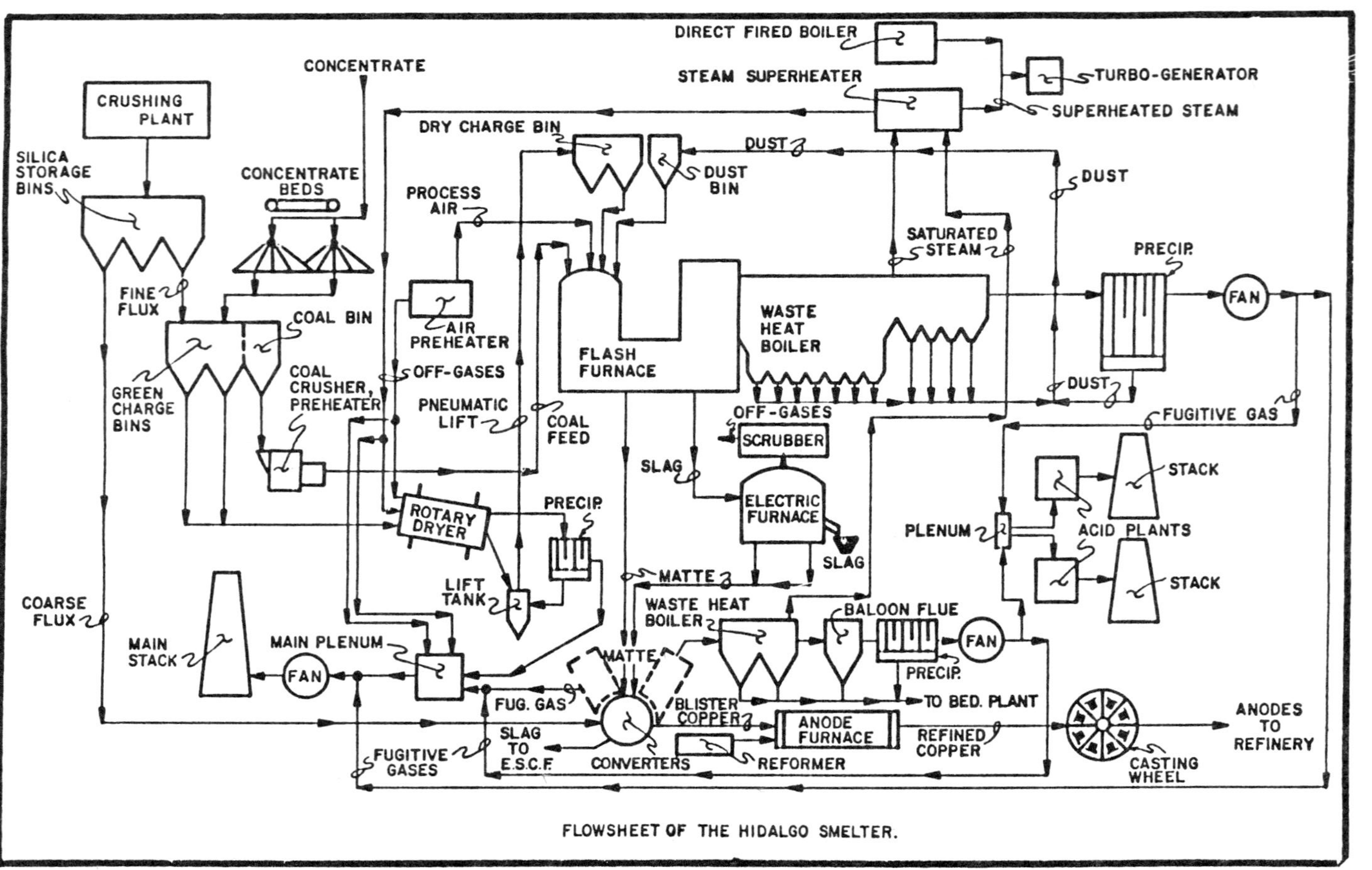

FLOWSHEET OF THE HIDALGO SMELTER.

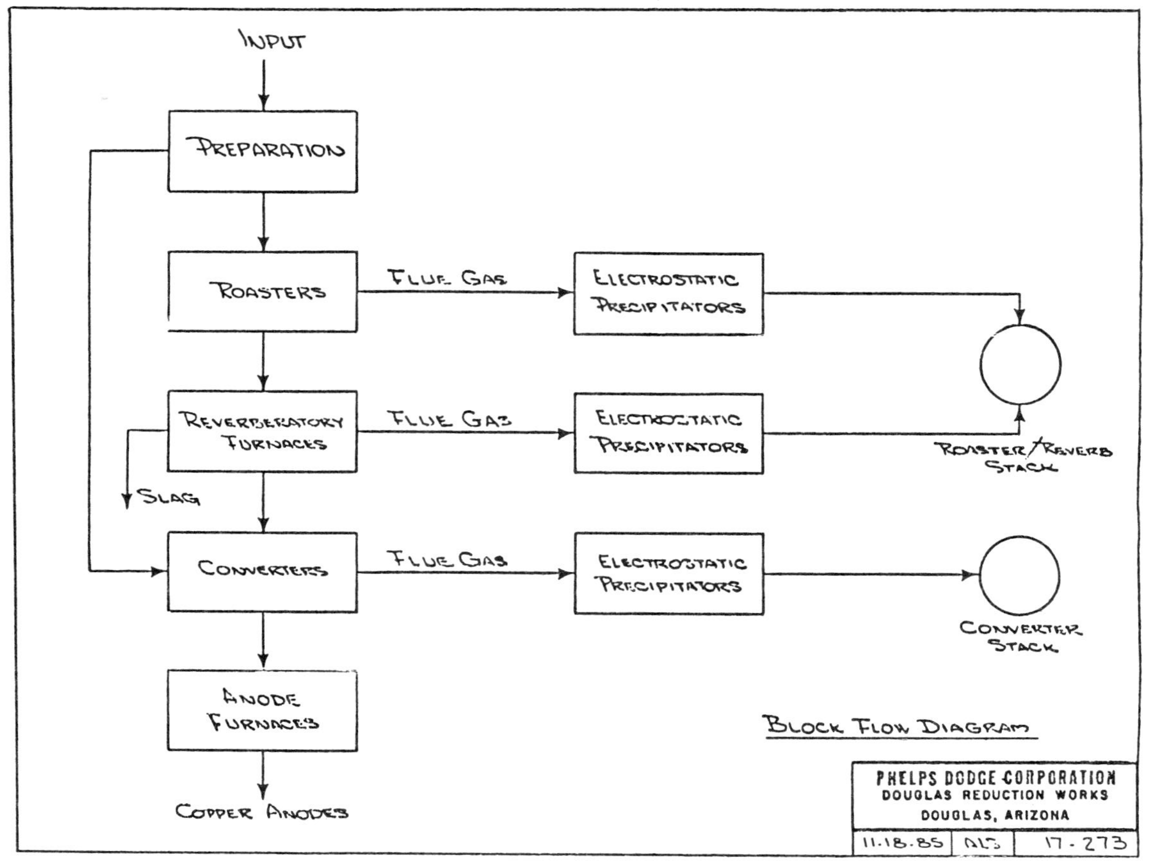

Input
Preparation
Roasters
Flue Gas
Electrostatic Precipitators
Reverberatory Furnaces
Flue Gas
Electrostatic Precipitators
Roaster/Reverb Stack
Slag
Converters
Flue Gas
Electrostatic Precipitators
Converter Stack
Anode Furnaces
Copper Anodes
Block Flow Diagram
PHELPS DODGE CORPORATION
DOUGLAS REDUCTION WORKS
DOUGLAS, ARIZONA
11.18.85
17-273

CHINO MINES COMPANY - COPPER FLASH SMELTER

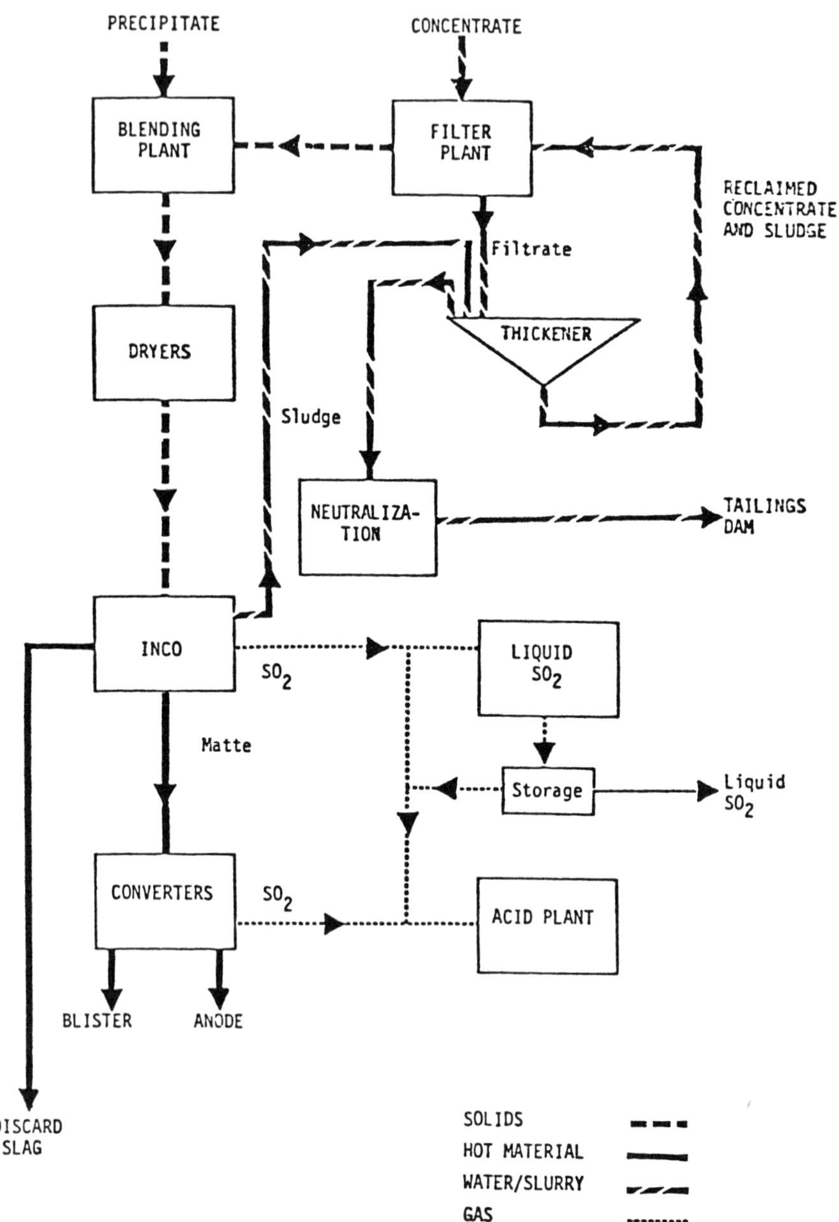

Nickel Smelters

Nickel Smelters - Survey Review

C. A. Landolt
Superintendent, Smelter Technical Services
Copper Cliff Smelter Complex
Inco Limited
Copper Cliff, Ontario, Canada
POM 1NO

John C. Taylor
Consulting Engineer
Jan H. Reimers and Associates Inc.
Oakville, Ontario, Canada
L6J 1H7

Introduction

All the major nickel producers of the Western World are included in TMS Pyrometallurgical Committee's survey on nickel smelters. Data for the main plant operations are given in the tables and supplemented by selected papers.

Primary nickel is extracted from two radically different types of ore, and this has a significant bearing on the type of process which can be used. Sulphide deposits account for about 36% of land-based reserves in the Western World(1), while laterite deposits account for the balance of 64%.

Sulphide ores typically contain about 1% Ni and can be upgraded to 6 to 12% Ni using conventional mineral dressing techniques, while laterites are only amenable to minor upgrading by screening out the coarser low grade fraction. Metal values in sulphide nickel concentrates range from 6 to 15% Ni + Cu + Co and frequently include economic quantities of precious metals. In addition, they have a fuel value in the form of iron sulphides, equivalent to 500 to 1,000 MCal/MT of concentrate. Alternatively, laterites contain 1.6 to 3% Ni, occasionally small amounts of cobalt, and have no fuel value. Typical analyses of the feed to sulphide and laterite smelters are given in Table I. (2)

Energy requirements and sulphur abatement are key factors in the comparison of pyrometallurgical processing of sulphide concentrates and of laterites. Sulphide concentrates require lower energy inputs than laterite ores(3), due to their higher grades and fuel contents. In addition, laterite smelting requires higher temperatures in view of the presence of significant quantities of refractory oxides, particularly MgO. However, laterite operations are not faced with the problem of fixing sulphur in order to meet environmental regulations. The sulphur-to-nickel ratios for sulphide concentrates range from 2 to 5; consequently, limits on sulphur emissions have a greater impact on nickel smelters than on those producing copper, lead or zinc, where the sulphur-to-metal range is only 0.2 to 1.6(4). About 30% of the sulphide smelters responding to the survey have some sulphur fixation facilities.

TABLE I. NICKEL SMELTER FEED COMPOSITION (2)

SMELTER	Weight Percent							
	Ni	Cu	Co	Fe	S	SiO_2	MgO	Moisture
Sulphide								
Falconbridge	6.3	5.5	0.26	41	30.5	-	-	-
Inco-Sudbury	11	2.7	0.3	40.8	30.9	-	-	-
-Thompson	10.5	0.4	0.16	38	27	-	-	12.5
Western Mining	11.4	0.85	0.24	38	32	8.5	4.5	-
BCL-Botswana	2.77	3.16	0.15	43.5	31	9	-	27
Outokumpu	8-9	2-4	-	24-27	19-22	15-17	-	9-10
Bindura	10.6	3.4	0.27	25	20	21.3	-	13-15
Laterite								
Falconbridge-Dominicana	1.6-1.9	-	-	15.6-18.3	<0.03	33-37	-	27
PT INCO	1.97	-	0.07	20.6	-	32.6	17.6	32
Pacific Metals	2.4	-	0.06-0.09	13-15.5	-	36	23.5-25.9	28
PT Aneka Tambang	2.25(Ni+Co)	-	-	13	-	45	24	-
Cerro Matoso	2.9	-	-	14	-	46	15	-
SLN-Doniambo	2.5-2.9	-	0.05 -0.10	14-15	-	35-40	20-28	20-30

OVERVIEW OF NICKEL ORE PROCESSING
SULPHIDE
LATERITE
Ore
MINING
NORMALLY UNDERGROUND
CONCENTRATION
Copper conc. to
Copper Smelter
SEPARATING
FLOTATION
ROASTING
ELECTRIC FURNACE,
REVERB. or FLASH
SMELTING
ACID or
AMMONIA LEACHING
SMELTING
CONVERTING
MATTE LEACHING
REFINING
OR
REDUCTION
REDUCTION
&
BRIQUETTING
REFINING
Copper to
Cu Circuit
Matte
FINAL PRODUCTS
CLASS I or II
Ore
SURFACE MINING
&
LIMITED BENIFICIATION
REDUCTION
ROASTING
ELECTRIC
SMELTING
AMMONIA
LEACHING
MATTE CONVERTING
LADLE REFINING
CARBONATE
PRECIPITATION
REDUCTION
&
BRIQUETTING
FERRONICKEL PRODUCTS
CLASS II
FINAL PRODUCT
CLASS I

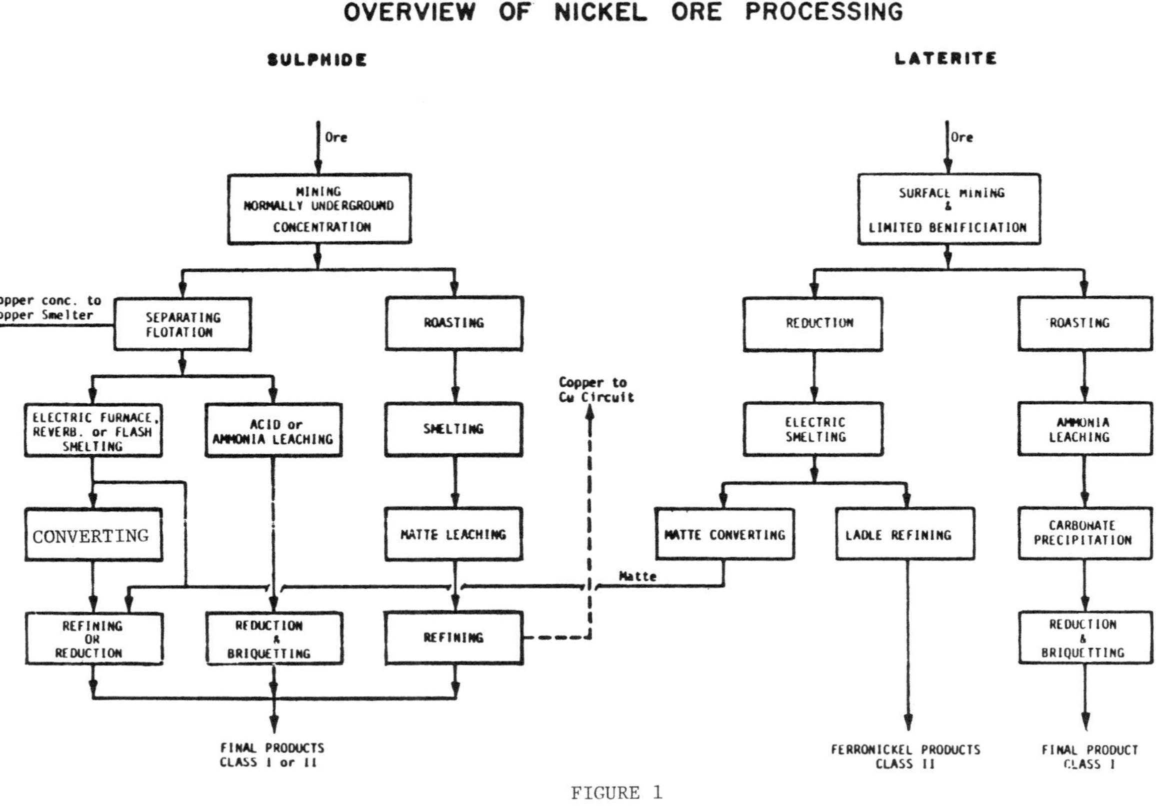

FIGURE 1

The processing of sulphide concentrates can include steps such as drying and/or roasting, smelting, and converting to a low iron sulphur-deficient matte. Further treatment follows a diversity of flowsheets, depending to some extent on the required final product. These can include mineral dressing, pyro, hydro or electrometallurgical steps.

The production of nickel from laterite ores typically includes drying and pre-reduction, followed by reduction smelting; the product is either a crude ferronickel or sulphur deficient (<10% S) mattes.

Alternative process routes for sulphides and laterites are presented schematically in Figure 1.

Smelter Operations

Feed Preparation

The mineralogy of most sulphide deposits is far more homogeneous than for laterites. Consequently, with grade control during mining and the blending which occurs during the production and transport of concentrate, the composition of the feed to most sulphide smelters is relatively constant in composition.

In comparison, laterite nickel deposits have been formed as a result of extensive weathering, in tropical climates, with the result that they exhibit a stratified structure with great variations both vertically and horizontally. Selection of the metallurgical process for these ores requires a thorough knowledge of the chemical and physical characteristics of the mineral deposits. Mining procedures are usually complex and designed to mix the various sections of the orebody in order to provide a relatively uniform grade of feed to the metallurgical plant. In addition to selective mining, some operations also include pre-screening to remove rock boulders which are remnants of the original geological formation and have a lower nickel content. In any laterite operation, the tonnage of ore which must be handled through the initial process steps, including smelting, is significant.

Pre-Treatment for Smelting

Feed preparation for sulphide smelting is relatively simple and typically includes drying, usually in rotary driers, or drying and roasting for control of furnace matte grade. The majority of sulphide concentrates contain less than 15% moisture, the exception being Botswana which receives a 27% H_2O concentrate and feeds a slurry to its spray drier. The rotary-flash drier combination, common in Japanese copper smelters, has not been adopted by the nickel industry. Fluid bed or multiple hearth roasters are used to produce a calcine containing from 12 to 15% S (including the flux). The fluid bed roasters at Falconbridge remove between 50% and 60% of the sulphur in feed as a high strength gas for the production of sulphuric acid. Inco's Thompson smelter employs similar roasters, but recent studies indicate that it is not economic to fix sulphur at a location so remote from established acid markets(5).

Laterite ores contain a substantial amount of water of hydration as well as free moisture. Drying is usually carried out in rotary driers and requires a substantial amount of fuel. However, even after drying, the product still contains up to 20% H_2O. Further processing may include a pre-reduction step, using a rotary kiln as at Cerro Matoso, or a shaft furnace as is the case at Falconbridge Dominicana. Inco's Soroako operation combines pre-reduction and sulphidization with addition of elemental sulphur in a kiln. Société Le Nickel at Doniambo adds gypsum to the hot sinter as it is charged to the blast furnace to produce a 10% S matte.

Smelting

Sulphide concentrates account for about 70% of the primary nickel production reported in the survey. Oxy-fuel reverberatory smelting at Inco's Copper Cliff Smelter is the largest single production at 26%, followed by electric smelting and Outokumpu flash smelting at about 22% each. The mattes produced by these operations range from 22 to 42% nickel.

Laterite ores account for the remaining 30% of the production reported; all smelters surveyed use electric furnaces, with the exception of Société Le Nickel's (SLN) blast furnaces at Doniambo, for smelting. Most of the nickel from laterites is in the form of ferronickel, containing between 18% and 43% Ni; the two exceptions are PT Inco Indonesia, which produces a 79% Ni and 19% S matte, and SLN which produces a 78% Ni and 20 to 22% S matte.

Reverberatory Smelting

At Inco's Copper Cliff Smelter, calcine from multiple hearth roasters is smelted in two reverberatory furnaces. These were originally of conventional design and fired with pulverized coal. They were then converted to oil firing and most recently have been equipped with roof mounted oxygen/natural gas burners(6,7). The matte grade produced is relatively low at about 23% Ni, 38% Fe. All converter slag is returned to the reverbs for cleaning; the furnace slag is discarded without further processing.

Flash Smelting

The Outokumpu flash smelting process is used in three of the nickel smelters in the survey and also at the Norilsk smelter in the U.S.S.R. The most recent project is the installation of a flash furnace (similar to Western Mining's) at Jinchuan, Gansu Province, in the People's Republic of China(4). Outokumpu's own installation at Harjavalta operates with 85 to 95% oxygen enrichment and produces a 34 to 38% Ni matte. The furnace slag is cleaned in a separate electric furnace(8) at all Outokumpu type installations. Inco has operated its oxygen flash furnace at Copper Cliff on a test basis using nickel concentrates(9); the matte produced during the test runs was considerably lower (at 20 to 32% Ni) than found in other nickel flash smelters; however, the slag could be discarded directly without a cleaning step. Both the Outokumpu and Inco processes produce a high strength SO_2 offgas suitable for sulphur fixation.

Electric Smelting

Electric smelting is used in two sulphide smelters and in most laterite operations.

Falconbridge smelts a 6.5% Ni, 12% S calcine from fluid bed roasters, in two 36 MVA furnaces; the matte contains 18% Ni and about 13% Cu. Inco (Thompson) smelts a 13.5% Ni, and 14% S calcine in 18 MVA or 30 MVA furnaces, to produce a 27% Ni matte. The furnace slag from both operations is low in metal values and is discarded without further treatment.

The production of ferronickel from oxide ores was developed by Elkem in the early 1950s for the Doniambo smelter in New Caledonia(10). Since then, this technology, including pre-reduction in rotary kilns, has been adopted for a number of laterite smelters producing ferronickel. Inco developed a different approach for its laterite projects in Indonesia and Guatemala (the latter mothballed early in 1982). The Inco process includes pre-reduction and sulphidization in a rotary kiln prior to electric furnace smelting; the furnace matte contains about 32% Ni and 10% S.

The presence of high quantities of MgO in the slags from laterite operations lead to operating temperatures in the range of 1,550 to 1,600°C, in marked contrast to the 1,230 to 1,250°C required for sulphide smelting. These high temperatures, combined with the corrosive nature of the laterite slags, can lead to severe refractory erosion; complex cooling systems are required to attain reasonable furnace campaigns.

Electric smelting of laterites to ferronickel or low sulphur matte requires highly reducing conditions; all the slags from these operations can be directly discarded.

Converting

The Peirce-Smith converter is the standard unit in all the nickel sulphide smelters surveyed. The operation of these converters is similar to the "slag blow" of conventional copper converters. Furnace mattes assaying 22 to 42% Ni and 24 to 43% Fe are converted to low iron mattes, usually containing about 1% Fe. Siliceous flux is used in all smelters; the converter slag will assay in the range of 24 to 26% SiO_2 with about 25% Fe_3O_4. The nickel content ranges from 1 to 2% Ni before slag cleaning and disposal.

Top blown rotary converters (TBRC or Kaldo) were originally installed by PT Inco at its Soroako plant in Indonesia. These converters are very versatile and have the capability of producing a wide range of products, ranging from low sulphur "charge" nickel to the more conventional high sulphur-low iron mattes. However, the controls and drive mechanisms of these units are quite complex and require very specialized maintenance. In addition, the capacity of the TBRCs at Soroako was limited to 50 MT of product per cycle. In 1985, PT Inco installed one conventional 3.96 m Ø x 8.53 m Peirce-Smith converter which handles the total production of furnace matte at Soroako. The TBRCs are now kept on standby.

The Kaldo or TBRC is most suitable for specialized processing where its ability to switch from oxidizing to reducing conditions and/or auxiliary fuel is required. This is the case, for example, at Inco's Copper Cliff Nickel Refinery(11). Here, a 4.2 m diameter by 6.7 m TBRC is used to convert smelter intermediate products to a form which is suitable for pressure carbonyl refining. During this operation, the converter removes sufficient sulphur to maximize the extraction of nickel and slag off the refractory oxides which would interfere with subsequent processing of the carbonyl residue.

Typically, the cycle consists of an oxidation and a reducing step. The oxidation is carried out by top blowing with technical oxygen blown through a lance onto the surface of the bath while the converter rotates. The temperature during the blow will reach 1,580°C without the addition of supplementary fuel(11); reduction of the nickel oxides formed during the oxidation step is carried out using coke and natural gas. The hot metal is then poured into pre-heated teeming ladles and transferred to a granulation station, where high pressure water jets produce 0.2 to 2.0 mm granules for processing in the pressure carbonyl reactors.

Slag Cleaning

Most of the chemical slag losses in nickel smelting occur as dissolved metal oxides. This precludes the recovery of metal values by slow cooling of the slags followed by flotation, as commonly practiced in the copper industry. Cleaning of nickel slags in electric furnaces has been the standard since first initiated by Outokumpu in 1959. Data from various operations were compiled recently (12,13).

An interesting combination of flash smelting and slag cleaning technology was developed in recent years by Western Mining(13). The Outokumpu flash furnace at Kalgoorlie was altered to include slag cleaning as an integral part of the furnace settler. The modifications included the addition of an appendage to the settler and installation of the electrodes from the original slag cleaning furnace. With this design there is a transition from the oxidizing conditions under the reaction shaft to reducing conditions at the slag cleaning end. Some coke is added to the appendage to ensure that strongly reducing conditions are maintained in this area. Converter slag is returned to the appendage, and experience to date indicates that there is no increase in the nickel build-up on the hearth.

Nickel Recovery

The TMS survey did not request data on overall smelter recovery; however, data from the 1986 CIM compilation(2) are included in Table II for reference. In general, the average overall recovery from laterite smelters is lower than in sulphide operations, in spite of the highly reducing conditions which prevail during the production of ferronickel or of sulphur deficient mattes. This is due to the lower grade of laterites compared to sulphide concentrates; this requires the processing of high tonnages, higher handling losses and very high slag-to-metal ratios.

TABLE II. NICKEL SMELTERS - AVERAGE RECOVERY (2)

SMELTER	PERCENT
Sulphide	
BSR Limited, Bindura	93
BCL Limited, Botswana	91.5
Outokumpu Oy, Harjavalta	94-95
Western Mining, Kalgoorlie	96
Inco Limited, Copper Cliff	97
Inco Limited, Thompson	97.5
Laterite	
P.T. Aneka Tambang, Pomalaa	90-92
Pacific Metals, Hachinohe	98.5
P.T. Inco, Soroako	90
Cerro Matoso, Montelibano	87.4

Further Processing

The ferronickel or matte from nickel smelters are further refined using a variety of processes.

Matte, whether granulated, in lumps or as anodes, is refined using diverse flowsheets, including mineral dressing, pyro, hydro, vapo or electrometallurgical processes. The refining of nickel matte permits the production of a variety of final nickel products with which to meet market requirements for Class I (high purity, over 99.5% Ni), or Class II nickel (normally 18 to 96% Ni).

Crude ferronickel usually requires de-sulphurization, typically carried out with the addition of soda ash in a stirring ladle, to reduce the sulphur content from the 0.3% range to less than 0.03% S. The carbon, phosphorous and silicon content are then reduced, using a shaking furnace or Thomas furnace, by lancing with oxygen enriched air. Fluxes, such as fluorspar, limestone, etc., are added to produce a fluid slag which can be discarded. The final ferronickel, a Class II product, is then cast into pigs or shot as required by the market.

Conclusion

While the nickel smelters surveyed in general employ modern and efficient processes, further improvements can be expected as efforts are made to reduce costs. In the case of laterite smelters, reduced energy consumption will be the main target. For sulphide smelters, sulphur fixation will be a requirement, particularly for Canadian producers.

It is to be expected, then, that the next smelter survey will present operating data considerably different from that included in the present tables.

References

1. D.L. Buchanan, "Nickel: A Commodity Review", IMM, 44 Portland Place, London W1N 4BR, 1982.

2. E. Ozberk, S.A. Gendron, G.H. Kaiura, "Review of Nickel Smelters - Responses to Questionnaire", 25th Annual Conference of Metallurgists, Toronto, August 1986.

3. H.J. Roorda, J.M.A. Hermans, "Energy Constraints in the Extraction of Nickel from Oxide Ores", Erzmetall 34, Nr. 2, 1981.

4. J.C. Taylor, "Non-Ferrous Pyrometallurgy - A Changing Industry", 89th Annual General Meeting of CIM, Toronto, May 1987.

5. Environmental Protection Service, Environment Canada, "Alternative Solutions for SO_2 Containment at the Thompson Smelter", unpublished report prepared by Jan H. Reimers and Associates Inc. under Contract No. 52SS.KE145-5-0472, March 1986.

6. J.A. Blanco, T.N. Antonioni, C.A. Landolt, G.J. Danyliw, "Oxy-Fuel Smelting in Reverberatory Furnaces at Inco's Copper Cliff Smelter", 50th Congress of the Chilean Institute of Mining & Metallurgical Engineers, Santiago, November 1980.

7. J.A. Blanco, T.N. Antonioni, C.A. Landolt, C.M. Mitchell, "Productivity Improvements at Inco's Copper Cliff Smelter", TMS-AIME, New Orleans, March 1986.

8. J. Asteljoki, J. Sulanto, T.T. Talonen, "Outokumpu Flash Smelting Method and its Application for Nickel and Lead Production", Mineral Processing and Extractive Metallurgy, M. J. Jones and P. Gill, eds., IMM Conference, Kunming, PRC, November 1984.

9. M.Y. Solar, R.J. Neal, T.N. Antonioni, M.C. Bell, "Smelting Nickel Concentrates in Inco's Oxygen Flash Furnace", Journal of Metals, January 1979, pp 26-32.

10. A.A. Dor, H. Skretting, "The Production of Ferronickel by the Rotary Kiln/Electric Furnace Process", International Laterite Symposium, SME-AIME, New Orleans, February 1979.

11. W.J. Thoburn, P.M. Tyroler, "Optimization of TBRC Operation and Control at Inco's Copper Cliff Nickel Refinery", 18th Annual CIM Conference of Metallurgists, Sudbury, August 1979.

12. J.M. Floyd, P.J. Mackey, "Developments in the Pyrometallurgical Treatment of Slag: A Review of Current Technology and Physical Chemistry, op cit (1), 1981.

13. C.W. Hastie, Dr. T. Hall, C.A. Hohnen, J.M. Limerick, "Kalgoorlie Nickel Smelter: Integration of Flash Smelting and Slag Cleaning Within One Process Unit", AIMM Extractive Metallurgy Symposium, Melbourne, November 1984.

NICKEL SMELTERS
of
AFRICA AND FINLAND

COMPANY NAME	BSR Limited	BCL Limited	Outokumpu Oy
PLANT	Bindura	Botswana	Harjavalta
Annual Production - Nickel MTPY	12,500	19,500	16,000
- Copper MTPY	5,900	21,500	7,000
Flowsheet Page No.	-	141	300
FEED ANALYSIS			
Concentrate			
% Ni	10.5 (3.4% Cu)	5.87 (Cu + Ni)	8-9
% Fe	26.0	43.5	24-27
% S	20.0	31.0	19-22
% SiO_2	21.3	9.0	15-17
Mesh	72% -200	70% -200	60-70% -270
% H_2O	13-15	27	9-10
Furnace Flux			
% SiO_2	90	82	86-89
% CaO	-	0.7	0.5-1.0
% Fe	-	0.5	1.0-1.5
Mesh	-3 mm	-3 mm	25-90% -50
Converter Flux			
% SiO_2	98	82	86-89
% CaO	-	0.7	0.5-1.0
% Fe	-	0.5	1.0-1.5
Mesh	-20 mm	+3 -35 mm	5-15% -50
FEED PREPARATION			
Blending System	Crane & Front End Loader	None	Proportioned to Belt Conveyor

PLANT - continued	Bindura	Botswana	Harjavalta
Dryer - Type	Rotary Drum	Niro Spray	Rotary
- Number	1	2	1
- Dimensions - m	3 Ø x 22	10 Ø	2.2 Ø x 24
Average Feed Rate - WMTPH	20-25	114	30-35
Feed Moisture - % H_2O	13-15	-	9-10
- % Solids	-	73	-
Fuel Used - Type	Coal	Coal	Heavy Oil
Calorific Value - kCal/kg	7,200	5,600	9,600
Retention Time - min.	-	-	30-35
Inlet Gas Temp. - °C	750	850-1,000	800-900
Outlet Gas Temp. - °C	132	135	140-150
Product Moisture - % H_2O	<4	0	0.1-0.3
Exhaust Gas Volume - Nm^3/hr	-	52,000	20,000-30,000
Cleaned by	Cyclones (6) & Scrubber	Electrostatic Precipitators (2)	Electrostatic Precipitator
SMELTING			
Furnace Type	15 MW Electric Fce.	Flash (OKO)	Flash (OKO)
Number of Units	1	1	1
Nominal Capacity - MTPH Concentrate	28	120	40
Dimensions - m	7.78 x 24.24 x 5.19	9 x 27	4.9 x 18.3 x 1.8
Oxygen Enrichment - % O_2	-	23	85-95
Auxiliary Fuel - Type	-	Coal	Heavy Oil
Calorific Value - kCal/kg	-	5,600	9,600
Operating Temperature - °C	-	1,350	1450-1550
Flux as % of Concentrate	7	18	7-10
Campaign Life - Years	9	6.6	4-6
Furnace Product - Type	Matte	Matte	Matte
Average Production - MTPD	190	425	120-150

PLANT - continued	Bindura	Botswana	Harjavalta
Analysis - % Ni	22	33	34-38
- % Cu	-	-	16-19
- % Fe	38	38	16-19
- % S	27	24	24.5-25.5
Temperature - °C	1,250	1,130	1,320 ±20°
Furnace Slag - MTPD	280	2,500	280-350
% Ni	0.33	1	1.0-1.5
% Fe	42	43	41-43
% CaO	1	2	1.0-1.5
% SiO_2	40	30	28-29
Temperature - °C	1,350	1,240	1,460 ±20°
Disposition	-	To Slag Cleaning	To Slag Cleaning
Offgas - Volume - Nm^3/hr	30,000	170,000	14,000-20,000
- Temp. - °C	200	1,460	1,400
- % SO_2	<2.0	8	20-30
Cooled by	-	Boiler	Boiler
Inlet Temp. - °C	-	1,460	1,400
Outlet Temp. - °C	-	350	280
Disposition of Dust	Recycled	Recycled	Recycled

CONVERTING

	Bindura		Botswana	Harjavalta	
Converter - Type	P-S		P-S	P-S	
Number - Hot	1		2	1	
- Standby	2		0	1	
Dimensions - m	2 @ 3.05 Ø x 7.0 1 @ 3.96 Ø x 7.32		3.96 Ø x 9.14	3.2 Ø x 5.8 & 3.6 Ø x 6.7	
Number of Tuyeres	32	36	48	25	32
Size - mm	51	51	51	-	-
Average Blowing Rate - Nm^3/hr	14,200	22,500	-	17,000-20,000	
Oxygen Enrichment - % O_2	None		None	21-28	
Air Rate - Nm^3/min.	-		580	280-300	
Scrap added as % of matte charged	5		0	2-5	
Flux as % of matte charged	24		26.5	13-15	

PLANT - continued	Bindura	Botswana	Harjavalta
Matte - Average MTPD	54	70	80-90
Analysis - % Ni	63	) 85	60-64
- % Cu	-	)	27-31
- % Fe	0.5	0.7	0.35-0.45
- % S	6	13.5	6.5-8.0
Slag - Analysis - % Ni	1.26	2.6	4-5
- % Fe	65	45	42-45
- % SiO_2	29.3	30	24-26
- Disposition	Returned to Smelting Furnace	Returned to Flash & Electric Fce.	To Slag Cleaning
SLAG CLEANING	None	Barnes Birlec Reduction Furnaces	
Slag Treated - Furnace MTPD		2,500	280-350
- Converter MTPD		50	90-110
- Total MTPD		2,550	370-460
Number of Electric Furnaces		2	1
KWH/MT Slag Charged		25	180-200
Retention Time - Hrs.		2 max.	2.5
Matte Grade - % Ni		35 Ni + Cu	47-52
Discard Slag - % Ni		0.6 Ni + Cu	0.15-0.20
- % Fe		45	42-44
- % SiO_2		29	31-32
% Ni Recovery (overall)		92.5 (Cu 85%)	97-98
PRODUCT CASTING			
Casting Machine - Type	-	Granulation	Granulation
Average Casting Rate - MTPH	-	-	60-120
Normal Casting Temp. - °C	-	-	1250-1300
Cast Product - Type	-	Granules	Granules
Rejects as % of new material cast	-	-	-
SULPHUR FIXATION	None	None	Carried out by a separate company

NICKEL SMELTERS
of
AUSTRALIA, JAPAN and SOUTH EAST ASIA

COMPANY NAME	Western Mining Corporation Ltd.	P.T. Aneka Tambang	Pacific Metals Co. Ltd.	Société Le Nickel
PLANT	Kalgoorlie	Pomalaa (1)	Hachinohe (1)	Doniambo (2)
Flowsheet Page No.	142	-	-	-
Annual Production - Nickel, MTPY	48,000	4,000	25,000	45,000
FEED ANALYSIS				
Concentrate	Sulphide	Laterite	Laterite	Laterite
% Ni	11.4	2.2	2.4	2.5-2.9
% Fe	38.0	12-14	13-15.5	-
% S	32.0	-	-	-
% SiO_2	8.5	44-46	36	35-40
% MgO	4.5	23-25	23.5-25.9	20-28
% H_2O	-	27-28	28	20-30
Furnace Flux				Limestone: 8 kg/kg Ni Gypsum: 3 kg/kg Ni
% SiO_2	55.9	-	-	
% CaO	0.8	53.0	-	
% Fe	10.7	-	-	-
Mesh	65%-200	-	-	-
Converter Flux				silica and sodium sulphate
% SiO_2	87.9	-	-	
% CaO	0.5	-	-	-
% Fe	2.3	-	-	-
Mesh	70%-28 micron	-	-	-

PLANT - continued	Kalgoorlie	Pomalaa	Hachinohe	Doniambo
FEED PREPARATION				
Blending System	Bucket from stockpile	Beds	-	-
Number of Beds		-	-	-
Bed Size - MT		1,000	-	-
Dryer - Type	-	-	Rotary & Impact Dryer	Sintering
- Number	-	-	1 / 2	-
- Dimensions - m	-	-	1.2 Ø x 13 / -	-
Average Feed Rate - WMTPH	-	-	65 / 100	-
Feed Moisture - % H_2O	-	-	28 / 28	10-12
PRETREATMENT				
Equipment - Type	-	Countercurrent Kiln	Rotary Kilns	Electric Fce Feed only: Rotary Kilns
- Number	-	1	3	-
Dimensions - m	-	4 Ø x 90	4.6 Ø, 4.7 Ø, 5.5 Ø	-
Nominal Capacity - MTPH	-	31	50 to 110	-
Air Rate - Nm^3/hr	-	-	-	-
Fuel Used - Type	-	Bunker C	-	Fuel Oil
Calorific Value	-	-	-	-
Discharge Temperature - °C	-	600	-	900
Offgas Volume - Nm^3/hr	-	30,000	-	-
- % SO_2	-	-	-	-
SMELTING				
Furnace Type	Flash (OKO)	20 MVA Elkem	2x40 & 1x60 MVA Fces.	Blast Furnaces, 11 & 30 MVA El.Fces
Number of Units	1	1	3	
Nominal Capacity		31	-	-
MTPH Concentrate	75	-	-	-
Dimensions - m	36 x 7	15 Ø x 5.6	-	-
Oxygen Enrichment - % O_2	up to 28	-	-	None
Auxiliary Fuel - Type	Oil & Coal	-	-	Coke (Blast Fce.)
Calorific Value	10,900 & 5,000 kCal/kg	-	-	-

PLANT - continued	Kalgoorlie	Pomalaa	Hachinohe	Doniambo	
Operating Temperature - °C	1280	1400	-	-	
Flux as % of Concentrate	27	-	-	-	
Campaign Life - years	5+	-	-	-	
Furnace Product - Type	Matte	FeNi	FeNi	Matte	Crude FeNi
Average Production - MTPD	36.0	-	-	-	-
Analysis - % Ni	42.2	-	-	22-27	22-25
- % Cu	-	-	-	Co 0.6-0.7	0.4-0.5
- % Fe	28.5	-	-	60-70	67-73
- % S	24.0	0.3	-	9-10	0.25
Temperature - °C	1150	1400	-	1,370	1500-1550
Furnace Slag - MTPD	1000	-	-	Blast Fce.	-
- % Ni	0.25	0.9	-	0.3	
- % Fe	44.6	6-9	-	-	
- % CaO	1.4	4-8	-	-	
- % SiO_2	31.5	50	-	-	
Temperature - °C	1300	-	-	-	
Disposition	Discard	Discard	-	Discard	
Offgas - Volume - Nm^3/hr	75,000	-	-	-	
- Temperature - °C	1,350	-	-	-	
- % SO_2	12	-	-	-	
- Cooled by	Boiler	-	-	-	
Inlet Temp. - °C	1,150	-	-	-	
Outlet Temp. - °C	400	-	-	-	
Disposition of Dust	Recycled or Bled	-	-	-	
CONVERTING					
Converter - Type	P.S.	-	-	Matte only: 2 P.S. operated in series	
Number - Hot	2	-	-		
- Standby	1	-	-	-	
Dimensions - m	3.66 Ø x 7.0	-	-	-	
Number of Tuyeres	28	-	-	-	
Size - mm	76.1	-	-	-	

PLANT - continued	Kalgoorlie	Pomalaa	Hachinohe	Doniambo
Average Blowing Rate - Nm^3/hr	-	-	-	-
Oxygen Enrichment - % O_2	None	-	-	1st stage only
Air Rate - $Nm^3/min.$	292	-	-	-
Scrap added as % of matte charged	16.7	-	-	-
Flux as % of matte charged	15.0	-	-	-
Matte - Average MTPD	-	-	-	1st Stage / 2nd Stage
- Analysis - % Ni	76.1	-	-	77.5 / 78
- % Cu	-	-	-	Co 2.5 / 0.4
- % Fe	1.03	-	-	2.0 / 0.1
- % S	20.4	-	-	18-20 / 20-22
Slag - Analysis - % Ni	1.95	-	-	-
- % Fe	53.5	-	-	-
- % SiO_2	23.8	-	-	-
- Disposition	To Furnace Appendage	Discard	Discard	Returned to BF & 1st Stage P.S.
SLAG CLEANING				
Slag Treated - Furnace MTPD	1000	-	-	-
- Converter MTPD	160	-	-	-
- Total MTPD	1160	-	-	-
Number of Electric Furnaces	Part of Flash	-	-	-
KWH/MT slag charged	180	-	-	-
Retention Time - Hrs.	6-7	-	-	-
Matte Grade - % Ni	Returns to Flash	-	-	-
Discard Slag - % Ni	0.25	-	-	-
- % Fe	44.6	-	-	-
- % SiO_2	31.5	-	-	-
% Ni Recovery	-	-	-	-
FIRE REFINING				Crude FeNi only:
Furnace - Type	-	Shaking Converter	2 Stirring Ladles 1 L.D. Converter	15 MT Thomas Fce.
- Number	-	1	-	
- Capacity, MT/charge	-	5	30	-
- Oxygen Rate, Nm^3/hr	-	400	-	0.07 to 0.08 m^3/MT FeNi

PLANT - continued	Kalgoorlie	Pomalaa	Hachinohe	Doniambo
PRODUCT CASTING				
Casting Machine - Type	Matte Granulation	In-line Caster	-	-
Average Casting Rate - MTPH	-	35	-	-
Normal Casting Temp. - °C	-	-	-	-
Cast Product - Type	-	Pig & Shot	-	Matte anodes, lump matte and FeNi
Dimensions - m	-	-	-	
Average weight - kg	-	-	-	
Rejects as % of new material cast	-	-	-	-
SULPHUR FIXATION	None	None	None	None

Reference (1) CIM 25th Annual Conference of Metallurgists - 1986 - Proceedings, Nickel Metallurgy Volume I, Editors E. Ozberk and S.W. Marcuson, page 311.

(2) R.J. Testut and P. Raffinot, Le Nickel, SME Mineral Processing Handbook, editor N. L. Weiss, pages 17-28 to 17-31.

LATERITE NICKEL SMELTERS of CENTRAL AMERICA AND SOUTH EAST ASIA

COMPANY NAME	PT International Nickel Indonesia	Falconbridge Dominicana C. por A.	Cerro Matoso S.A. Colombia
PLANT	Soroako	Santo Domingo	Montelibano (3)
Annual Production - MTPY	30,000 Nickel	28,750 Ferronickel	22,680 Ferronickel
Flowsheet Page Number	143	144	-
FEED ANALYSIS			
Concentrate	Laterite	Laterite	Laterite
% Ni	1.69-1.89	1.6-1.9	3.0
% Fe	15 -15.9	15.6-18.3	15.0
% S	0	0.013-0.03	-
% SiO_2	43-35	33-37	46.0
% MgO	26-24	-	15.0
Mesh	-150mm	85%-4	65mm
Furnace Flux	Nil	Nil	Nil
% SiO_2	-	-	-
% CaO	-	-	-
% Fe	-	-	-
Mesh	-	-	-
Converter Flux		Nil	
% SiO_2	75-80	-	-
% CaO	0	-	-
% Fe	5	-	-
% MgO	7-10	-	-
Mesh	-25 mm	-	-

PLANT - continued	Soroako	Santo Domingo	Montelibano
FEED PREPARATION			
Blending System	Rotary Dryers	Trippers	Stacker
Number of Beds	-	3	-
Bed Size - MT	-	50,000	25,000
Method of Reclaim.	-	Front End Loader	Front End Loader
Dryer - Type	Rotary	Rotary	Rotary
- Number	2	2	1
- Dimensions - m	5 I.D. x 50 5.5 I.D. x 50	4.27 Ø x 24.4	-
Average Feed Rate - WMTPH	300 & 500	250	200 dry
Feed Moisture - % H_2O	32	27	-
Retention Time - min.	20-25	20	-
Fuel Used - Type	High S Fuel Oil	Naphtha	-
- Calorific Value kCal/kg	10,300	10,960	-
Inlet Gas Temp. - °C	1,000	800	-
Outlet Gas Temp. - °C	225	90	-
Product Moisture - % H_2O	20	18.3-18.9	12
Exhaust Gas Volume - Nm^3/hr	250,000 & 420,000	75,000	
Cleaned by	Multiclone	Cyclones (4)	
REDUCTION			
Equipment	Countercurrent Kiln	Vertical Shaft Fce.	Countercurrent Kiln
Number of Units	3	12	1
Dimensions - m	5.5 Ø x 100	1.37x5.49x8.38	6.9 Ø x 185 m
Nominal Capacity - WMTPH	125	22	100 dry
Air Rate - Nm^3/hr	60,000	12,000	-
Fuel	High S Oil	Naphtha	Gas
Calorific Value - kCal/kg	10,300	10,960	-

PLANT - continued	Soroako	Santo Domingo	Montelibano
Calcine - Discharge Temp. °C	770	825	900-950
- Analysis - % Ni	2.3	1.9	-
- % Fe	22	19.0	-
- % S	0.5-1.0	0.02	-
Offgas Volume - Nm^3/hr	210,000	270,000	-
Strength - % SO_2	0.1	-	-
SMELTING			
Furnace Type	45 MVA Elkem	Shielded Arc	50 MW Electric Fce.
Number of Units	3	3	1
Nominal Capacity - MTPH	-	100	90
Concentrate	75	-	-
Dimensions - m	1.8 Ø x 7.4	2.43x8.8x7.3	22 Ø
Power Consumption - KWH/MT dry ore	-	-	430
Auxiliary Fuel - Type	None	None	-
Calorific Value	-	-	-
Operating Temp. - °C	700-900	-	-
Flux as % of Concentrate	0	Self fluxing	-
Campaign Life - Years	5	-	-
Furnace Product - MTPD	100 per furnace	240	-
Analysis - % Ni	32	36	42-47
- % Fe	57	63	-
- % S	10	0.15	-
Temperature - °C	1,360	1,485	1,420-1,440
Furnace Slag - MTPD	1,400 per furnace	4,500	1,900
Analysis - % Ni	0.16	0.15	0.2
- % Fe	20.0	17	14.5
- % CaO	1.0	-	-
- % SiO_2	45.0	46	57.0
- % MgO	23.5	-	19.0
Temperature - °C	1,550	1,575	1,600-1,630
Disposition	To Dump	-	To Dump

PLANT - continued	Soroako	Santo Domingo	Montelibano
Offgas - Volume - Nm^3/hr	12,100	-	-
Temperature - °C	800	800	-
% SO_2	<0.1	-	-
Cooled by	Nil	Tramp Air	-
Inlet Temp. - °C	-	-	-
Outlet Temp. - °C	-	-	-
Disposition of Dust	Bled from system	Recycled	-
CONVERTING			
Converter Type	PS (1), TBRC (2)	Not Applicable	Not Applicable
Number - Hot	3		
- Standby	-		
Dimensions - m	PS 3.96 Ø x 8.53 TBRC 4.22 Ø		
Number of Tuyeres	26		
Size - mm	50 Ø, 300 Ø lance		
Average Blowing Rate - Nm^3/hr	24,000		
Oxygen Enrichment - % O_2	Nil		
Normal TBRC Rotation - rpm	15		
Scrap added as % of matte charged	40	-	-
Flux added as % of matte charged	45	-	-
Matte Converted - MT/day	PS 300, TBRC 150	-	-
Total Production - MTPD	450	-	-
Analysis - % Ni	79	-	-
- % Fe	0.5	-	-
- % S	19.5	-	-
Slag - Analysis - % Ni	2-3	-	-
- % Fe	50-56	-	-
- % SiO_2	24	-	-
- Disposition	70% recycled to kilns	-	-

PLANT - continued	Soroako	Santo Domingo	Montelibano
FIRE REFINING			
Furnaces - Type	-	Electric Arc Ladle	ASEA/SKF Ladle
Number of Units	-	2	1
Dimensions - m	-	2.3 Ø x 3.2	-
PRODUCT CASTING			
Casting Machine - Type	Direct Granulation	Belt Caster	Pig Caster or Shot
Number of Moulds	-	230	-
Average Casting Rate - MTPH	45	55	-
Normal Casting Temp. - °C	1,200	1,520	-
Cast Product - Type	Granulated Matte	Ferronickel	Pigs or Shot
Dimensions - m	20%-100M	750 x 9 x 50	Shot 3 to 50 mm
Average Weight - kg	-	17	Pigs 22
Rejects as % of new material cast	2-3	6	-

Reference (3) S.C.C. Barnett, I. Patino, F.A. Perez, J. G. Schofield, "Smelting of Laterite Ore at Cerro Matoso S.A., Columbia", Extraction Metallurgy '85, IMM, September 1985, pages 877 to 890.

NICKEL SMELTERS of NORTH AMERICA (CANADA)

COMPANY NAME	INCO Limited		Falconbridge Ltd.
PLANT	Copper Cliff	Thompson	Falconbridge
Annual Production - Nickel MTPY	103,000	45,000	32,000
Flowsheet Page No.	107 & 334	145	355
FEED ANALYSIS			
Concentrate			
% Ni	12 (2.5% Cu)	10.4	12 Cu+Ni
% Fe	40	36.5	45
% S	31.5	27.5	30
% SiO_2	6	10.5	8
Mesh	95% -100	60% -325	60% -325
% H_2O	-	12.5	-
Furnace Flux			
% SiO_2	80	62	74
% CaO	1-2	6.3	8 Al_2O_3
% Fe	1.9	2.7	4
Mesh	70% -48	100% -3	-10 +65
Converter Flux			
% SiO_2	70		74
% CaO	2		8 Al_2O_3
% Fe	3		4
Mesh	100% -32 mm		-50 mm
FEED PREPARATION			
Blending System	None	None	None

PLANT - continued	Copper Cliff	Thompson	Falconbridge
PRETREATMENT			
Roasters - Type	Herreshoff	Fluid Bed	Fluid Bed
- Number	24	5 (2 operating)	2
Dimensions - m	6.6 Ø x 9.15	2 @ 5.49 Ø & 3 @ 3.66 Ø	5.6 Ø x 10
Nominal Capacity - MTPH	11.5	60	46
Air Rate - Nm^3/hr	8,490	2 @ 34,000 & 3 @ 17,000	35,000
Fuel Used - Type	Natural Gas	-	None
Calorific Value	8900 kCal/m^3	-	-
Calcine Composition - % Ni	10.5	13.5	6.5
- % Fe	35	36	30
- % S	14-15	14	12
Discharge Temp. - °C	500	650	680
Offgas Volume - Nm^3/hr	15,000	-	40,000
- % O_2	2-6	10	9
SMELTING			
Furnace Type	Reverb (Oxy-Fuel)	2 @ 30 MVA, 3 @ 18 MVA Submerged Arc Electric	36 MVA Elkem Electric Fce.
Number of Units	3 (1 standby)	5 (2 operating)	2
Nominal Capacity MTPH Concentrate	70	2 @ 50 and 3 @ 25	-
Dimensions - m	3.8 x 10 x 34.8	-	10 x 30
Oxygen Enrichment - % O_2	-	None	None
Auxiliary Fuel - Type	Natural Gas	-	Petroleum Coke
Calorific Value	8900 kCal/m^3	-	8000 kCal/kg
Operating Temp. - °C	1,230	-	1,230
Flux as % of Concentrate	22	18	40
Campaign Life - Years	2	4	over 7
Furnace Product - Type	CuNi Matte	Matte	Matte
Average Production - MTPD	900	500	800
Analysis - % Ni	23	27	18
- % Cu	8	-	-
- % Fe	36	42	38
- S	27.5	27	25
Temperature - °C	1,180	1,140	1,130

PLANT - continued	Copper Cliff	Thompson	Falconbridge
Furnace Slag - MTPD	-	1,240	-
- % Ni	0.4 (0.2% Cu)	0.22	0.1
- % Fe	41	34.5	35
- % CaO	1.5	4.0	2
- % SiO_2	34	35.7	40
Temperature - °C	1,230	1,250	1,230-1,250
Disposition	Discard	Discard	Discard
Offgas - Volume - Nm^3/hr	30,000	50,000	28,300
- Temp. - °C	1,250	400	780
- % SO_2	3-6	∿1	1
- Cooled by	-	-	Air Infiltration
Inlet Temp. - °C	-	-	780
Outlet Temp. - °C	-	-	300
Disposition of Dust	Recycled	Recycled	Recycled
CONVERTING			
Converter - Type	P-S	P-S	P-S
Number - Hot	5	3	2
- Standby	3	3	1
Dimensions - m	2 @ 3.97 Ø x 13.7 3 @ 3.97 Ø x 10.7	3.96 Ø x 10.67	3.96 Ø x 9.75
Number of Tuyeres	52 & 42	28-37	48
Size - mm	60.2	51	50
Average Blowing Rate - Nm^3/hr	31,800-41,400	-	-
Oxygen Enrichment - % O_2	up to 27	None	None
Air Rate - Nm^3/min	520-680	400-600	500
Scrap added as % of matte charged	15-20	15	5
Flux as % of matte charged	26	25	30
Matte - Average MTPD	850	210	-
- Analysis - % Ni	50	73.5	40
- % Cu	25	-	33
- % Fe	0.7	0.6	2.5
- % S	22	20.5	23

PLANT - continued	Copper Cliff	Thompson	Falconbridge
Slag - Analysis - % Ni	4 (2% Cu)	1.2	1
- % Fe	47	50	46
- % SiO_2	24	26	25
- Disposition	Returned to Reverb	Returned to Fce (420 MTPD)	Returned to Furnace
PRODUCT CASTING			
Casting Machine - Type	Slow Cooling in Pits	Stationary Molds	Cast Steel Molds
Number of Molds		54	20 (10 MT capacity)
Average Casting Rate - MTPH		12-15	-
Normal Casting Temp. - °C		930	1,150
Cast Product - Type	Cu-Ni Matte Ingots	Sulphide Anodes	Cu-Ni Matte Cakes
Dimensions - m		-	1.22 x 1.22 x 0.2
Average Weight - kg		250	317
Rejects as % of new material cast	None	<2	None
SULPHUR FIXATION	None	None	
Source of Gas			Roasters
Plant Type			Acid Plant
Rated Capacity - MTPD			1,180
Volume - Nm^3/hr			70,000
Average % SO_2			6.8
Average Conversion Efficiency %			98 +
Product Grade - % H_2SO_4			93.5
KWH/MT Sulphur Fixed in Product			150
Auxiliary Fuel			None
Tailgas Scrubbing			None

BCL LIMITED - BOTSWANA NICKEL SMELTER

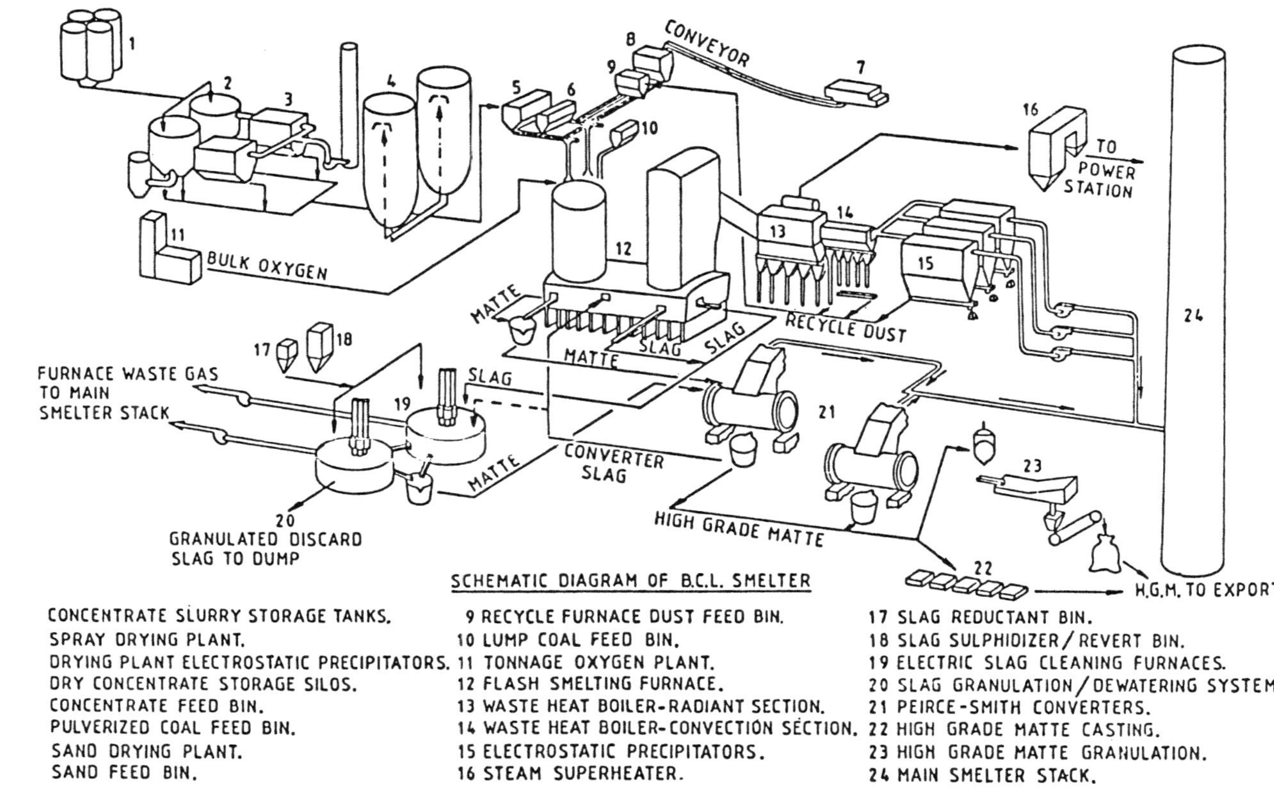

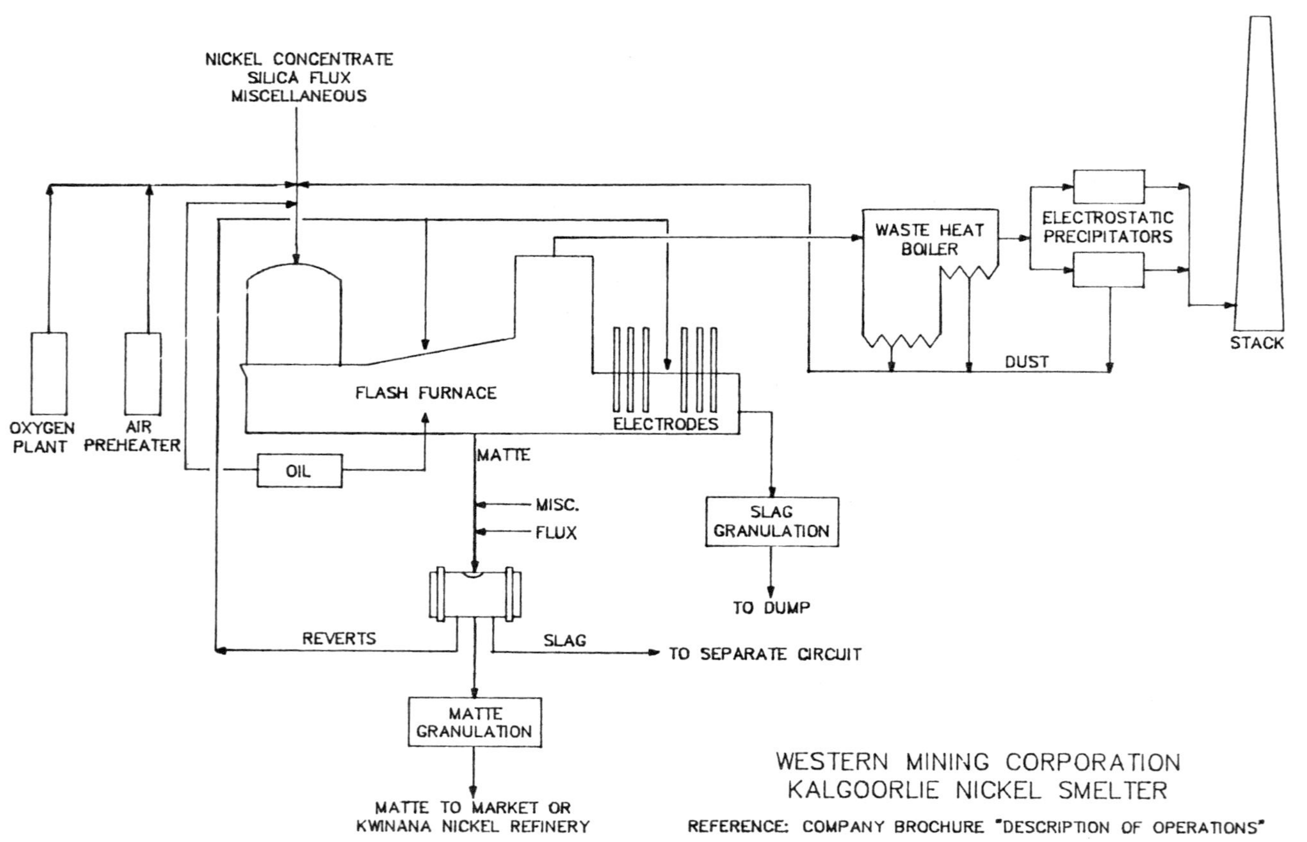

WESTERN MINING CORPORATION
KALGOORLIE NICKEL SMELTER

REFERENCE: COMPANY BROCHURE "DESCRIPTION OF OPERATIONS"

P.T. INCO INDONESIA - SIMPLIFIED FLOWSHEET

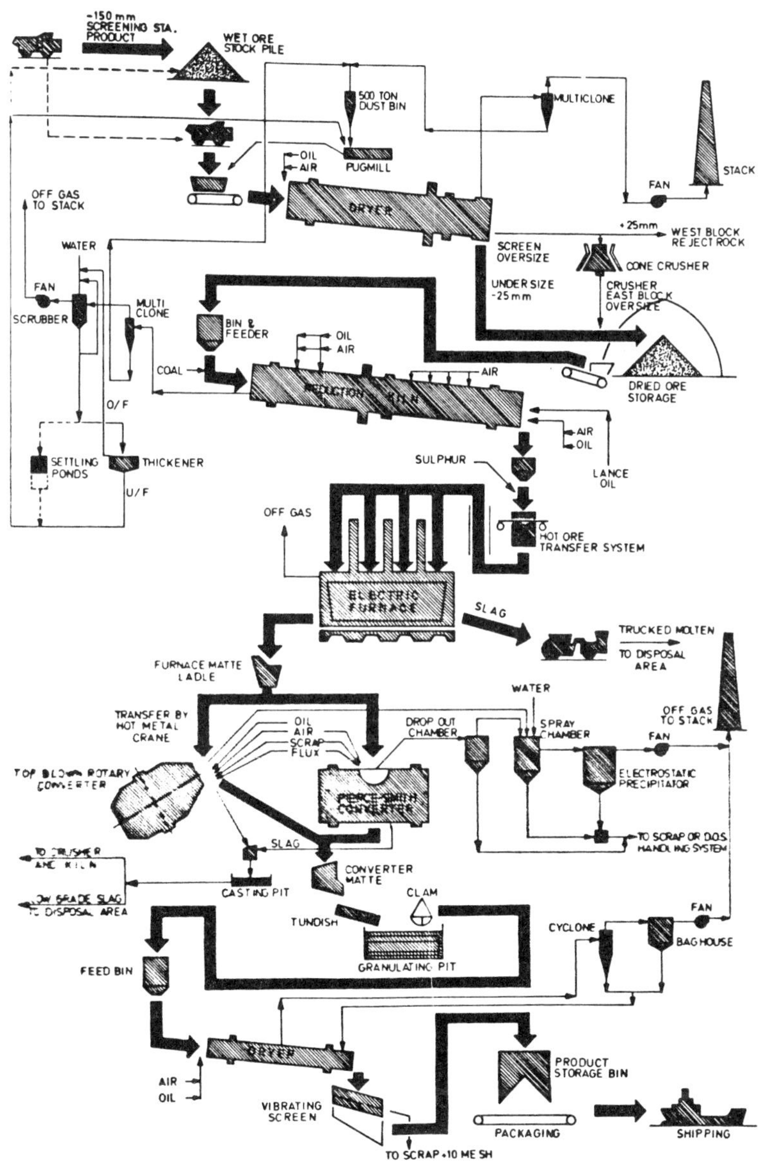

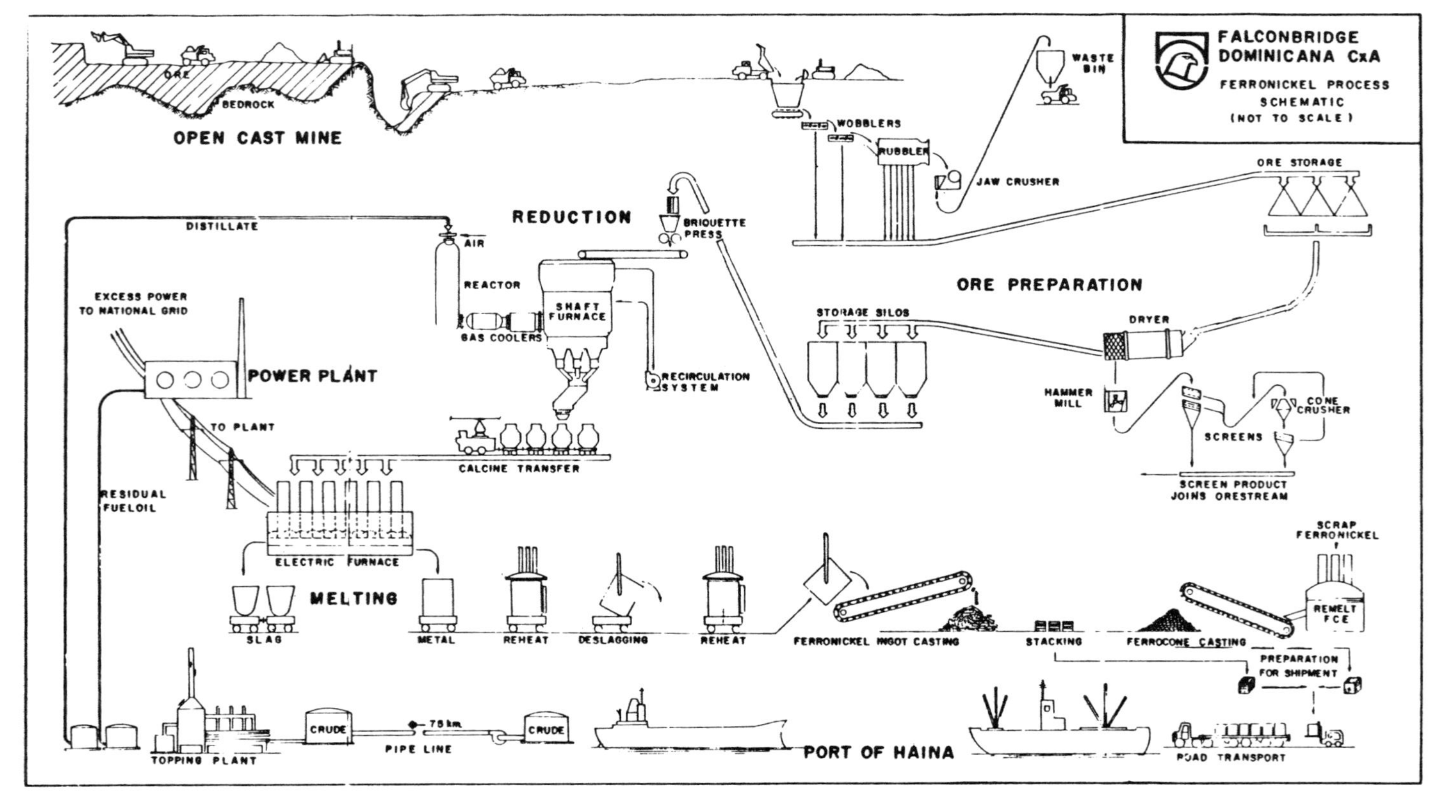
FALCONBRIDGE DOMINICANA CxA
FERRONICKEL PROCESS
SCHEMATIC
(NOT TO SCALE)
OPEN CAST MINE
ORE
BEDROCK
WASTE BIN
WOBBLERS
RUBBLER
JAW CRUSHER
ORE STORAGE
REDUCTION
DISTILLATE
AIR
REACTOR
GAS COOLERS
SHAFT FURNACE
RECIRCULATION SYSTEM
BRIQUETTE PRESS
ORE PREPARATION
STORAGE SILOS
DRYER
HAMMER MILL
CONE CRUSHER
SCREENS
SCREEN PRODUCT JOINS ORESTREAM
EXCESS POWER TO NATIONAL GRID
POWER PLANT
TO PLANT
CALCINE TRANSFER
RESIDUAL FUELOIL
ELECTRIC FURNACE
MELTING
SLAG
METAL
REHEAT
DESLAGGING
REHEAT
FERRONICKEL INGOT CASTING
STACKING
FERROCONE CASTING
SCRAP FERRONICKEL
REMELT FCE
PREPARATION FOR SHIPMENT
TOPPING PLANT
CRUDE
75 km
PIPE LINE
CRUDE
PORT OF HAINA
ROAD TRANSPORT

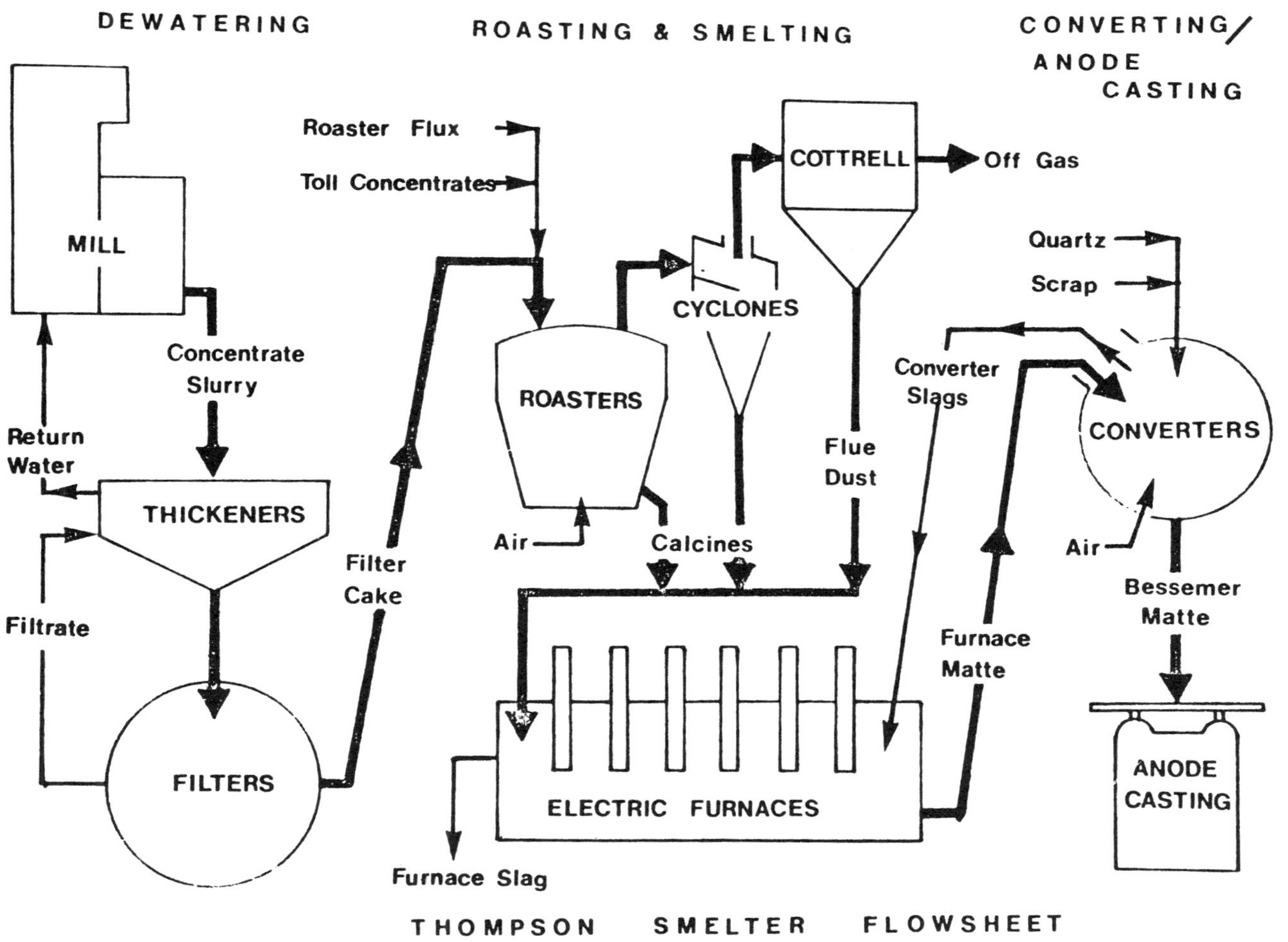

THOMPSON SMELTER FLOWSHEET

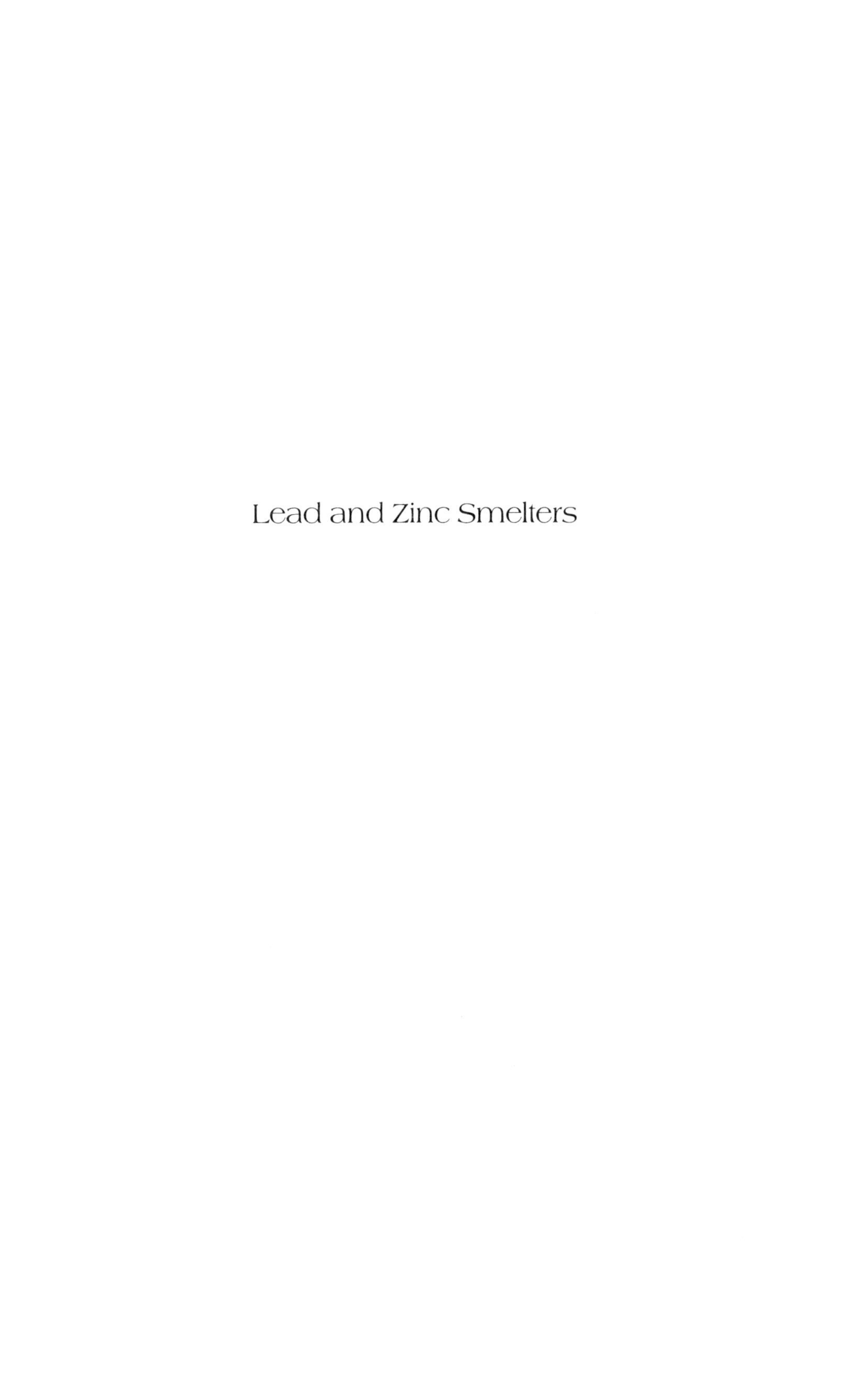

Lead and Zinc Smelters

LEAD SMELTERS - SURVEY REVIEW

J. E. Hoffmann
Consulting Metallurgist
Jan H. Reimers and Associates USA Inc.
Houston, Texas

Introduction

A total of nineteen lead smelters responded to the TMS Pyrometallurgical Committee's survey. Of these, the geographic distribution is:

Country	Number of Smelters
Japan	3
Germany	3
Sweden	3
Canada	2
United Kingdom	2
United States	2
Africa	2
Australia	1
Belgium	1
	19

Despite the geographic diversity, all the primary lead smelters reporting use the conventional blast furnace technology, with the exception of Boliden who employ a novel electric furnace smelting process. Over 85% of the lead production reported is produced by primary smelters, the remainder from secondary smelters.

Somewhat more diversity of process is apparent in the secondary smelters, the majority of which treat scrap batteries. One-half use the short rotary furnace technology and the remainder employ blast furnace technology. Boliden use a Kaldo to smelt dust and reverts, and have processed scrap batteries through this unit.

Lead Smelting

Blast Furnace

Consistent with the use of blast furnace technology, sinter composition is similar at all the primary operations and averages about 41% lead and 2.5% sulphur. The greatest variation was in the iron content of the sinter; this varied from 11 to 19%. However, the amount of blast furnace slag which may have been recycled was not specified. This could account for variations in the iron content of the sinter.

Blast furnace slags from primary lead smelters averaged about 3% lead for primary operation, with the exception of the much more strongly reducing Imperial Smelting Furnaces where the lead tenor in the slag was about 0.5%.

Oxygen enrichment of the blast air was practiced at six primary smelters. However, the degree of enrichment was generally low, up to a maximum of 25% oxygen. Its effect on productivity must be viewed as marginal, say a maximum of 10-15%. In all likelihood it is used more as a method of compensating for adverse operating conditions which may occur from time to time at any smelter.

Of the primary lead smelters which responded to the questionnaire, all but three practice some form of sulphur capture (sulphuric acid production in every case but one, Trail). Actual overall recovery figures of sulphur from concentrates was not provided by any smelting operations. This is understandable in light of the difficulties with achieving high recovery of sulphur dioxide from sinter belt operations. In comparison to the gas strength from other nonferrous smelting processes, sinter plant gases are relatively low, normally in the range of 2.5 to 7.0% SO_2.

New Processes

In considering the data developed from this survey, it should be recognized that the lead industry is at a crossroad where newly developed processes will gradually supplant the sinter plant-blast furnace which has held sway for so long in the lead industry (1). Preeminent among the new technologies is the QSL (Queneau, Schuhmann, Lurgi) process, developed at the Berzelius smelter in Germany where a commercial unit is presently being installed. This oxygen aspirated, primarily bath smelting process for lead concentrates is presently scheduled for installation at the Cominco lead smelter in Trail, B.C. (2), as well as an installation in mainland China.

As part of the Cominco plant, a separate gas offtake will be installed on the slag reduction end of the reactor which will increase zinc processing capacity and reduce the volume of the SO_2 gas stream (2). Cominco's plant, then, will include an added feature. In the near future the data tabulated in this survey for Cominco will change significantly.

The Kivcet process (1), essentially an oxygen flash smelting process for lead concentrates, is operating in the U.S.S.R. and Italy. The installation at the SAMIN smelter in Italy is due to start up this year and will be the first Western smelter to use this process.

The highly flexible Kaldo lead process was developed by Boliden Metall AB and has operated for a number of years at Rönnskär. It is used primarily for revert materials, but production tests have been carried out using concentrate (1) and a primary Kaldo lead smelter is being constructed in Iran. Scrap batteries have been processed through the Rönnskär Kaldo and with the environmental pressure on secondary smelters, particularly in

North America, the process may well be adopted in the future in this area.

Other processes under consideration for new lead smelting operations include the Outokumpu flash process (1) and the Isasmelt, developed by Mount Isa in Australia.

Conclusion

Based on the review of the data provided by the respondents to the questionnaire, little progress appears to have been made in lead smelting technology, when compared to developments in the copper, nickel or zinc industries.

In the past few years new processes have been developed for lead and it can be confidently projected that a similar survey, ten years from now, will show the sinter-blast furnace technology largely supplanted by processes employing the direct oxygen smelting of lead sulfide concentrates.

With these new technologies will come the attendent benefits of high energy efficiency, compact process equipment and virtually complete capture of sulphur dioxide and lead emissions.

References

1. Reimers, J.H. and Taylor, J.C., "The Future of Lead Smelting", Advances in Sulfide Smelting, edited by H.Y. Sohn, D.B. George and A.D. Zunkel, TMS-AIME, Warrendale, PA, November 1983, pp 529-551.

2. Sutherland, C.A., "Modernization of Cominco's Zinc Plant and Lead Smelter at Trail, British Columbia", 89th Annual General Meeting of CIM, Toronto, Ontario, May 1987.

Zinc and IS-Smelters - Survey Review

Heinrich Traulsen

Lurgi GmbH, D-6000 Frankfurt (Main), F. R. Germany

1. Introduction

The TMS smelter survey includes the pyrometallurgical operation lines of the two main zinc producing processes, industrially applied:

- roasting/electrowinning
- ISF-process.

The survey reports data of the major portion of the IS-furnace operations, while the response from operations applying roasting/electrowinning was not as good and is limited to some of the important producers.

This review will focus separately on aspects of the two process alternatives.

2. Roasting/Electrowinning

The pyrometallurgical operations within the typical flowsheet of an electrolytic zinc refinery is limited to the roasting of concentrates, mainly directed towards sulfur elimination prior to leaching and to remelting of zinc cathodes and zinc slab casting. Themelis and Freeman (1) have studied the fluid bed behavior in zinc roasters and present a review of operations and theoretical background information. The operating range of fluid bed roasters with regard to average retention time and particle size are indicated in Figure 1 (2). The figure also gives a comparison of fluid bed roaster and other reaction systems applied in non-ferrous industry.

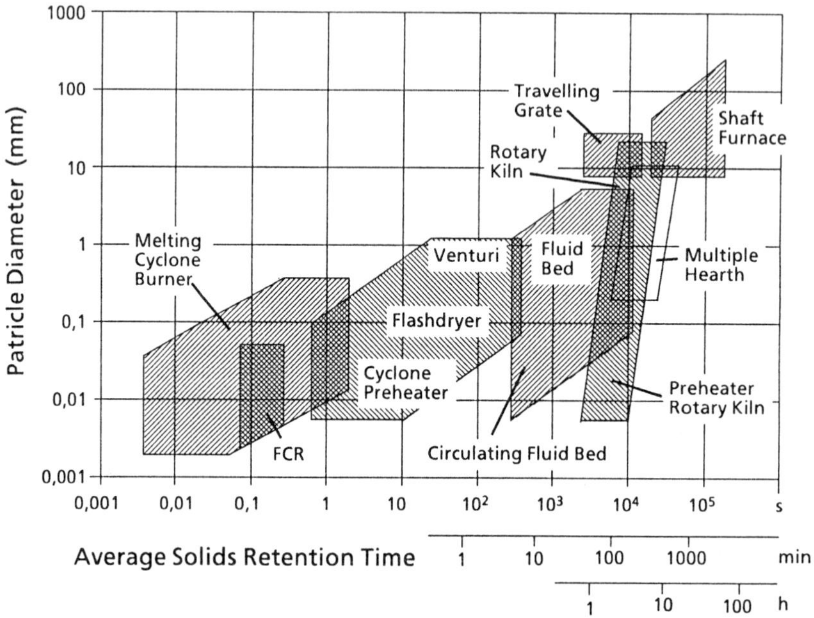

Figure 1: Operating Range of Industrial Gas Solid Reactors

The majority of the zinc roasting operations is based on a flowsheet as shown in Figure 2 (3). Normally zinc concentrates with moisture contents up to 10 % are directly charged to the roasting unit without pretreatment, i.e. by drying.

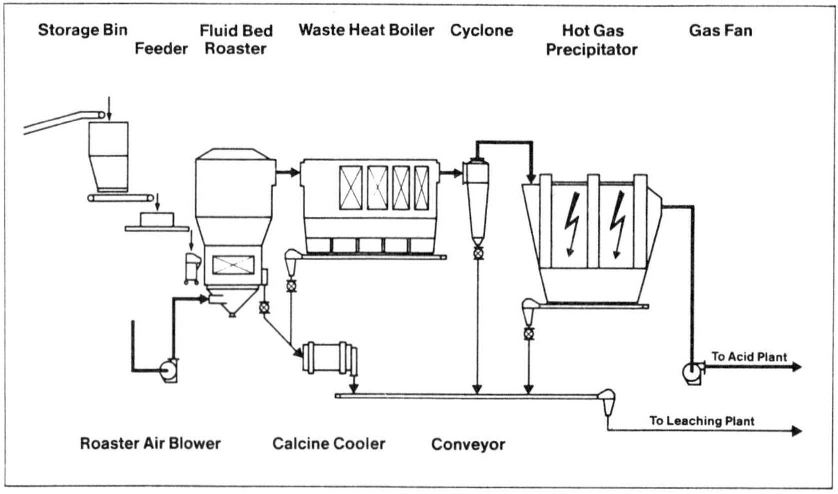

Figure 2: Fluid Bed Roasting Plant for Roasting of Zinc Concentrate (3)

In Table 1 some of the specific fluid bed roaster data of operations included in this survey have been compiled.

Table 1: Fluid Bed Roaster Data

Plant	Preussag Nordenham	EdZ	Norzinc	St. Joe Monaca	St. Joe Bartlesville
Type of feeding	dry	dry	dry	dry	slurry
Grate area, sq.m.	90	35	37.5/55		
Bed diameter, m				6.9	5.5/6.75
Reactor height				10.7	
Bed temperature, °C	950	900	925	850	900
Sulfur in calcine, %	2-3	2.1	2.2	2.0	2.5
Feed rate, WMTPH	24.2	9.2	10/14.5	15	3.8/7.6
Air flow, Nm3/h	52,000	19,500	18,000/29,000	20,400	8,400/17,000
Load factor, MTPH feed per sq.m. grate area	0.27	0.26	0.27/0.26		

Plant	Kidd Creek	Amax Zinc	Cominco Trail	Union Zinc Clarksville
Type of feeding	dry	dry	dry	dry
Grate area, sq.m.	46	32.2		64.7
Bed diameter, m	8.4	9.7	13.8/16.5	
Reactor height	16			
Bed temperature, °C			930	980
Sulfur in calcine, %			2.5	2.5-3.0
Feed rate, WMTPH	18	12.7	26	17-22
Air flow, Nm3/h	26,000	17,000	43,000	42,725
Load factor, MTPH feed per sq.m. grate area	0.39	0.39		0.26-0.34

The roaster units shown are to be considered as smaller to medium size roasters. The biggest unit built by Lurgi for Electrolytic Zinc of Australasia at Risdon has a grate area of 123 sq.m. and a nominal capacity of 33 MTPH concentrate feed with a load factor around 0.28 MTPH feed per sq.m. of grate area. Two plants are showing higher than the average load factors. Major changes in the basic design parameters mainly relate to the development of bigger roaster units as it can be taken from Figure 3.

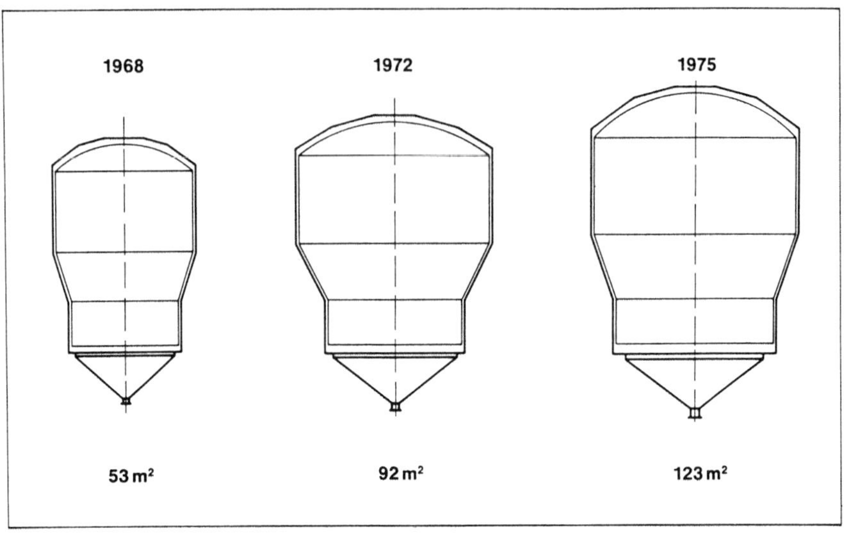

Figure 3: Increase of Performance of Lurgi Fluid Bed Roasters for Zinc Concentrates (3)

Oxygen application and changes in feed composition in regard to impurity levels, however, may influence the development of roasting units in future. The utilization of oxygen may increase the specific throughput and may influence the residual sulfur content. Normally approx. 15-20 % of the calcine sulfur content totalling 2-3 % reports as sulfides and 80-85 % as sulfate. Reduction of sulfates in calcine is a concern in zinc hydrometallurgy as the sulfur balance of the system is affected. The final outlet of sulfate surplus is the leach residue.

Lead contents in concentrates beyond 3 % are considered to be hazardous for roaster operations, specifically in presence of higher silica amounts. Feed preparation forming micro pellets as roaster input has shown, that this problem may be overcome also at high lead contents.

The circulating fluid bed system may offer some advantages over conventional systems (see Figure 1).It may allow a somewhat longer retention time, closer operating temperature control and therefore may overcome some of the constraints briefly outlined for conventional systems and the effects on downstream processing.

3. ISF-Process

ISF based smelters, competing with electrolytic zinc plants and conventional lead smelters, have undergone various experiments, changes and improvements with the objective to increase productivity and economics:

- The standard IS-furnace with about 17 sq.m. shaft area has achieved a substantial capacity increase:

	Typical 1963 furnace	Today achieved
MTPY zinc	35,000	85,000
MTPY lead	17,000	40,000

The increase in furnace capacity resulted from an increased blast rate, improved charge preparation, increased blast and coke preheat temperature, improved condenser efficiency and use of immersible coolers in the launder system (4).

Table 2: IS-Furnace Operating Data (4)

	Avonmouth	Duisburg	Kabwe	Hachinohe
Type of feed	sinter	sinter/briquettes from secondary materials	sinter	sinter
Feed rate, MTPH				27.9
Production, MTPD Zn	300-350	250		220
MTPD Pb	160-190	108		130
Dimensions, m				9.5x3x6.4
Shaft area, sq.m.	27.1	17	19.6	17
No. of tuyeres		16		
Fuel used	coke	coke (700°C)	coke	coke
Blast temperature, °C		1150		1000
Operating temp., °C	1300	1350	1300	
Annual design capacity				
- MTPY Zn	100,000	85,000	35,000	76,000
- MTPY Pb	45,000	40,000	20,000	35,000
Campaign life, years	2-3		3	5

- Furnace configuration: The original water jacketed furnace has been mostly replaced by spray cooled shafts in parallel with redesign of furnace shape. Ongoing experimenting is reported from various smelters in regard to tuyere number, angle and position.

- It is understood, that experiments with oxygen enrichment of the blast have not resulted in industrial application. Oxygen enrichment must be compared economically with increased blast air preheat. When surplus of low calorific gases from the furnace is available to increase blast air temperature, oxygen enrichment is not viable. Oxygen enrichment around 23 % today is only practiced on sinter machines resulting in increased sinter machine capacity.

- Drying the blast air to reduce coke requirement has been practiced by Japanese ISF-operators, leading to a considerable decrease in Pb/Zn-ratios. This procedure has justification where faced with constantly high humidity levels in the ambient air (4, 5).

- Increased heat recovery from launders producing high pressure saturated steam by utilization of a special low melting point lead and bismuth containing alloys is applied in Japan and has been experimented with at Duisburg.

The essential data on the ISF operations included in this survey are compiled in Table 2.

As it can be seen from the data, the Duisburg ISF is processing besides the sinter, briquettes produced from secondary materials or Waelz oxide resulting from steel industry dust. The ratio of briquettes in total feed is reported to be up to 40-50 % (6). This indicates that a future trend in the ISF-process may be seen in increasing ratios of processing low grade, complex and secondary materials (4, 7, 8).

References

(1) N. J. Themelis, G. M. Freeman: Fluid Bed Behavior in Zinc Roasters, Journal of Metals, August 1984, P. 52-57

(2) L. Reh, Auswahlkriterien für nichtkatalytische Gas-/Feststoffreaktoren, Chemie-Ingenieur-Technik, 49 (1977), P. 786-795

(3) Lurgi GmbH, Roasting - Process and Technology, Internal Reference 1985

(4) C. F. Harris, A. W. Richards, A. W. Robson, The Zinc-Lead Blast Furnace into the 1980s, Proceedings Lead-Zinc-Tin '80, TMS-AIME World Symposium, February 1980, Las Vegas, P. 247-260

(5) H. Nakagawe, Y. Sugawara, K. Nakayama, Energy Savings in the Zinc-Lead Blast Furnace Complex, Proceedings Lead-Zinc-Tin '80, TME-AIME World Symposium, February 1980, Las Vegas, P. 261-277

(6) R. Kola, Sekundärrohstoffe im IS-Schachtofen der Berzelius Metallhütten GmbH, Duisburg, Erzmetall 35 (1982), P. 132-137

(7) P. R. Mead, Treatment of a Diverse Range of Low Grade Feed Materials in ISF, Proceedings Lead-Zinc-Tin '80, TMS-AIME World Symposium, February 1980, Las Vegas, P. 278-289

(8) A. O. Adami, G. R. Firkin, A. W. Robson, Treatment of Complex Materials and Residues in the Imperial Smelting Process, Paper presented at joint G.D.M.B. and I.M.M. Meeting, Bad Harzburg, 1978

LEAD SMELTERS
of
AUSTRALIA AND JAPAN

COMPANY NAME	Mitsui Mining & Smelting Co. Ltd.		Toho Zinc Co. Ltd.	Mount Isa Mines Ltd.
PLANT	Kamioka	Takehara (1)	Chigirishima(2)	Mount Isa
Annual Production - Lead MTPY	30,000	36,000	72,000	160,000
- Zinc MTPY	3,000	-	-	-
Flowsheet Page No.	-	-	-	-
FEED ANALYSIS				
Concentrate		Secondaries		
% Pb	-	60	40.6	48
% Fe	-	-	-	12.0
% S	-	-	6.3	23.8
% SiO_2	-	-	-	2.9
Mesh	-	-	-	90%-400
% H_2O	-	-	-	-
Furnace Flux				
SiO_2-kg/MT bullion	-	-	-	20%
Lime- " "	290 @ 55% CaO	-	140	35.6%
Iron- " "	104 @ 36% Fe	-	4	4.5%
Mesh	100%-30mm	-	-	98%-6 mm
FEED PREPARATION				
Blending System	Rod Mill	-	Bin Feeders Mixing Drum	Drum Filter
SINTERING				
Number of Machines	1	-	1	1
Dimensions - m	1.385x17.9	-	2 x 15	3.05 x 30.5
Hearth Area - sq.m	24.4	-	30	93
Nominal Capacity - MTPH	24	-	45	58

PLANT - continued	Kamioka	Takehara	Chigirishima	Mount Isa
Fuel Used - Type	Heavy Oil	-	Heavy Oil	Light Oil
Calorific Value - kCal/kg	10,510	-	-	10,700
Feed Moisture - % H_2O	10-15	-	-	14
Sinter Composition				
% Pb	39.6	-	41.9	42
% Fe	11.8	-	11.9	11
% S	2.4	-	2.0	1.7
Exhaust Gas - Volume - Nm^3/hr	24,000	-	18,000	-
% SO_2	2.9	-	7-7.5	2.5
SMELTING				
Number of Furnaces	1	1	1	1
Type	Blast Fce(Round Ends)	Blast Fce(Oval)	Blast Fce.	Blast Fce.
Nominal Capacity - Sinter MTPH	11.4	3 Secondaries	16	55
Dimensions - m	5.2 m^2x5.036 m	2.4 x 1.6	5.5 x 1.7	1.83x7.02x5.8
Oxygen Enrichment - % O_2	-	22	21	21
Auxiliary Fuel - Type	Coke	Coke	Coke	Coke
Calorific Value - kCal/kg	-	7,000	-	6,450
Preheat Air Temp. - °C	-	400	300	-
Operating Temp. - °C	1,200	-	-	1,200
Flux as % of feed	-	Iron 8%, Lime 1%	-	Silica 18%, Lime 2
Campaign Life - Years	10	1		3
Furnace Products (average)				
Temperature - °C	900	900	-	1,200
Lead Bullion - MTPD	86.5	40	-	470
% Pb	98.6	93	-	98
% Fe	Trace	-	-	-
% S	Trace	-	-	-
Matte - MTPD	4.2	-	-	-
% Cu	60.2	2.3	-	-
% Fe	0.97	42.3	-	-
% S	13.8	-	-	-

PLANT - continued	Kamioka	Takehara	Chigirishima	Mount Isa
Slag - MTPD	115	9	-	650
- Temperature - °C	1,100	1,100	-	1,200
- Analysis - % Pb	3.6	1	3.6	3.5
- % Fe	24.2	30	22.4	22
- % CaO	13.7	10	17.1	23
- % SiO_2	20.1	20	20.5	18.5
Furnace Offgas				
Volume - Nm^3/hr	10,800	36,000	-	-
Temperature - °C	420	400	-	230
% SO_2	0.42	100-200 ppm	-	0.2
Gas Cooling by	Tube Coolers	Jacket Cooling	-	Spray Cooler
Number of Units	2	1	-	1
Inlet Temp. - °C	350	400	-	230
Outlet Temp. - °C	50	100	-	130
Disposition of Dust	Recycled to Sinter Machine	Recycled to Furnace	-	Recycled
SLAG CLEANING				
Slag Fumed - MTPD	84.7	-	-	-
Analysis - % Cu	0.41	-	-	-
- % Pb	3.6	-	3.6	-
- % Zn	18.0	-	18.2	-
- % SiO_2	20.1	-	20.5	-
Furnace Hearth Area - sq.m	30	-	-	-
Reductant - Type	Coke Breeze	-	-	-
- MTPD	2.8	-	-	-
Discard Slag Analysis - % Cu	0.2	-	-	-
- % Pb	0.2	-	0.3	-
- % Zn	8.0	-	6.5	-
- % SiO_2	23.1	-	27.8	-
Fume Produced - MTPD	-	-	-	-
PRODUCT CASTING				
Casting Machine - Type	Straight Line	-	-	Fixed Moulds
Number of Moulds	13	-	-	44

PLANT - continued	Kamioka	Takehara	Chigirishima	Mount Isa
Average Casting Rate - MTPH	30	-	-	88
Normal Casting Temp. - °C	360	-	-	370
Cast Product - Type	Anodes	-	Anodes	-
Dimensions - m	1.4x.99x.025	-	1.2x0.8x.024	-
Weight - kg	350	-	250	4,000
Rejects as % of new Pb cast	2.0	-	-	1
SULPHUR FIXATION				
Source of Gas	Sinter Machine	Not Applicable	Sinter Machine	
Plant - Type	Single & Double Contact		-	
- Rated Capacity, MTPD	65		98	
- Volume, Nm^3/hr	24,000		-	
- Average % SO_2	2.9		5	
- Average Conversion Efficiency, %	97.0		-	
Product Grade - % H_2SO_4	98		98	
KWH/MT Sulphur Fixed in Product	330		-	
Auxiliary Fuel	-		-	
Tailgas Scrubbing	Turbulent Contact Absorber (lime scrubbing)		-	

Reference (1) S.Hirakawa et al, "Lead Smelting at Takehara Refinery", Joint MMIJ-AIME, Tokyo, 1980.

(2) K. Tzuzuki, "Operation of Lead Smelting and Refining at Chigirishima Smelter", op.cit. ref.(1)

LEAD SMELTERS of EUROPE

COMPANY NAME	Berzelius Metallhuetten GmbH	Preussag-Boliden-Blei GmbH	Metallurgie Hoboken - Overpelt	Boliden Metall AB	
PLANT	Binsfeldhammer	Nordenham	Hoboken	Roennskaer	
				Lead Plant	Lead Kaldo
Flowsheet Page No	-	-	205	410	425
Annual Production					
Lead - MTPY	90 - 92,000	102,300	50,000	60,000	10,000
Byproduct - MTPY	Dore M. with Au: 0.13 - 02 Ag: 140 - 160 Cu-Matte with Cu: 1,500-2,000	Cu Dross Sb Slag As Dross Ag Scum Litharge			
FEED MATERIAL					
Primary Metal Feed					
Type	Concentrate	Concentrate	Various Concentrate:	Concentrate	Cu-Conv.Dust Slimes
Analysis - % Pb			Pb:40 ; Cu: 4	71.5	43
- % Fe			10	1.5	Zn: 11
- % S			23	15 - 23	10
- % SiO2			Concept:	6	-
Furnace Flux			Complex metal-lurgy operation		
Silica - % SiO2		85	with Pb and Cu		
Lime - % CaO		55	as collectors.	54.5	90
Other -Type		Pyrite	Pb circuit		Fumed Slag
% Fe	FeO: 13 - 17	65	comprising		35
% SiO2	8 - 9		roasting,		36.8
% CaO	8 - 9		sintering,		5
Size - mm	>25	> 3	blast furnace, el.slag cleaning.	5 - 12	

PLANT - continued	Binnsfeldhammer	Nordenham	Hoboken	Roennskaer	Roennskaer
FEED PREPARATION					
Dryer					
Type	-	-	-	Kiln	Kiln
Number				1	1
Dimensions - m				2.2 dia x 16	1.8 dia x 14
					Conv.Dust+Slime:
Feed Rate - MTPH				30	15
Inlet - % H2O				6 - 7	13 .5
Outlet - % H2O				0.5	< 1
Fuel - Type				Fuel Oil	Fuel Oil
Cal.Value-kcal/kg				9,000	9,000
Gas Temperature					
Dryer Inlet - oC				600	600 - 800
Dryer Outlet -oC				110	120
Gas Volume - Nm3/hr				15 - 20,000	11,000
Discharge Temp.oC				35	60
Roasters - Type	-	-	Wedge, 7 Hearths	-	-
Number			2		
Dimensions - m			7.3 dia x 19.525		
Capacity - MTPH			3 - 5		
Air Rate - Nm3/hr			6,300		
Calcine					
Analysis - % Pb			10 - 40		
- % Fe			5 - 30		
- % S			8		
Discharge T. - oC			150		
Offgas - Nm3/hr			6,700		
- % SO2			3 - 6		
Sinter Machines					
Number	1	1	2		
Dimensions	25 m2, 1.5 m W.	36 m2	2 x 28 m		
Capacity	0.072 MT S/m2xH	0.071 MT S/m2xH	8-10 MTPH Sinter		

PLANT - continued	Binnsfeldhammer	Nordenham	Hoboken	Roennskaer	Roennskaer
Fuel Used - Type	Natural Gas	Natural Gas	Natural Gas	-	-
Calorific Val.	9,700 kcal/Nm3	7,600 kcal/Nm3	7,600 kcal/Nm3		
Sinter Compos.				-	-
- % Pb	42	42 - 44	30 - 40		
- % Fe	17	10	6 - 12		
- % S	2	1.6 - 2.1	2 - 2.5		
Exhaust Gas					
- Vol: Nm3/hr	19,000	45,000	19,800	-	-
- % SO2	5 - 6.5	4 - 5	5.5		
SMELTING					
Furnaces					
Type	Blast Furnace	Blast Furnace	Blast Furnace	Elect. Furnace	Kaldo
Number	2	2 (1 Standby)	1 (48 Tuyeres)	1	1
Capacity - MTPH		700 - 900 Sinter	36 Charge:40% Sinter 20% Conv.Slag, plus Drosses,Cu-Matte,Dusts,Scrap.	12.5	130 (13m3) Dust Pellets
Dimensions	10 / 7 m2	12 m2 x 7.6 m	1.3 x 7.5 m	4.5 x 13.3 m	3.6 dia x 6.5 m Inclination Vessel: 18 degr.
Oxygen Enr. - % O2	23	23-24 at Times	24.5	25	-
Fuel Used - Type	Coke	Coke,40-100mm	Natural Gas	Elec. Power	Oil
Cal.Value:kcal/U.	7,000	7,000	7.600	8 MVA	9,700
Operating Temp.- oC	<1,100	1,300 - 1,500	400 (Blast)	1,350	1,150
Campaign Life	2 Months	1 - 2 Years	3 Years	Walls 1 Year, Bottom 10 Years	8 Weeks
Flux as % of Concentrate Feed	8		-	16 Coke Breeze:2.7%	Fumed Slag + Burnt Limestone
Metal Production - MTPD	150	270 - 300	16 (3Phases: Pb, Speiss, Matte)	210	

PLANT - continued	Binnsfeldhammer	Nordenham	Hoboken	Roennskaer	Roennskaer
Matte /Metal - MTPD	150	270 - 300	3 Phases: 408	210	Varying
Temp. - oC	950 - 1,050	1,000	900 - 1,000	1,000	1,000
Analysis - % Pb	97	98.5	Lead: 20 - 30 %	97	Pb+Sb+Bi: 99
- % Cu			Speiss: 5 - 10 %		< 0.1
- % As			CuPb Matte:5-10%		< 0.1
- % S	< 1	0.15	of 408 MT	1 - 2	< 0.5
Furnace Slag - MTPD	85 - 110	150 - 200	350	65	Varying
Analysis - % Pb	1 - 1.5	1.5 - 2	Pb:1.5 / Cu:0.5	4.3	1.5
- % FeO	32 - 34	32 - 33	30 - 33	15.5	28.3
- % CaO	16 - 19	20 - 21	15 - 18	32.5	23
- % SiO2	19 - 22	19 - 20	25 - 30	24	25
- % Zn			7 - 10		15
Temperature - oC	975 - 1,050	1,250	1,150	1,350	1,100
Furnace Offgas - Nm3/hr	40,000	40,000	25,000	20,000	25,000
% SO2	0.2	0.6	-	5.3	2
Temperature - oC	120	200 - 300	200	1,050	1,200
Cooling Method		Air Cooler	Tub.Air Cooler	Boiler + Convection Cooler	By Air to 800 oC + Venturi Scrub
No of Units:		1	3	1	1
Inlet Temp. - oC		200 - 300	150 - 450		800
Outlet Temp. - oC		50 - 70	80 - 100	110	65
Gas Cleaning				EP + Baghouse	Venturi:1,800 mm
Dust Disposition	Recycled	Recycled	Recycled: Transformed to Cake.	Recycled	Recycled
CONVERTING					
Type	-	-	-	P.S	-
Number				2	
Operating				1	
Standby				1	
Dimensions - m				2.7 Dia x 5.6	
Tuyeres - Number				18	
- Size, mm				30	

PLANT - continued	Binnsfeldhammer	Nordenham	Hoboken	Roennskaer	Roennskaer
Blowing Rate - Nm3/min.				100	
% Oxygen				21	
Lead per Converter -MTPD				200	
SLAG CLEANING					
Method	-	-	Electric Furnace Number: 1 & Electrodes in Line, 7 MW. Startup End 1986	Slag fuming. Refer to Copper Smelter.	-
Description					
FIRE REFINING					
Furnaces			No Fire Refining. Pb processed in Harris Refinery.		
Number	3	3			
Dimensions	1 x 150 MT Pb 2 x 250 MT Pb	5 x 3.1 x 1.7 m			
Number of Tuyeres	3	Lances: 2 or 3			
Tuyere Size - mm	19	12			
Oxidation Time per Charge - hrs.	8 - 10	4 - 6			
PRODUCT CASTING					
Casting Machine:					
- Type	Belt Casting	Sheppard			
- No Moulds	240	234			
- Rate, MTPH	30 - 40	30			
- Casting Temp. oC	400	420			
- Ingot Size	600x120x95 mm				
- Ingot Weight, kg	45 - 50	50			
- Rejects as % of new Material Cast	0.2	1			

PLANT - continued	Binnsfeldhammer	Nordenham	Hoboken	Roennskaer	Roennskaer
SULPHUR FIXATION					
Source of Gas	Sinter Machine	Sinter Machine	Sinter Machine	See Copper Smelter Data.	
Plant:					
- Type	Single Contact	Double Contact	Lead Chamber		
- Rated Capacity as Acid, MTPH	125 - 148	290	400		
- Volume, Nm3/hr	19,800	45,000	60,000		
- Average % SO2	5 - 6.5	4 - 5	4 - 5		
- Average Conversion Efficiency, %	97 - 98	99.2	> 99.8		
Product Grade - % H2SO4	96	96 and 78, resp.	78		
KWh/MT Sulphur Fixed in Product	220	375	230		
Auxiliary Fuel	-	Light Oil: 65 MTPY	-		
Tailgas Scrubbing	-	-	-		

LEAD SMELTERS
of
NORTH AMERICA (CANADA & UNITED STATES)

COMPANY NAME	Brunswick Mining and Smelting Corp. Ltd.	Cominco Ltd.	AMAX Lead Company of Missouri	St. Joe Lead Company
PLANT	Belledune	Trail	AMAX	Herculaneum
Annual Production - Lead MTPY	60,000	140,000	130,000	200,000
Flowsheet Page No.	206	-	-	-
FEED ANALYSIS				
% Pb	32-35	51	-	72
% Fe	22-24	10	-	3.3
% S	31-34	18	-	16.1
% SiO_2	0.5	10	-	0.7
Mesh	75% -400	80% -200	-	-400
% H_2O	8-12	8-10	-	7.9
Furnace Flux				
Sand % SiO_2	90	94	90	-
Limestone % CaO	56	54	56	-
Hematite % Fe	-	-	-	-
Mesh	-	6 to 12 mm	Sand 90% -16 Limestone 95% -100 Hematite < 6 mm	-
FEED PREPARATION				
Blending System	None	Metering Bins	Drum Mixer	Belt Mixing
Dryer - Type	-	Rotary	-	-
- Number	-	3	-	-
- Dimensions - m	-	2.5 Ø x 24	-	-
- Retention Time - min.	-	10	-	-
Fuel Used - Type	-	Natural Gas	-	-
Calorific Value	-	9500 $kCal/Nm^3$	-	-

PLANT - continued	Belledune	Trail	AMAX	Herculaneum
Average Feed Rate - WMTPH	-	28	-	-
Feed Moisture - % H_2O	-	8-10	-	-
Inlet Gas Temp. - °C	-	-	-	-
Outlet Gas Temp. - °C	-	200	-	-
Product Moisture - % H_2O	-	6	-	-
Exhaust Gas Volume - Nm^3/hr	-	145,000	-	-
SINTERING				
Number of Machines	1	2	1	1
Dimensions - m	-	3 x 32	2.5 x 24	3 x 30
Hearth Area - sq. m	120	-	-	-
Nominal Capacity - MTPH	32-38	27/machine	45-55 sinter	75
Fuel Used - Type	Propane	Natural Gas	Propane	Natural Gas
Calorific Value - kCal/-	6,100/L	9,500/Nm^3	-	9,000/Nm^3
Sinter Composition - % Pb	34	30-38	50	45.1
- % Fe	19	13-14	10	17.3
- % S	1.2	1-2	<2	1.7
Exhaust Gas - Volume - Nm^3/hr	140,000	300,000	138,000	230,000
% SO_2	4.8-5.2	1	-	4
SMELTING				
Number of Furnaces	1	3 (2 operating)	2 (1 operating)	3 (2 operating)
Type	Converted ISF	Blast Furnace	Blast Furnace	Blast Furnace
Nominal Capacity - Sinter MTPH	40	53 (total)	21-23	30 per Fce.
Dimensions - m	6 x 2 x 10	2 x 5.5 x 7	1.5 x 6.4 x 6.1	1.67 x 8.5
Oxygen Enrichment - % O_2	1:5	24	23	23
Auxiliary Fuel - Type	Coke	None	Coke	Coke
Calorific Value - kCal/kg	-	-	-	-
Tuyeres	40	-	-	-
Operating Temperature - °C	1,300	1,250	1,400	1,500
Flux as % of Concentrate Feed	-	20	-	7
Campaign Life - Years	1	5	1-3	2

PLANT - continued	Belledune	Trail	AMAX	Herculaneum
Furnace Products (average)				
Temperature - °C	1,050	-	1,200	1,200
Lead Bullion - MTPD	250	-	440	650
% Pb	96	-	96	97+
% Fe	-	-	-	<0.5
% S	0.5	-	<1	<2
Slag - MTPD	600	720 (total)	430	760
- Temperature - °C	1,050	1,250	1300-1400	1,200
- Analysis - % Pb	3.5	3-4	2-3	2.7
- % Fe	31	26.5	22-24	33.5
- % CaO	16	12.5	9-11	10.8
- % SiO_2	21	21.1	22-24	24.2
Furnace Offgas				
Volume - Nm^3/hr	255,000	170,000 (total)	93,000	170,000/Fce.
Temperature - °C	200	250	100	300
% SO_2	0.01	0.1	<0.01	0.025
Gas Cooled by	Air Infiltration	Water Cooled Tower	Water Spray	Dilution
Number of Units	-	1	1	-
Inlet Temp. - ,C	-	160	-	-
Outlet Temp. - ,C	-	110	150 max.	-
Disposition	Recycled to Sintering	Recycled to Sintering	Recycled to Sintering	Recycled to Sintering
SLAG FUMING				
Slag Treated - MTPD	-	720	-	-
Analysis - % Cu	-	0.4	-	-
- % Pb	-	3-4	-	-
- % Zn	-	17	-	-
- % SiO_2	-	21.1	-	-
Furnace Dimensions - m	-	3.5 x 8	-	-
Reductant - Type	-	Coal	-	-

PLANT - continued	Belledune	Trail	AMAX		Herculaneum	
Final Slag Analysis - % Cu	-	0.3	-		-	
- % Pb	-	0	-		-	
- % Zn	-	2-2.5	-		-	
- % SiO_2	-	30	-		-	
Fume Production - MTPD	-	185	-		-	
Analysis - % Pb	-	22	-		-	
- % Zn	-	53	-		-	
PRODUCT CASTING						
Casting Machine - Type	Sheppard	-	Straight Line	Block Caster	Treadwell	Sheppard
Number of Molds	250	Pig 150-Jumbo 16	200	24	438	300
Average Casting Rate - MTPH	28	25 17	22	30	55	36
Normal Casting Temp. - °C	-	375	420	420	400-450	
Cast Product - Type	Pigs	Pigs & Jumbo	Pigs	Jumbo	-	-
Dimensions - m	-	-	0.1x0.1x0.5	0.61x0.61x0.3	-	-
Weight - kg	25, 45 & 1000	-	45	907	45	27
Rejects as % of new Pb cast	-	1	1	<1	<1	<1
SULPHUR FIXATION						
Source of Gas	Sinter Machine	-	Sinter Machine		Sinter Machine	
Plant - Type	Acid Plant	Ammonia Absorption	Acid Plant		Acid Plant	
Rated Capacity - MTPD	600	-	250		270	
Volume - Nm^3/hr	146,000	340,000	42,000		60,000	
Average % SO_2	5	0.9	4.5		4	
Average Conversion Efficiency %	96	91	97.5		99.9	
Product Grade - % H_2SO_4	93.5	-	93		93	
KWH/MT Sulphur Fixed in Product	-	-	360		650	
Auxiliary Fuel	No. 2 & 4 Fuel Oil	None	Propane		Natural Gas	
Tailgas Scrubbing	None	None	None		None	

IMPERIAL AND LEAD SMELTERS of AFRICA AND THE UNITED KINGDOM

COMPANY NAME	Zambia Consolidated Copper Mines Limited		Commonwealth Smelting Ltd.	Associated Lead Manufacturers Ltd.	Tsumeb Corporation Limited
PLANT	Kabwe		Avonmouth	Elswick	Tsumeb
Annual Production - Lead MTPY	-		36,000	60,000	44,000
Annual Production - Zinc MTPY	19,300		90,000	-	-
Flowsheet Page No.	207-208		209	-	-
FEED ANALYSIS					
Concentrate	Lead	Zinc			
% Pb	25	5	-	Secondary Sources	See Description Pages 233-238
% Fe	18	18	-		
% S	26	25	-	-	
% Zn	22	50	-	-	
Mesh	60% -325		-	-	
% H_2O	18.8		-	-	
Furnace Flux					
% SiO_2	51.9		-	-	
% CaO	2.6		-	-	
% Fe	0.9		-	-	
Mesh	62.8% + 1.7 mm		-	-	
FEED PREPARATION					
Blending System	-		16 Weigh Bins	-	
Equipment	Rotary Dryer		-	-	Rotary Dryer
Number	1		-	-	1
Dimensions	2.25 Ø x 18.3		-	-	2.3 Ø x 18.3
Retention Time - min.	18		-	-	8
Fuel Used - Type	Heavy Oil		-	-	Producer Gas
Inlet Gas Temp. - °C	800		-	-	750
Outlet Gas Temp. - °C	60		-	-	-

PLANT - continued	Kabwe	Avonmouth	Elswick	Tsumeb
Exhaust Gas Volume - Nm^3/hr	500	-	-	-
Cleaned by	Rotoclone	-	-	-
SINTERING				
Number of Machines	2	1	-	1
Dimensions - m	1.5 x 18	132 m^2	-	2.44 x 47.8
Nominal Capacity - MTPH	-	180 incl. 140 MTPH return fines	-	100
Fuel Used - Type	Light Oil & Kerosene	Natural Gas	-	Producer Gas
Calorific Value	60:40 by vol.	9200 $kCal/m^3$	-	1600 $kCal/m^3$
Sinter Composition				
% Pb	20	20	-	28
% Fe	12.8	10	-	-
% S	0.7	0.5	-	-
% Zn	24.5	38	-	4
Exhaust Gas - Volume - Nm^3/hr	80,000	103,000	-	165,000
% SO_2	-	6.5	-	2.5
SMELTING				
Number of Furnaces	1	1	2	1
Type	ISF	ISF	Short Rotary	Blast Furnace
Nominal Capacity				
Feed - MTPH	100 MTPD Carbon	-	4	-
Dimensions - m	-	-	3.5 Ø x 4.5	-
Oxygen Enrichment - % O_2	None	None	None	None
Auxiliary Fuel - Type	Coke	Coke	Recovered Oil	-
Calorific Value - kCal/kg	Analysis 85% C/1.7% VM/13% Ash	-	10,500	-
Operating Temperature - °C	1,300	1,300	1,000	-
Flux as % of feed	-	5% limestone to sinter	15	-
Campaign Life - Years	3	2-3	1.5	1.1

PLANT - continued	Kabwe	Avonmouth	Elswick	Tsumeb
Furnace Products (average)				
Temperature - °C	800	1,030	950	-
Lead Bullion - MTPD	80	160-190	55/Fce.	-
% Pb	94.1	95	97.5	77
% Fe	0.5	-	-	-
% Zn	1.3	-	-	-
% Cu	2.9	-	-	-
Zinc Vapour - MTPD	0	300-350	-	-
Slag - MTPD	230	300	15/Fce.	-
Temperature - °C	1,270	1,400	950-1000	-
Analysis - % Pb	0.5	0.7	2-7	2
- % Fe	30.4	30	20-25	22
- % CaO	21.7	15	1-2	23
- % SiO_2	22.3	18	5-10	27
- % Zn	7.4	7	-	-
Furnace Offgas - Volume - Nm^3/hr	44,070	65,000	15,000	70,000
Temperature - °C	1,020	1,030	600-1000	-
% SO_2	-	None	100 ppm	-
Gas Cooled by	Water Scrubbing	Pb Splash Condenser	Water Jacket & Dilution	-
Number of Units	3	2		-
Inlet Temp. - ,C	450	1,050	-	-
Outlet Temp. - ,C	50	450	-	-
Disposition	Recycled (18% CO & 12% CO_2)	-	Recycled	Recycled to Sintering
SLAG CLEANING	None	None	None	None
PRODUCT CASTING				
Casting Machine - Type	Transverse Mould Conveyor	2 Metpro Casters	Worswick Carousel	Sheppard Caster
Number of Molds	120	160	-	300
Average Casting Rate - MTPH	-	12-18	20	40
Normal Casting Temp. - °C	450	470	380	450
Cast Product - Type	Ingot	Various	Ingot	Pigs
Dimensions - m	0.44 x 0.22 x 0.05	"	-	-
Weight - kg	25	25, 800, 1000, 1400	25	-
Rejects as % of new Pb cast	<5	2	-	1

PLANT - continued	Kabwe	Avonmouth	Elswick	Tsumeb
SULPHUR FIXATION	None		None	None
Source of Gas		Sinter Machine		
Plant - Type		Single Contact Acid Plant		
Rated Capacity - MTPD		600		
Volume - Nm^3/hr		103,000		
Average % SO_2		6.5		
Average Conversion Efficiency %		98.5		
Product Grade - % H_2SO_4		98.5		
KWH/MT Sulphur Fixed in Product		440		
Auxiliary Fuel		Natural Gas		
Tailgas Scrubbing		None		

PYROMETALLURGICAL ZINC OPERATIONS of EUROPE

COMPANY NAME	Preussag-Weser-Zink GmbH	Espanola de Zinc SA Zincsa	Norzink AS
PLANT	Nordenham	Cartagena	Odda
Annual Production			
Metal - MTPY	SHG-Zn/Alloys: 120,000	40,000	90,000
Sulfuric Acid - MTPY	76,000		
Flowsheet Page No	-	-	210
FEED PREPARATION			
Feed Materials	Concentrates	Concentrates	Concentrates
Analysis - % Zn			54
- % Fe			8
- % S			32
- % SiO2			2
- Mesh			100 % <100
- % H2O			7
ROASTING			
Type	Lurgi Fluid Bed	Lurgi Fluid Bed	Lurgi Fluid Bed
Number	1	1	2
Dimensions			
Grate Area - m2	90	35	37.2 / 55
Capacity - MTPH	24.2	9.2	10 / 14.5
Air Rate - Nm3/hr	52,000	19,500	18,000 / 29,000
Calcine Analysis - % Zn	55 - 65	58	60
- % Fe	4 - 10	12.5	9
- % S	2 - 3	2.1	2.2
Temperature - oC	950	250	925

PLANT - continued	Nordenham	Cartagena	Odda
Offgas - Volume, Nm3/hr	52,000	20,500	22,000 / 35,000
SO2-Content - %	10	10.5	10
PRODUCT CASTING	SHG-Zn Alloy		
Casting Machine - Type	Wheel Belt	Casting Belt	Norzink Casting Belt
Number of Moulds	144 544	80	509
Casting Rate - MTPH	18 20	10	12
Casting Temperature - oC	500	500	450
Slab Size - kg	Slab: 28 Ingot: 6	25	25/500/1,000/2,000
SULPHUR FIXATION			
Plant Type	Acid Plant	Double Catalysis Acid P.	2 Acid Plants
Capacity - MTPD Acid		300	220 / 330
Volume - Nm3/hr		35,000	42,000 / 44,000
Average SO2-Content - %		7.6	5 / 8
Conversion Efficiency - %		99.5	96.5 / 99.5
Product Grade - % H2SO4		98.5	96 - 98
Tailgas Scrubbing		Candle Filter	Limestone + Seawater

ZINC AND ISF SMELTERS of EUROPE

CAMPANY NANE	Berzelius Metall-hutten GmbH	Preussag AG	Cinko Kursun Metal Sanayii SA (Cinkur)
PLANT	Duisburg	Hüttemwerk Harz	Kayseri
Basic Process	ISF Process	Clinker Kiln, New Jersey Retorts	Waelz and Clinker Kiln, Electrolysis
Annual Production - Zn, MTPY	85,000	60,000	40,000
- Pb, MTPY	40,000		6,000
- Cd, MTPY			125
Flowsheet Page No	211	212	213
FEED ANALYSIS			
Smelter Feed - Type	Concentrates: Zn / Pb / Bulk	Secondary Materials	Ore
% Zn	45 / 6 / 34		21.82
% Pb	3 / 60 / 16		2.1
% Cu	1.4/ 1.7/ 1.6		
% Fe			20
% S	32 / 20 / 30		-
% CaO			3
% SiO2			13
FEED PREPARATION			
Crushing			
Equipment			1 Jaw,2 Cone Crushers
Feed Ore Size - mm			- 10
Capacity - MTPH			150

PLANT - continued	Duisburg	Hüttenwerk Harz	Cinkur
Separation for Zn Containing Material		Low Cl / High Cl Materials	
Equipment		2 Impact Mills / 1 Double Roll Breaker 1 Morgensen Sizer Screens / Vibro Screen / 4 Ball Mills	
Capacity - MTPH (wet)		10 - 20 / 4	
Products			
- % Metallic Fraction		30 /	
- % Oxide Fraction		70 /	
Ventilation Gas Cleaning			
Filter Type		1 Baghouse / 2 Baghouses	
Filter Area - m2		764 / 350	
Gas Temperature - oC		Ambient	
Volume - Nm3/hr		47,000 / 21,000	
Kiln Plant		(Also Used in Pb Circuit)	
Type of Process	Waelz Kiln	Clinkering	Waelz / Clinkering
Type Of Feed Used	Dust of Steel Ind.	Oxide Fraction + Zn-Oxide	Ore / Waelz Oxide
Dimensions			
Diameter - m		3.1	4.34 / 2.6
Length - m		40	70 / 36
Number of Units	1	1	2 / 2
Capacity - MTPH and Unit		10 (Wet Feed)	385.5 / 110
Feed Composition - MTPD			771 Ore / 220 Oxide 122 Coke Breeze / 56 Limestone
Analysis - % Zn			21.82 /
- % Pb			2.1 /
- % Fe			20 /
- % CaO			3 /
- % SiO2			13 /
Ore Type			Smithomite+Zincite
Fuel Used - Type			Oil No 6

PLANT - continued	Duisburg	Hüttenwerk Harz	Cinkur
Operating Temp. - oC			1,100-1,200/ 950
Products			
Oxide - Type	Waelz Oxide	Clinker	Waelz Oxide / Clinker
Analysis - % Zn			65,6 /
- % Pb			6.4 /
- % Cd			0.2 /
Production - MTPD			220 / 220
Slag			
Analysis - % Zn			3,5 /
- % Fe			26 /
- % CaO			14 /
- % Al2O3			11 /
- % SiO2			22 /
Gas Cleaning Method		Scrubbing	
Equipment		1 Scrubber, 1 Neutralisation Tank, 1 Thickener, 1 Filter Press, 2 Filters for Water.	
Temperature - oC		90	
Volume - Nm3/hr		50,000	
Briquetting Plant			
Feed Preparation			
Equipment	1 Heating Drum	1 Impact Mill, 3 Heated Ball Mills with Air Classifier, 2 Impact Mills for Coal, 1 Eirich Mixer.	
Capacity - MTPH		Impact Mill: 40 Ball Mill: 7.5 Coal Impact Mill: 7.5	
Feed Material - Type	Ashes, Pb/Zn Dust	Clinker Oxide, Oxide Zn Ore , Oxide Recycle Material, Bituminous Coal and Sulfite Liquor.	
- Size, micron		40 % < 75	

PLANT - continued	Duisburg	Hüttenwerk Harz	Cinkur
Gas Cleaning by		Baghouse	
Used for		Zn Circuit / Coal Circuit	
Number of Filters		3 / 2	
Filter Area - m2		815 / 361	
Temperature - o C		120 / 120	
Volume - Nm3/hr		50,000 / 26,000	
Briquetting Installations			
Method	Hot Briquetting	Cold Briquetting	-
Equipment		2 Presses for Prepressing, Briquet Diam: 650 mm 1 Press for Final Pressing, Briquet Diam: 1,300 mm.	
Capacity - MTPH		20	
Feed Analysis			
- % Zn		36 - 39	
- % Pb		2 - 3	
- % C		20 - 22	
- % H2O		4	
Briquette Coking			
Equipment	-	4 Cokers (Autogenous Operation)	-
Capacity - MTPH		5 , with 3 Units in Operation	
Gas Cleaning by		Baghouse	
Number of Units		2	
Filter Area - m2		1,863	
Temperature - oC		120	
Volume - Nm3/hr		80,000	
Sintering			
Type of Process	Updraught	-	-
Number of Units	1		
Area - m2	75		
Number of Windboxes	15		
Capacity - MTPH	23		

PLANT - continued	Duisburg	Hüttenwerk Harz	Cinkur
Ignition			
Number of Boxes	1		
Area - m2	2.5		
Ignition Layer - mm	30 -50		
Total Layer - mm	340		
Speed of Machine - m/min	1 - 1.2		
Sinter Capacity - MTPD	360		
Fuel Used	Light Oil		
Calorific Value	9,800 kcal/kg		
Comsumption - m3/hr	0.13		
Feed			
Concentrates - Type	Zn / Pb / Bulk		
Feed Rate - MTPD	150 / 30 / 210		
Analysis - % Zn	45 / 6 / 34		
- % Pb	3 / 60 / 16		
- % Cu	1.4 / 1.7 / 1.6		
- % S	32 / 20 / 30		
Various - Type	Dross + /Various/Sinter/Flux		
	Blue / / Re- /		
	Powder / /cycled/		
Feed Rate - MTPD	72 / 120 / 2,100 / 18		
Analysis - % Zn	34 /		
- % Pb	36 /		
- % Cu	0.2 /		
- % S	2 / / 2.5 /		
Sinter Feed Preparation			
Proportioning Bins - No	7		
Capacity - m3	50		
Recycled Sinter Bins			
Number	4		
Capacity - m3	100		
Mixing Drum - No	1		
Dimensions - m	2.5 Dia x 5		
Conditioning Drum - No	1		
Dimensions -	2.5 Dia x 5		

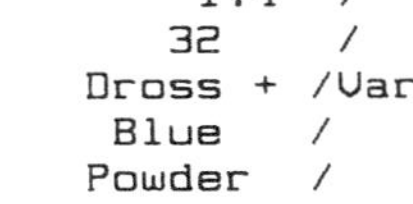

PLANT - continued	Duisburg	Hüttenwerk Harz	Cinkur
Sinter Analysis - % Zn	40		
- % Pb	20		
- % Cu	1.5		
- % S	0.7		
- % FeO	12		
- % CaO	4		
- % SiO_2	3.3		
Sinter Gas - Cleaned by	E.P.		
Volume - Nm3/hr	66,000		
SO_2 - Content - %	7		
SMELTING			
Furnace - Type	ISF with Lead Splash Condenser	New Jersey Vertical Retorts with Recuperators for Air Preheating	-
Number	1	24	
Dimensions	17 m2 Area at Tuyere Zone	SiC Lined Section Height: 9.7 m Heated Area Height: 9 m Internal Length: 2.2 m Internal Width: 0.25 m Heated Area: 39.6 m	
Number of Tuyeres	16 (H2O Cooled), 120mm Dia.		
Capacity	85,000 TPY Zn 40,000 TPY Pb	15.5 TPD Charge per Unit	
Feed Analysis - Zn	See Sinter.	42 - 44	
- % Pb		2.5 - 18	
- % C		17 - 18	
Operating Temp. - oC		Combustion Chamber: 1,280-1,320 Preheated Air: 500 - 750 Offgas Recup. Inlet: 1,000-1,200 Offgas Recup. Outlet: 500	
Fuel - Type	Hot Coke (700 oC)	Natural Gas (Impulse Burners)	
Calorific Value		7,300 kcal/Nm3	
Comsumption		2,407 Mcal/ MT Zn	

PLANT - continued	Duisburg		Hüttenwerk Harz		Cinkur
Blast Air - Nm3/hr	37,000				
Temperature - oC	1,150				
Pressure - mmWG	5,000				
Furnace Metal - MTPD	Zn : 250 Pb: 108		Zn per Unit: 7.5		
Analysis	Zinc	Bullion	Briquettes	Secondaries	
- % Zn	98.6	0.4	99.5 - 99.7	80 - 90	
- % Pb	1.1	91	0.2 - 0.4		
- % Cd			0.02- 0.045		
- % Cu	0.06	6			
Temperature - oC	500	1,150			
Furnace Slag/Residue	Slag		Residue		
Production - MTPD	155				
Analysis - % Zn	6.5		3.5 - 4		
- % Pb	1		17		
- % C					
- % FeO	44				
- % CaO	13				
- % SiO2	18				
Temperature - oC	1,350				
Furnace Offgas					
Cooled by	Condenser: Inlet 1,050 oC Outlet 460 oC Cooling Tower: Outlet 80 oC		Condenser and Waste Heat Boiler		
Gas Cleaning by	Theisen Disintegrator		Baghouse: Area 4.800 m2		
Temperature - oC	42		120		
Volume - Nm3/hr	54,500		160,000		
Composition - % Zn	7.5				
- % CO	23				
- % CO2	9.3				
- % H2	1.4				
Electric Furnace - Type	-		Induction		
Feed			Zn and Zn Alloy Scrap		
Capacity - MTPH			2.5		

PLANT - continued	Duisburg	Hüttenwerk Harz	Cinkur
Rating - kW		1,800	
Consumption		160 kWh/MT Charge	
Product to		New Jersey Refinery or Zn Oxide Production Line	
REFINERY			
Basic Process	Distillation Columns	Distillation Columns	-
Pb Column - Number	3		
Number of Trays	59		
Tray Size - mm	1,371 x 762		
Capacity - MTPD	30.5 Condensed Product		
Cd Column - Number	1		
Number of Trays	61		
Capacity - MTPD SHG-Zn	70		
Reboiler Column - Number	1		
Number of Trays	39		
Capacity - MTPD HG-Zn	45		
Fuel - Type	Heavy Fuel Oil		
Consumption - kg/MT Zn	180		
CASTING			
Casting Machine - Type	Continuous Belt	Continuous Belt	
Number of Machines	2	1	
Number of Moulds	PW-Zn: 160 SHG-Zn: 120		
Casting Rate - MTPH	16 / 12		
Casting Temperature - oC	480		
Slab Size - mm	455 x 220 x 50		
Weight	25 kg / Slab	Up to 1 and 2 MT	
SULPHUR FIXATION			
Type of Plant	Contact Acid Plant	-	
Source of Gas	Sinter Machine		

ZINC AND IMPERIAL SMELTERS of JAPAN

COMPANY NAME	Mitsui Miike Smelting Co.	Nikko Zinc Co. Ltd.	Hachinohe Smelting Company Limited	
PLANT	Miike	Mikkaichi	Hachinohe	
Annual Production - Zinc, MTPY	100,000	103,400	64,800	
- Lead, MTPY	-	-	29,500	
Flowsheet Page No.	-	214	215	
FEED ANALYSIS				
Concentrate			Zinc	Lead
% Zn	11	53.2	48.0	10.7
% Pb	-	-	4.7	51.2
% Fe	16	6.4	-	-
% S	2.7	31.5	30.5	20.6
% SiO_2	11.1	-	-	-
Mesh	-	90%-100	-	-
% H_2O	25		-	-
Furnace Flux				
% SiO_2	93	-	-	
% CaO	3	-	-	
% Fe	-	-	-	
Mesh	-	-	-	
FEED PREPARATION				
Blending System	-	-	Mixing Drum	

PLANT (cont'd)	Miike	Mikkaichi	Hachinohe
Equipment	Rotary Dryer	Fluosolids Roaster	-
Number	1	3	-
Dimensions - m	3.0 Ø x 25	5.7 Ø x 9.8	-
	-	6.5 Ø x 10.0	-
	-	9.6 Ø x 13.75	-
Calcine, % Zn/Fe/S	-	61.7/7.7/2.1	-
Average Feed Rate - WMTPH	21.3	5.6/7.1/15.2	-
Feed Moisture, % H_2O	25	-	-
Sinter Machines	-	2	1
Nominal Capacity - MTPH	-	28	-
Dimensions - m	-	-	2.5 x 36
Fuel Used - Type	Heavy Oil	Coke Breeze	Bunker C
Calorific Value - kCal/ℓ	9,800	-	9,900
Sinter Composition % Zn/Fe	-	54.2/9.8	38.3 Zn/9.1 Fe/31.4 Pb
Inlet Gas Temp - °C	800	-	-
Outlet Gas Temp - °C	70	-	-
Product Moisture - % H_2O	12	-	-
Retention Time - min.	48	-	-
Exhaust Gas - Volume - Nm^3/hr	36,000	-	60,000
Cleaned by	TCA	Absorbing Tower	Acid Plant
SMELTING			
Number of Furnaces	2	9	1
Type	Half Shaft	Electrothermic	ISF
Nominal Capacity - Feed MTPH	7.6 and 5.0	2 and 3	27.9
Dimensions - m	7x1 and 5.4x1	2.2 x 11.8 (8)	9.5x3.0x6.4
		2.9 x 13.7 (1)	
Auxiliary Fuel - Type	Steam	Coke	Coke
Calorific Value - kCal/kg	0.8	-	-
Operating Temperature - °C	300	-	950
Flux as % of conc. feed	4.6	-	Silica - 1-2%
Campaign Life - years	17	-	5

PLANT (cont'd)	Miike	Mikkaichi	Hachinohe
Furnace Products (average)			
Temperature - °C	1,000	-	-
Zinc - MTPD	Oxide 70	Metal 8.9	Metal 220
% Zn	51	98.86	98.7
% Pb	12	-	1.1
% Fe	-	-	0.02
Matte - MTPD	25	-	Lead Bullion 130
% Cu	10	-	0.11
% Fe	51	-	% Pb 91.0
% S	18	-	-
Slag - MTPD	160	Residue to magnetic separation	200
- Temp. - °C	1250		1200-1250
- Analysis - % Zn	3.5		7.0
- % Fe	29		32.0
- % CaO	6		16.0
- % SiO_2	25		20.0 % Pb 0.9
Furnace Offgas - Volume - Nm^3/hr	60,000 & 45,000	-	42,000
- Temp. - °C	1000	-	445
- % SO_2	0.04	-	-
Gas Cooled by	Boiler & Hairpin Cooler	-	Water Cooling
Number of Units	1	-	3
Inlet Temp. - °C	1000	-	445
Outlet Temp. - °C	100	-	35-40
Disposition of Dust	-	-	Recycled to Sintering

REFINING

	Miike	Mikkaichi	Hachinohe
Columns - Number	-	-	4
Fuel Used	-	-	LPG & LCV
Calorific Value - $kCal/Nm^3$	-	-	3,500
SHG Zinc - MTPD	-	-	70
Analysis - % Zn	-	-	99.997
- % Pb	-	-	0.002
- % Cd	-	-	0.0015

PLANT (cont'd)	Miike	Mikkaichi	Hachinohe
PRODUCT CASTING			
Casting Machine - Type			Chain Link
Number of Moulds			96
Average Casting Rate - MTPH			11
Normal Casting Temp. - °C			530
Rejects as % of new material cast			5-6
SULPHUR FIXATION			
Source of Gas	-	Zinc Roaster	Sinter Machine
Plant - Type	Basic Aluminum Sulphate Scrubbing	Acid Plant	Double Contact
- Rated Capacity, MTPD	3 (5 net)	550	445
- Volume, Nm^3/hr	130,000	-	60,000
- Average % SO_2	5 ppm	10.5	7
- Average Conversion Efficiency %	98.5	-	99.6
Product Grade - % H_2SO_4	Gypsum	98% H_2SO_4	98% H_2SO_4 (80% Prod.) 77% " (20% ")
KWH/MT Sulphur Fixed in Product	4.5	-	100
Auxiliary Fuel	-	-	Oil
Tailgas Scrubbing	None	-	Yes

PYROMETALLURGICAL ZINC OPERATIONS
of
NORTH AMERICA

COMPANY NAME	AMAX Zinc Co.	Canadian Electrolytic Zinc Ltd.	Cominco Ltd.	Union Zinc Inc.
PLANT	Sauget	Valleyfield	Trail	Clarksville
Annual Production - Zinc MTPY	73,000	230,000	272,000	82,000
- Cadmium MTPY	730	-	570	-
Flowsheet Page No.	216	-	-	-
FEED ANALYSIS				
Concentrate				
% Zn	57.7	-	55-56	62
% Fe	2.96	-	5.0	1.5
% S	29.57	-	31.0	32.0
% SiO_2	1.12	-	1.4	-
Mesh	90% -325	-	78% -200	90% -150
% H_2O	-	-	8-9	-
FEED PREPARATION				
Blending System	Front End Loader	-	Stockpiles	-
Pretreatment	Preleach	-	None	None
ROASTING				
Type	Lurgi Fluid Bed	Lurgi Fluid Bed	Lurgi Fluid Bed	Lurgi Fluid Bed
Number	2	4	2	1
Dimensions - m	9.7 Ø	-	13.8 Ø x 16.5	-
Grate Area - m^2	32.2	2x32, 1x72, 1x76	-	64.7
Nominal Capacity - MTPH	12.7	55 total	26	17-20
Air Rate - Nm^3/hr	17,000	106,200 total	43,000	42,725
Start-up Fuel Used - Type	Natural Gas	-	Oil	Oil
Calorific Value	-	-	-	-

PLANT - continued	Sauget	Valleyfield	Trail	Clarksville
Calcine Composition				
% Zn	64.3	52	61-62	73-74
% Fe	3.4	10	6.0	0.8-1.2
% S	3.18	32	2.5	2.5-3.0
% Sulphide S	-	-	-	0.25-0.65
Discharge Temp. - °C	980	950	930	960
Offgas - Volume - Nm^3/hr	56,000	-	76,600	-
- % SO_2	7	10-11.5	8	10.5
PRODUCT CASTING				
Casting Machine - Type	Continuous	1 x E+J 2 x Sheppard	2 Continuous 1 Jumbo 3 Slab Casters	Metpro
Number of Molds	432	120 & 80	28 & 200	160
Average Casting Rate - MTPH	13	15	120, 9, 20	20
Normal Casting Temp. - °C	-	-	-	-
Slab Size - kg	500	500	500	525
Rejects as % of new Zn cast	-	-	2	<0.6
SULPHUR FIXATION				
Plant Type	Acid Plant	3 Acid Plants	3 Monsanto Acid Plants	Acid Plant
Rated Capacity - MTPD	-	-	1,250 total	450
Volume - Nm^3/hr	56,000	-	160,000	-
Average % SO_2	7	10-11.5	7	7
Average Conversion Efficiency %	98	2 @ 97.4, 1 @ 99.7	96	99.6
Product Grade - % H_2SO_4	-	93,96,98, 20° Oleum	93	93 & 103
Auxiliary Fuel	None	-	-	-
Tailgas Scrubbing	Single Absorption Tower	None	Ammonia Absorption	None

PYROMETALLURGICAL ZINC OPERATIONS
of
NORTH AMERICA

COMPANY NAME	St. Joe Resources Company		Falconbridge Ltd.
PLANT	Monaca	Bartlesville	Kidd Creek (3)
Annual Production - Zinc MTPY	95,000	51,500	107,000
Flowsheet Page No.	-	-	-
FEED ANALYSIS			
Concentrate			
% Zn	58	52	55
% Fe	5	6	10
% S	32	32	32
% SiO_2	2	2	-
Mesh	70% -150	70% -100	98% -325
% H_2O	10	-	8-9
FEED PREPARATION			
Blending System	-	Slurry Tanks	-
Pretreatment	-	None	None
Dryer Type	Rotary		
Number of Units	1		
Dimensions - m	3.5 Ø x 15.2		
Fuel Used	Natural Gas & CO		
Calorific Value - $kCal/Nm^3$	8,900 & 2,200		
Average Feed Rate - WMTPH	15		
Product Moisture - % H_2O	6		
Exhaust Gas Volume - Nm^3/hr	4,420		
Temperature - °C	150		
Cleaned by	Wet Scrubber		

PLANT - continued	Monaca	Bartlesville	Kidd Creek
ROASTING			
Type	Fluid Bed	Slurry Fed Fluosolids	Fluid Bed
Number of Units	1	2	
Dimensions - m	6.9 Ø x 10.7	5.5 Ø x 6.75	8.4 Ø x 16
Grate Area - m^2	-	-	46
Nominal Capacity - MTPH	15	3.8 & 7.6	-
Average Feed Rate - MTPH	-	-	18
Air Rate - Nm^3/hr	20,400	8,500 & 17,000	26,000
Fuel Used - Type	None	Natural Gas for Start-up	-
Calcine Composition			
% Zn	65	60	-
% Fe	6	7	-
% S	2	2.5	-
% Sulphate S	-	-	-
Discharge Temp. - °C	850	900	-
Offgas - Volume - Nm^3/hr	21,240	32,000	-
- % SO_2	12	7.5 at Acid Plant	-
Sinter Machines - Number	3	N/A	N/A
Nominal Capacity - MTPH	30		
Dimensions - m	2 @ 1.5 x 13.4 1 @ 1.8 x 13.4		
Fuel Used - Type	Coke Breeze		
Sinter Composition - % Zn/Fe/S	49/7/0.1		
Exhaust Gas Volume - Nm^3/hr	382,320		
SMELTING			
Number of Furnaces	5	N/A	N/A
Type	Electrothermic		
Dimensions - m	2.4 Ø x 12.2		
Operating Temp. - °C	1,200		

PLANT - continued	Monaca	Bartlesville	Kidd Creek
Furnace Products (average)			
Temperature - °C	500		
Zinc - MTPD	68/Furnace		
% Zn	98.5		
% Pb	-		
% Fe	0.01		
Slag - MTPD	100		
% Zn	2		
% Fe	4		
% CaO	7		
% SiO_2	52		
Furnace Offgas - Volume - Nm^3/hr	4,250		
- Temperature - °C	850		
Cooled by	Zinc Condenser & Wet Scrubbing		
Inlet Temp. - °C	500		
Outlet Temp. - °C	25		
Disposition of Dust	Recycled		
PRODUCT CASTING			
Casting Machine - Type	Slab Conveyor	Morganite Slab Caster	2 E+J Casters
Number of Molds	124	-	-
Average Casting Rate - MTPH	7	-	-
Normal Casting Temp. - °C	500	-	-
Slab Size - kg	25	26, 272, 1090	227, 1090
Rejects as % of new Zn cast	Low	-	-
SULPHUR FIXATION			
Plant Type	2 Contact Acid Plants in Series	Single Contact Acid Plant	Single Contact Acid Plant
Rated Capacity - MTPD	330	250	-
Volume - Nm^3/hr	42,000	32,000	-
Average % SO_2	6	7.5	-
Average Conversion Efficiency %	99.5	98.1	-
Product Grade - % H_2SO_4	93	93, 98 and Oleum	-

PLANT - continued	Monaca	Bartlesville	Kidd Creek
Auxiliary Fuel	Sulphur 7,500 MTPY	-	-
Tailgas Scrubbing	Second Acid Plant	None	None

Reference (3) N.J. Themelis and G.M. Freeman, "Fluid Bed Behaviour in Zinc Roasters", Journal of Metals, August 1984, pages 52 to 57.

LEAD SMELTERS
of
EUROPE

COPAMY NAME	Boliden Bergsoe AB
PLANT	Landskrona
Flowsheet Page No	217
FEED MATERIAL	
Primary Feed	
Type	Battery Scrap
Analysis	
- % Pb	43
Furnace Flux	
Type	Recirculated Slag
Fuel Used	
Type	Coke
Size - mm	80 - 120
SMELTING	
Furnace	
Type	Blast Furnace
Number	1
Capacity	25,000 MTPY Pb
Dimensions	
- Shaft	4.25 x 1.4 x 5.3
- Tuyere Zone	4.0 x 1.0 x 4.3
Tuyeres	
- Diameter , mm	80
- Number	18
Blast	
- Flow, Nm3 /hr	4,500
- % O2	23
- Temperature, oC	350
Furnace Gas	
- Flow, Nm3/hr	7,000
- Temp. at Shaft, oC	200
- Afterburning Temp. oC	1,000
Gas Cooling	Boiler
Gas Cleaning	Baghouse
Metal	
- Production, MTPH	3
- Temperature, oC	900
Slag	
- Production, MTPH	3 - 4
- Temperature, oC	1,300
METAL TREATMENT	Blast furnace Pb is refined and/or alloyed followed by ingot casting.

SECONDARY LEAD SMELTER
OF
F.R.Germany

COMPANY NAME	Preussag AG
PLANT	Hüttenwerk Harz
Annual Production - Lead , MTPY	
Flowsheet Page No	218
FEED MATERIALS	
Type	Lead Oxide, Battery Scrap, Recycled Lead.
FEED PREPARATION	
Battery Scrap Separation	
Equipment:	
Grinding	Double Roll Crusher, Hammer Mill
Classifying	3 Oxide Classifiers 2 Screens 1 Metal Classifier 1 Plastic Classifier
Feed Rate - MTPH	30 (Wet Material)
Feed Composition	
- % Pb	58
- % Case Material	16
Products	
Pb Containing Products	
Metal Fraction	
- % of Dry Feed	40
Oxide Fraction	
- % of Dry Feed	60
Non-Metallic Products	Polypropylene, Hard Rubber
Equipment Ventilation	
Cleaning Method	Wet Scrubbing
Volume - Nm3/hr	25,000
Water Teatment	
Type	Neutralisation
Reagent	Lime
Equipment	2 Thickener, 2 Sand/Coal Filters
SMELTING	
Kiln Plant	
Furnace - Type	Rotary Kiln
Number	2
Dimensions - m	3.1 dia. x 40 / 2.1 Dia. x 40
Capacity - MTPH	230 (Metal Frac.)/ 120 (Oxide)
Fuel - Type	Natural Gas
- Rate , Nm3/hr	max. 1,200 / max. 420
- Consumption, Nm3/MT Pb	70 / 90
Flux - Type	Soda Ash and Iron Chips

PLANT - continued	Hüttenwerk Harz
Reductant	Coke Fines
Products	
Metal - Type	Crude Lead
Slag	
Production - % of Feed	14
Analysis - % Pb	< 4
- % Fe	18
- % S	22
Offgas	
Volume - Nm3/hr	70,000 / 35,000
Cleaning Method	Baghouse
Filter Area - m2	1,800 / 830
SRF Plant	
Furnace - Type	Short Rotary Furnace
Number	1
Dimensions - m	3.5 dia. x 3.5
Fuel - Type	Natural Gas
Consumption - Nm3/MT Feed	70
Capacity - MTPD Metallic Feed	150
- MTPD Oxide Feed	60
Flux - Type	Soda Ash and Iron Chips
Reductant	Coke Fines
Products	
Metal - Type	Crude Lead
Slag	
Production - % of Feed	15 - 20
Analysis - % Pb	4
- % Fe	up to 20
- % S	up to 15
Offgas	
Volume - Nm3/hr	10,000
Cleaning Method	Baghouse
Filter Area - m2	1,920
Ventilation System	
Volume - Nm3/hr	56,000
REFINING	
Kettles	With Stirrer, Skimming, Vent Hood
Number	11
Holding Capacity - MT	90
Fuel - Type	Natural Gas
Consumption - Nm3/MT Pb	70
Offgas	
Volume - Nm3/Hr	60,000
Cleaning Method	Baghouse
Filter Area - m2	1,680
Refinery Capacity - MTPY	60,000 (Refined Lead)

TIN SMELTER
OF
EUROPE

COMPANY NAME	Berzelius Metallhuetten GmbH
PLANT	Duisburg
Flowsheet Page No	219
Annual Production	
Primary Metal - MTPY	7,000 Sn/Pb Alloy (3,500 Sn)
Byproducts - MTPY	2,000 ZnO
	100 As2O3
	450 Cu
	300 Sn/Pb/Sb Alloy
	(Including slime from anode alloy)
FEED MATERIAL	Mainly Sn/Pb secondary materials are used.
SMELTING	
Furnace	Smelting Route 1:
Type	Short Rotary Furnace
Number	1
Nominal Capacity - MTPH	1 - 1.5 (Sn/Pb Feed)
Dimensions - m	3 dia. x 4 (Outside Dimensions)
Oxygen Enrichment - %	Oxy-Fuel Burners: 95 %
Auxiliary Fuel	
- Type	Fuel Oil
- Calorific Val. kcal/kg	9,600
Operating Temperature, oC	1,100 - 1,300
Campaign Life - months	4
Flux as % of Feed	2 - 4
Furnace Metal	
Analysis	
- % Sn	20 - 95
- % Pb	5 -10
- % Fe	0.1 - 1
- % S	< 0.1
Temperature - oC	950
Production - MTPD	12.5
Furnace Slag	
Analysis	
- % Sn	5 -15
- % Pb	< 5
- % Fe	5 -20
- % CaO	3
- % SiO2	10
- % NaCO3	5 -10
Temperature - oC	1,000
Production - MTPD	12

PLANT - continued	<u>Duisburg</u>
Furnace Slag	
Analysis	
- % Sn	5 -15
- % Pb	< 5
- % Fe	5 -20
- % CaO	3
- % SiO_2	10
- % $NaCO_3$	5 -10
Temperature - oC	1,000
Production - MTPD	12
Furnace Offgas	
Volume - Nm^3/hr	5,000 (+ 10,000 Infiltration Air)
Temperature - oC	1,000
SO_2 Content - ppm	300
Gas Cooled by	Air Infiltration + Evaporative Cooling
Number of Units	1
Inlet Temp - oC	800
Outlet Temp - oC	220 (Inlet Baghouse)
Disposition of Gas	Recycled to short rotary furnace or kiln.
Furnace	Smelting Route 2:
Type	Rotary Kiln
Number	1
Capacity - MTPH	1.2 - 1.4 (Mixed Sn/Pb/Zn Oxides)
Dimensions - m	2 dia. x 25
Auxiliary Fuel	
Type	Light Oil
Calorific Value-kcal/kg	9,600
Campaign Life - months	6
Temperature - oC	1,250 - 1,350
Flux as % of Feed	40
Furnace Metal	
Analysis - % Sn	30 - 60
- % Pb	40 - 60
- % Fe	0.1 - 0.3
- % S	< o.1
Temperature - oC	950
Production - MTPD	10.5
Furnace Slag	
Analysis - % Sn	0.4 - 0.7
- % Pb	0.5
- % Fe	1 - 2
- % CaO	0.3 - 1
- % SiO_2	6 - 8
Temperature - oC	1,050
Production - MTPD	12.5
Furnace Offgas	
Temperature - oC	400 - 600
Volume - Nm^3/hre	6,000
SO_2 - Content - ppm	< 100

PLANT - continued	Duisburg
Cooling Method	Evaporative Cooling
Number of Units	1
Inlet Temp - oC	400 - 600
Outlet Temp - oC	350
Disposition of Dust	ZnO - dust for shipment.
SLAG CLEANING	
Slag Processed	Sb/Pb containing slag from short rotary furnace.
Furnace	
Type	Kiln
Number of Units	1
Capacity - MTPH	1,3 - 1.5
Dimensions - m	2 dia. x 25
Oxygen Enrichment	-
Auxiliary Fuel	
Type	Light Oil
Calorific Value -kcal/kg	9,800
Temperature - oC	1,250 - 1,350
Campaign Life - months	6
Flux as % of Feed	-
Furnace Metal	
Analysis - % Sn	20 - 40
- % Pb	60 - 80
- % Fe	0.1 - 0.5
- % S	< 0.1
Temperature - oC	950 - 1,000
Production - MTPD	9.5
Furnace Slag	
Analysis - % Sn	0.2 - 0.5
- % Pb	0.2 - 0.5
- % Fe	5 - 8
- % CaO	2 - 5
- % SiO2	10 - 25
- % Na2CO3	1 - 3
Temperature - oC	1,150
Production - MTPD	22
Disposition	To Dump.
Furnace Offgas	
Temperature - oC	400 - 500
Volume - Nm3/hr	6,00
SO2-Content - ppm	2,000
Cooling Method	Evaporative Cooling
Number of Units	1
Inlet Temp. - oC	400 - 600
Outlet Temp. - oC	350
Disposition of Dust	Recycled to kiln processing secondary feed material.

PLANT - continued

	Duisburg
PRODUCT CASTING	
Casting Machine	
Type	Own Design
Number of Moulds	100
Casting Rate - MTPH	1 - 3
Casting Temp. - oC	350 - 450
Ingot Size - mm	205 x 60 x 30 and 400 x 120 x 80
Weight - kg	3 - 25
SULPHUR FIXATION	Minimal sulphur in feed.

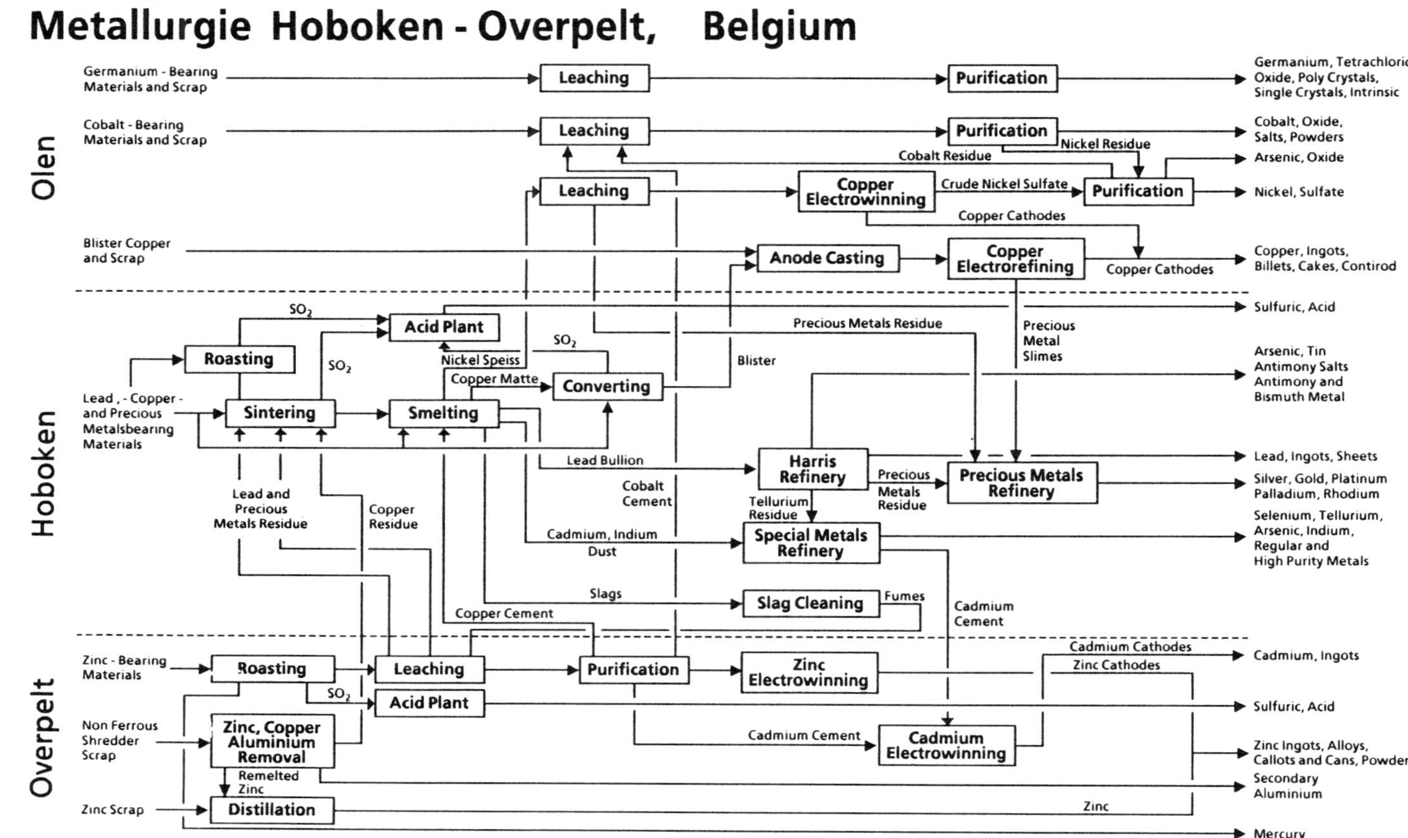
Flowsheet
Metallurgie Hoboken - Overpelt, Belgium
Olen
Hoboken
Overpelt
Germanium - Bearing Materials and Scrap
Cobalt - Bearing Materials and Scrap
Blister Copper and Scrap
Lead, - Copper - and Precious Metalsbearing Materials
Zinc - Bearing Materials
Non Ferrous Shredder Scrap
Zinc Scrap
Leaching
Leaching
Leaching
Purification
Purification
Purification
Copper Electrowinning
Anode Casting
Copper Electrorefining
Nickel Residue
Cobalt Residue
Crude Nickel Sulfate
Copper Cathodes
Copper Cathodes
Roasting
Sintering
Acid Plant
Smelting
Converting
SO_2
SO_2
SO_2
Nickel Speiss
Copper Matte
Blister
Precious Metals Residue
Precious Metal Slimes
Lead Bullion
Cobalt Cement
Harris Refinery
Precious Metals Refinery
Precious Metals Residue
Tellurium Residue
Special Metals Refinery
Cadmium, Indium Dust
Lead and Precious Metals Residue
Copper Residue
Slags
Slag Cleaning
Fumes
Copper Cement
Cadmium Cement
Roasting
Leaching
Purification
Zinc Electrowinning
Acid Plant
SO_2
Zinc, Copper Aluminium Removal
Remelted Zinc
Distillation
Cadmium Cement
Cadmium Electrowinning
Cadmium Cathodes
Zinc Cathodes
Zinc
Germanium, Tetrachloride, Oxide, Poly Crystals, Single Crystals, Intrinsic
Cobalt, Oxide, Salts, Powders
Arsenic, Oxide
Nickel, Sulfate
Copper, Ingots, Billets, Cakes, Contirod
Sulfuric, Acid
Arsenic, Tin Antimony Salts Antimony and Bismuth Metal
Lead, Ingots, Sheets
Silver, Gold, Platinum Palladium, Rhodium
Selenium, Tellurium, Arsenic, Indium, Regular and High Purity Metals
Cadmium, Ingots
Sulfuric, Acid
Zinc Ingots, Alloys, Callots and Cans, Powder
Secondary Aluminium
Mercury

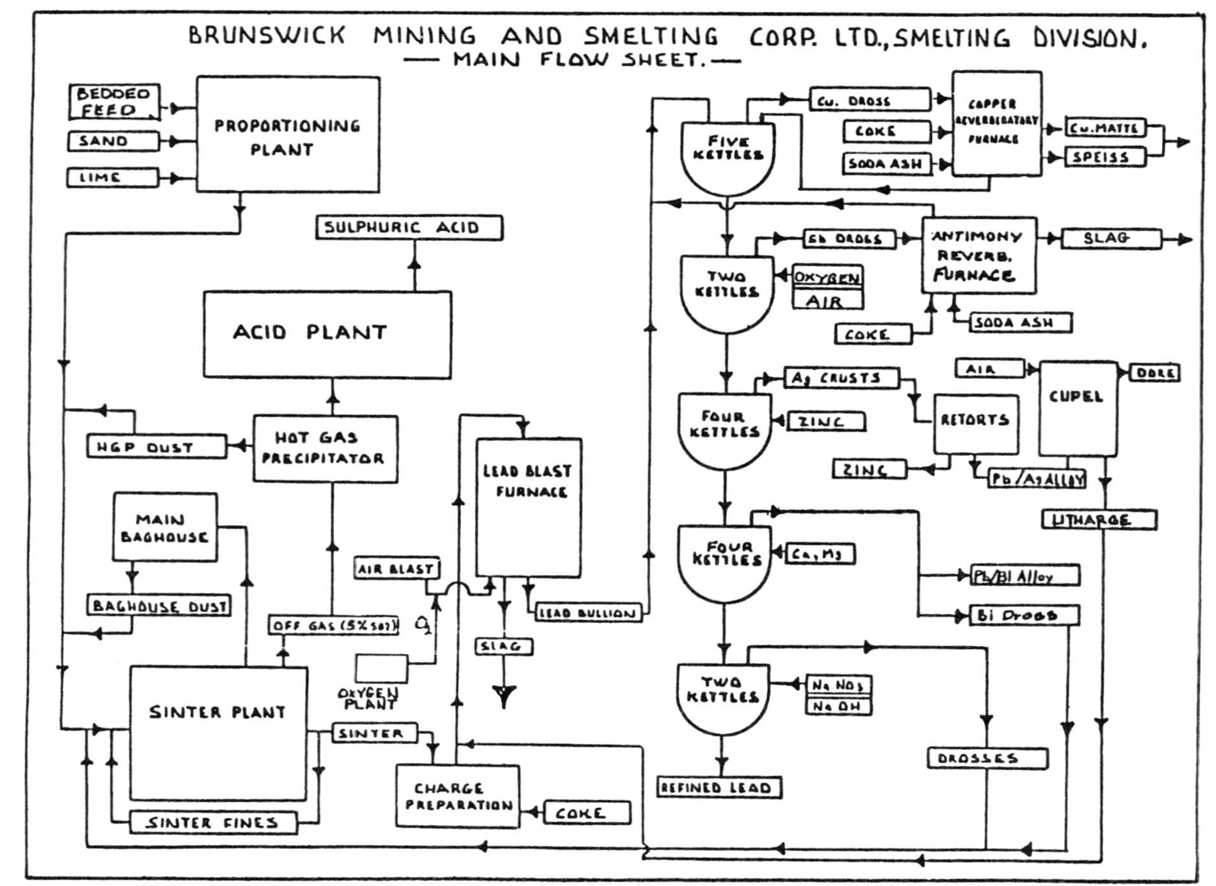
BRUNSWICK MINING AND SMELTING CORP. LTD., SMELTING DIVISION.
— MAIN FLOW SHEET. —
BEDDED FEED
SAND
LIME
PROPORTIONING PLANT
SULPHURIC ACID
ACID PLANT
HGP DUST
HOT GAS PRECIPITATOR
MAIN BAGHOUSE
BAGHOUSE DUST
AIR BLAST
OFF GAS (5% SO2)
O2
OXYGEN PLANT
SINTER PLANT
SINTER
SINTER FINES
CHARGE PREPARATION
COKE
LEAD BLAST FURNACE
LEAD BULLION
SLAG
FIVE KETTLES
Cu. DROSS
COKE
SODA ASH
COPPER REVERBERATORY FURNACE
Cu. MATTE
SPEISS
TWO KETTLES
Sb DROSS
OXYGEN
AIR
ANTIMONY REVERB. FURNACE
SLAG
COKE
SODA ASH
FOUR KETTLES
Ag CRUSTS
ZINC
RETORTS
ZINC
AIR
CUPEL
DORE
Pb/Ag ALLOY
LITHARGE
FOUR KETTLES
Ca, Mg
Pb/Bi Alloy
Bi Dross
TWO KETTLES
Na NO3
Na OH
DROSSES
REFINED LEAD

Flowsheet KABWE
Sinter Plant

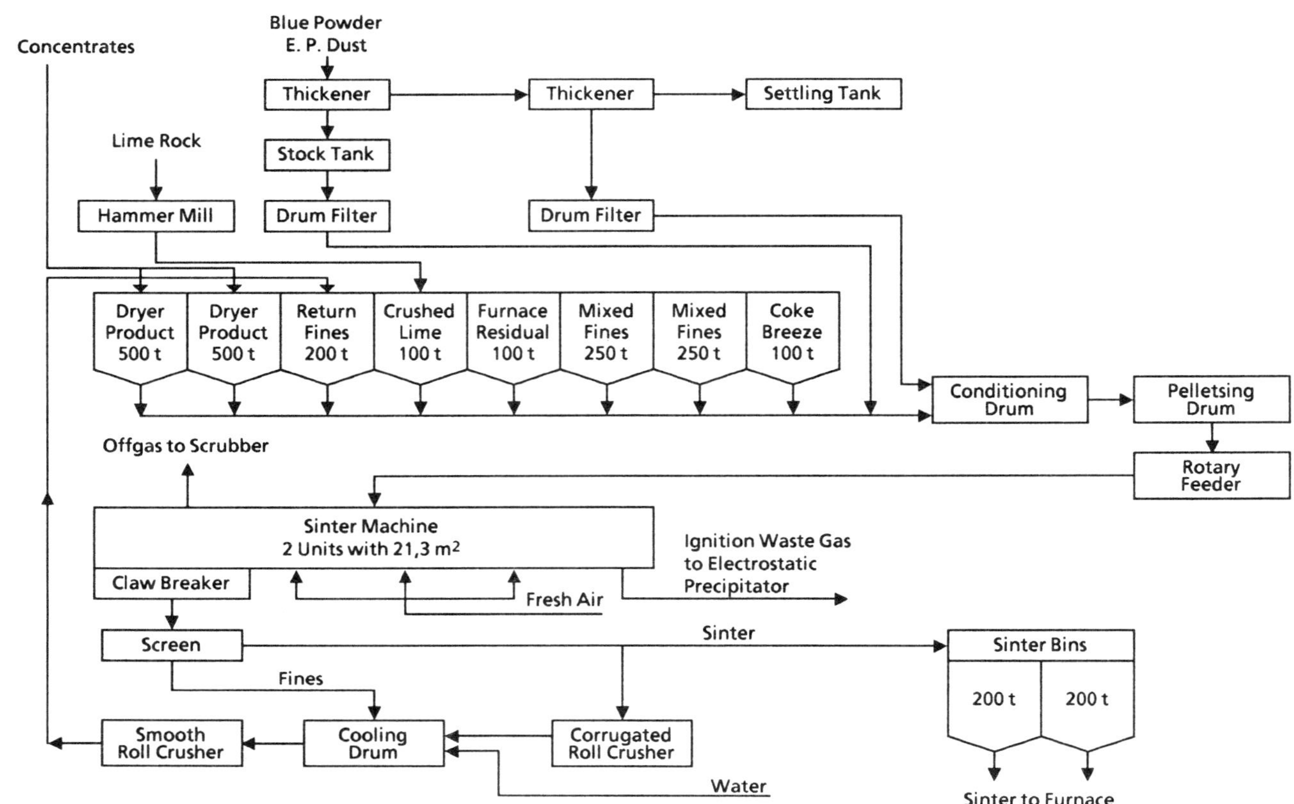

Flowsheet KABWE
Imperial Smelting Furnace

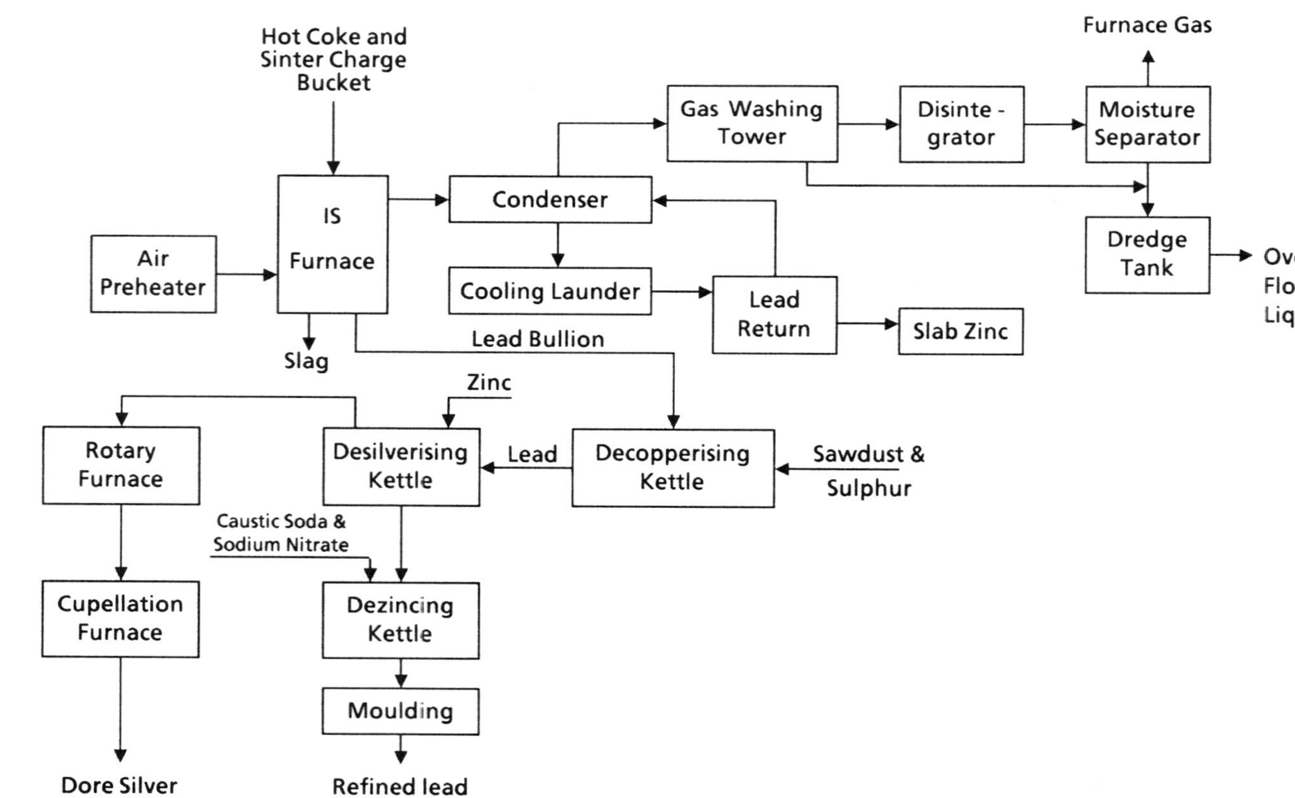

COMMONWEALTH SMELTING LIMITED

PROCESS FLOWSHEET OF "AVONMOUTH SMELTER"

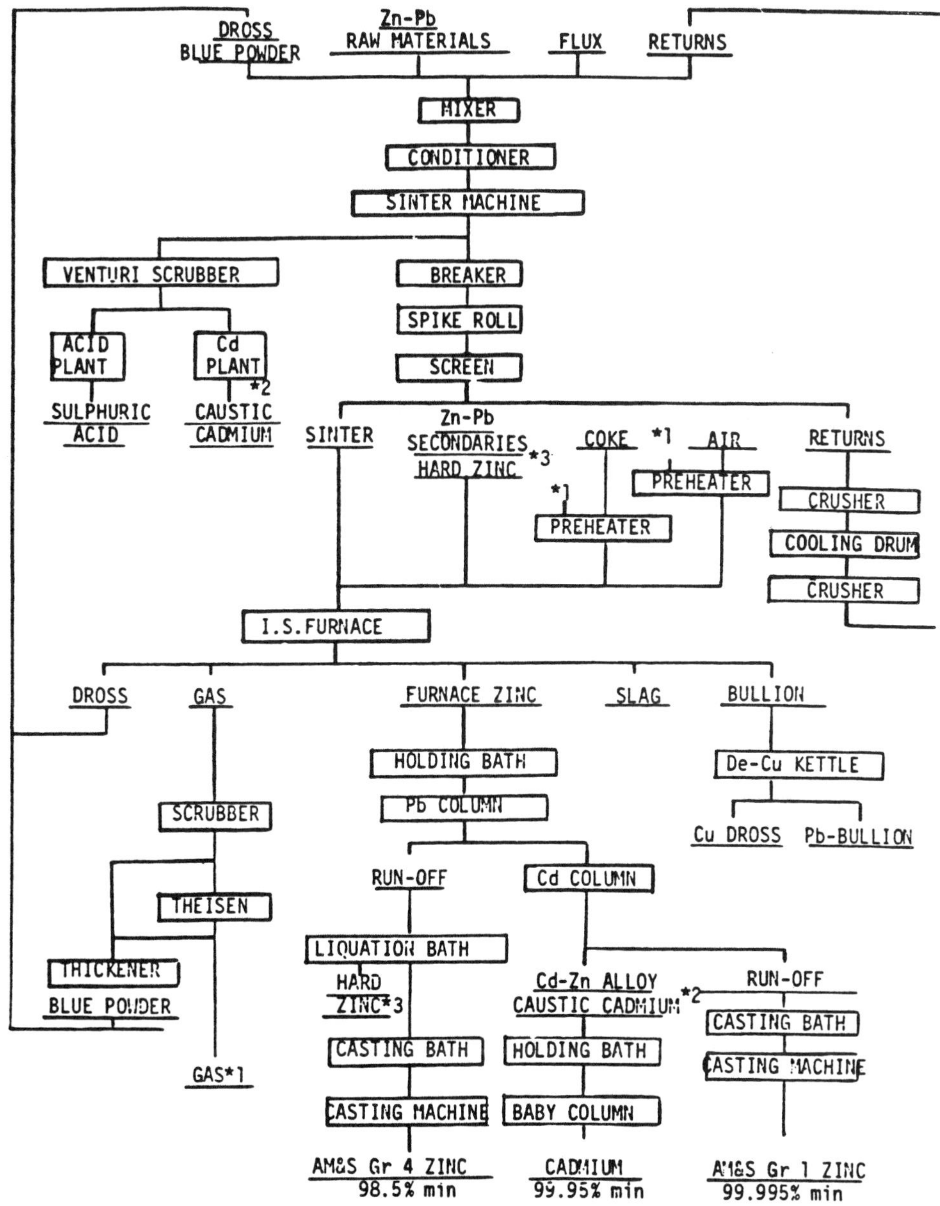

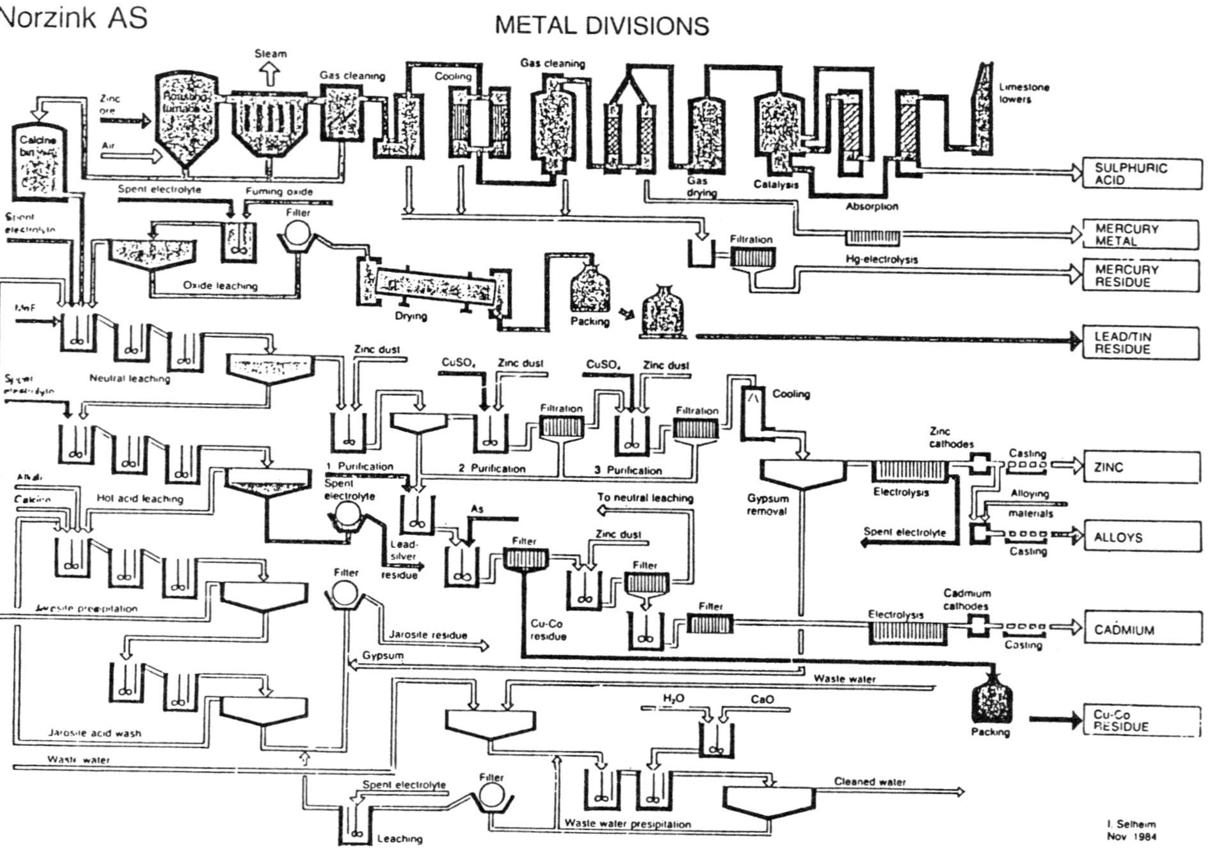

Norzink AS
METAL DIVISIONS
Steam
Gas cleaning
Cooling
Gas cleaning
Limestone towers
Zinc ore
Air
Calcine bin
Gas drying
Catalysis
Absorption
SULPHURIC ACID
Spent electrolyte
Fuming oxide
Filter
Spent electrolyte
Oxide leaching
Filtration
Hg-electrolysis
MERCURY METAL
MERCURY RESIDUE
Drying
Packing
LEAD/TIN RESIDUE
Neutral leaching
Spent electrolyte
Zinc dust
CuSO4
Zinc dust
CuSO4
Zinc dust
Filtration
Filtration
Cooling
1 Purification
2 Purification
3 Purification
Zinc cathodes
Casting
ZINC
Gypsum removal
Electrolysis
Alloying materials
Spent electrolyte
Casting
ALLOYS
Hot acid leaching
Spent electrolyte
To neutral leaching
As
Filter
Zinc dust
Lead-silver residue
Filter
Filter
Cu-Co residue
Filter
Jarosite precipitation
Jarosite residue
Electrolysis
Cadmium cathodes
Casting
CADMIUM
Gypsum
Waste water
H2O
CaO
Packing
Cu-Co RESIDUE
Jarosite acid wash
Waste water
Spent electrolyte
Filter
Cleaned water
Leaching
Waste water presipitation
I. Selheim
Nov 1984

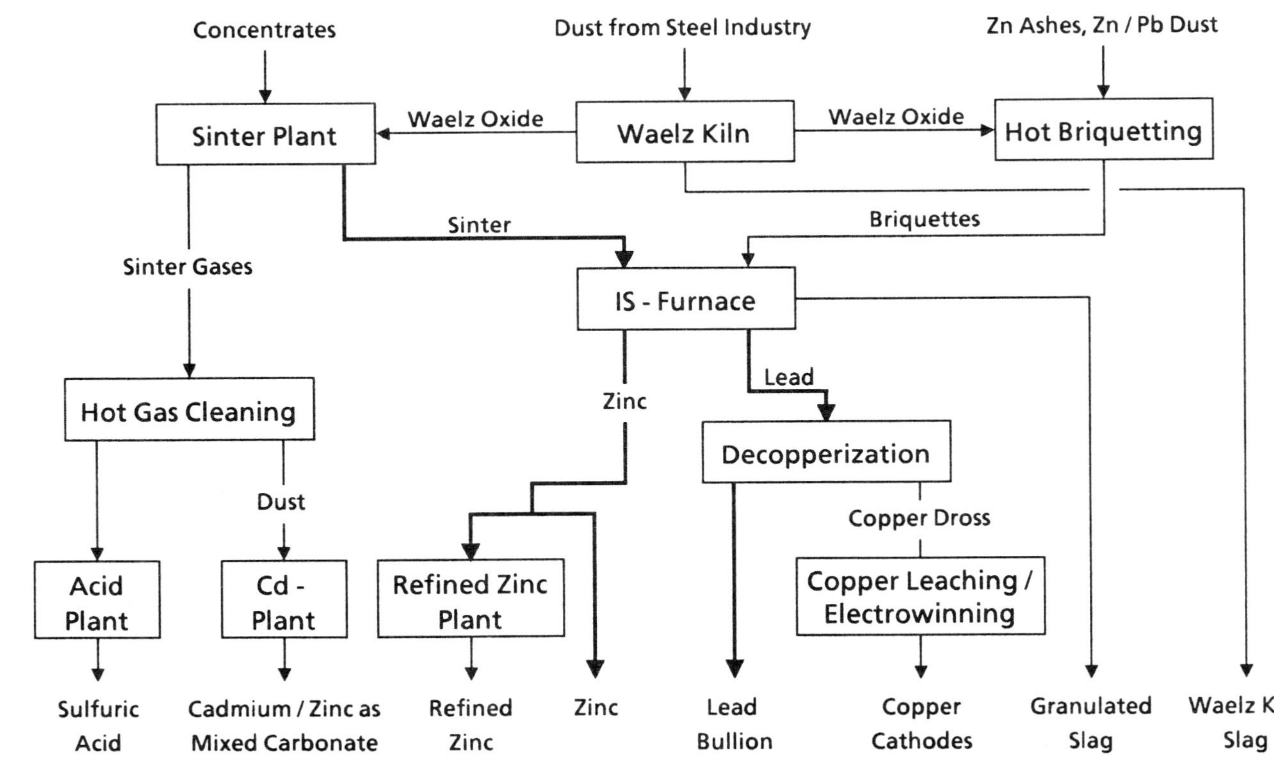
Flowsheet ISF Smelter
Berzelius Metallhütten GmbH, Duisburg, F.R. Germany
Concentrates
Dust from Steel Industry
Zn Ashes, Zn / Pb Dust
Sinter Plant
Waelz Oxide
Waelz Kiln
Waelz Oxide
Hot Briquetting
Sinter
Briquettes
Sinter Gases
IS - Furnace
Lead
Zinc
Hot Gas Cleaning
Decopperization
Dust
Copper Dross
Acid Plant
Cd - Plant
Refined Zinc Plant
Copper Leaching / Electrowinning
Sulfuric Acid
Cadmium / Zinc as Mixed Carbonate
Refined Zinc
Zinc
Lead Bullion
Copper Cathodes
Granulated Slag
Waelz Kiln Slag

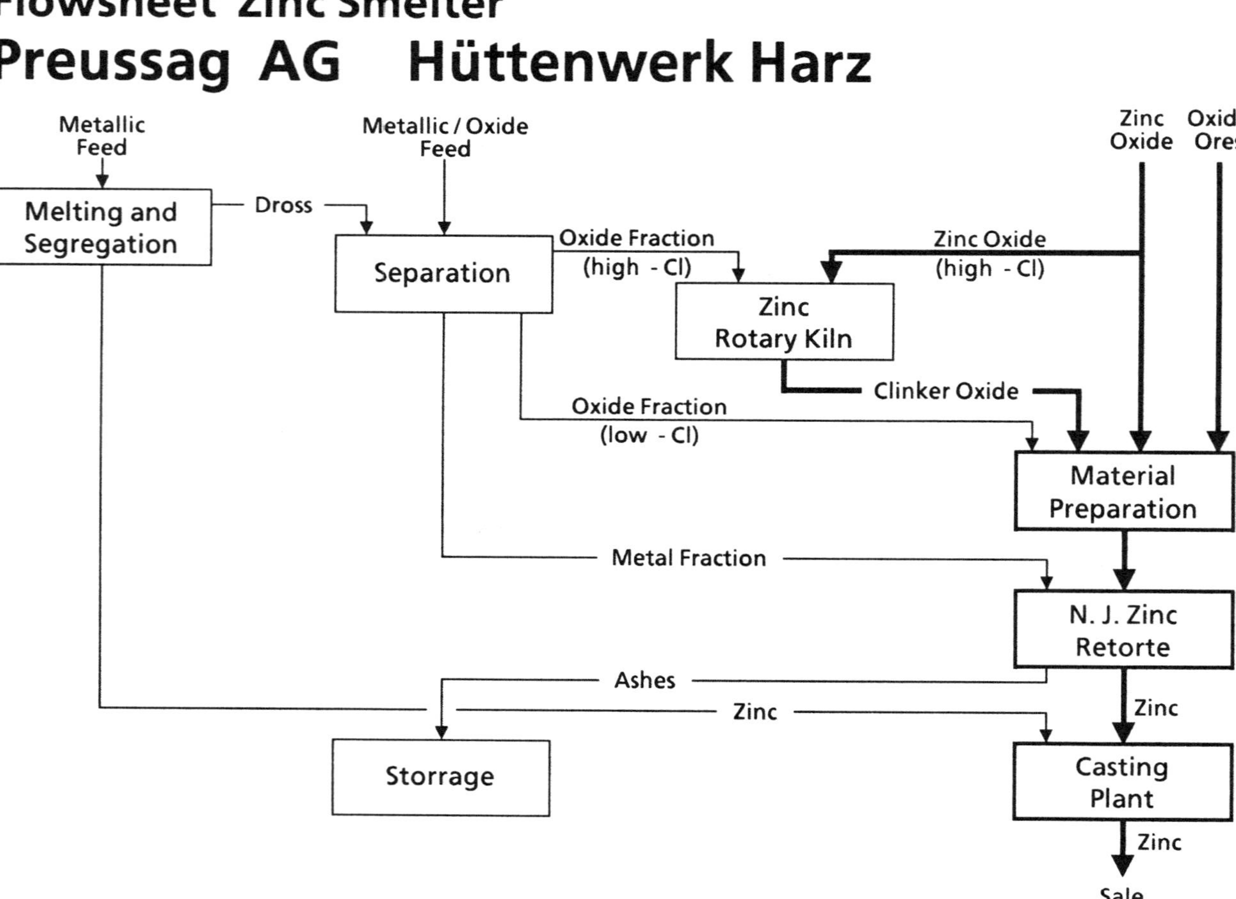
Flowsheet Zinc Smelter
Preussag AG Hüttenwerk Harz
Metallic Feed
Melting and Segregation
Dross
Metallic / Oxide Feed
Separation
Oxide Fraction (high - Cl)
Zinc Rotary Kiln
Zinc Oxide (high - Cl)
Zinc Oxide
Oxide Ores
Clinker Oxide
Oxide Fraction (low - Cl)
Material Preparation
Metal Fraction
N. J. Zinc Retorte
Ashes
Zinc
Zinc
Storrage
Casting Plant
Zinc
Sale

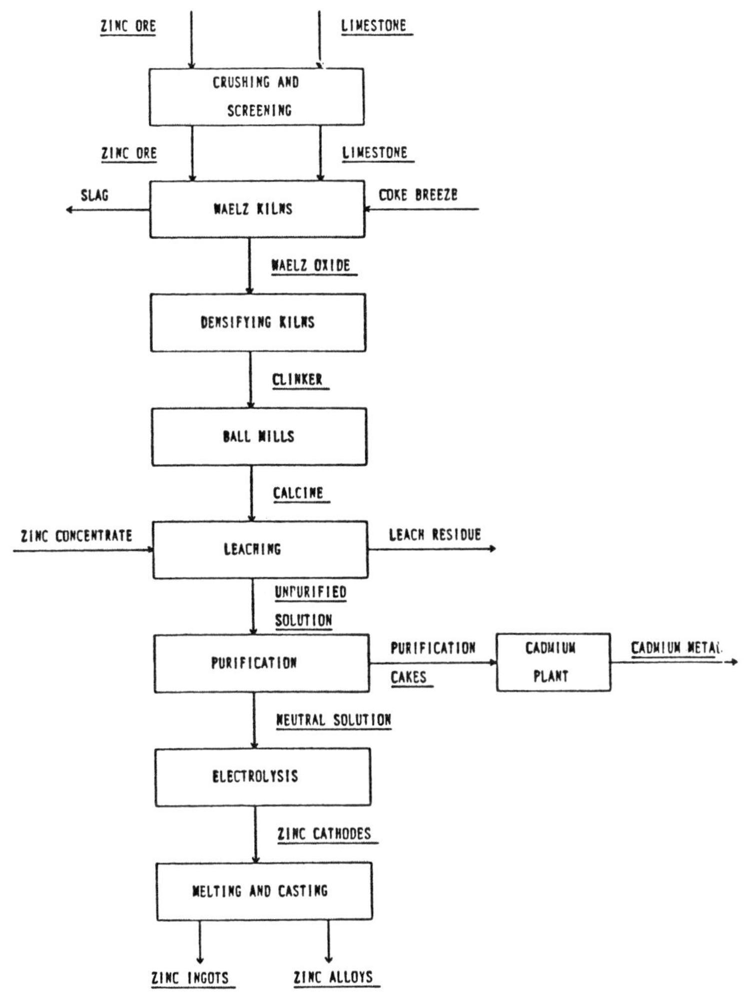

PROCESS FLOW DIAGRAM
for
CINKO KURSAN METAL SANAYII S.A.

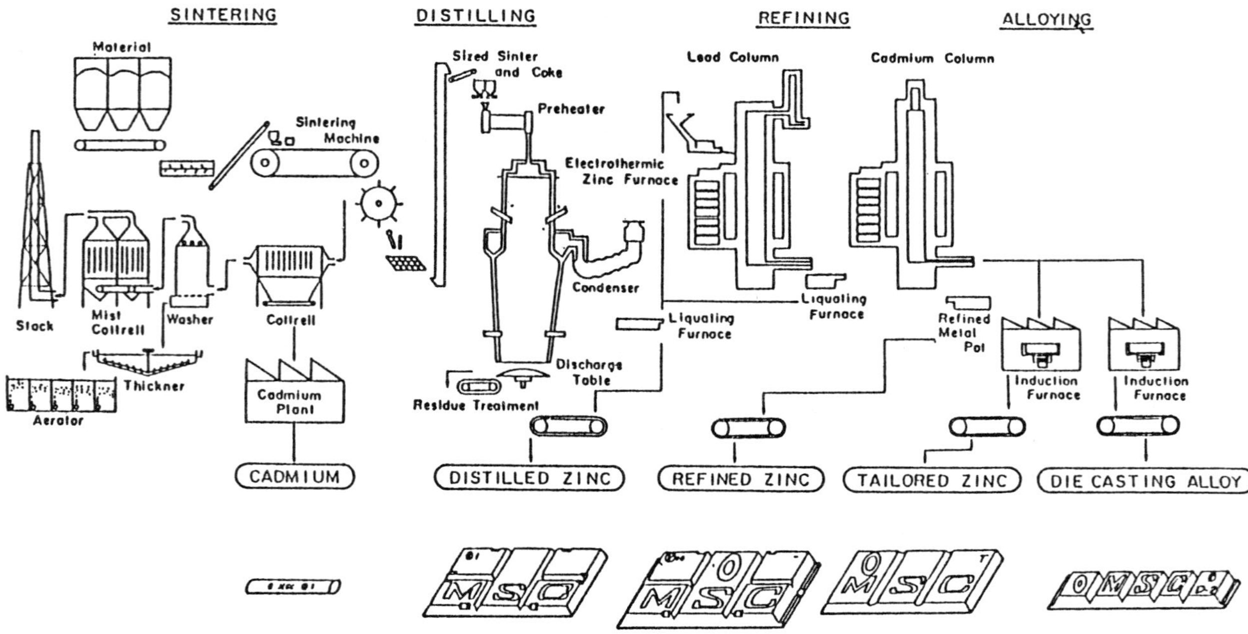

NIKKO ZINC CO. LTD. - MIKKAICHI SMELTER

The Imperial Smelting Process

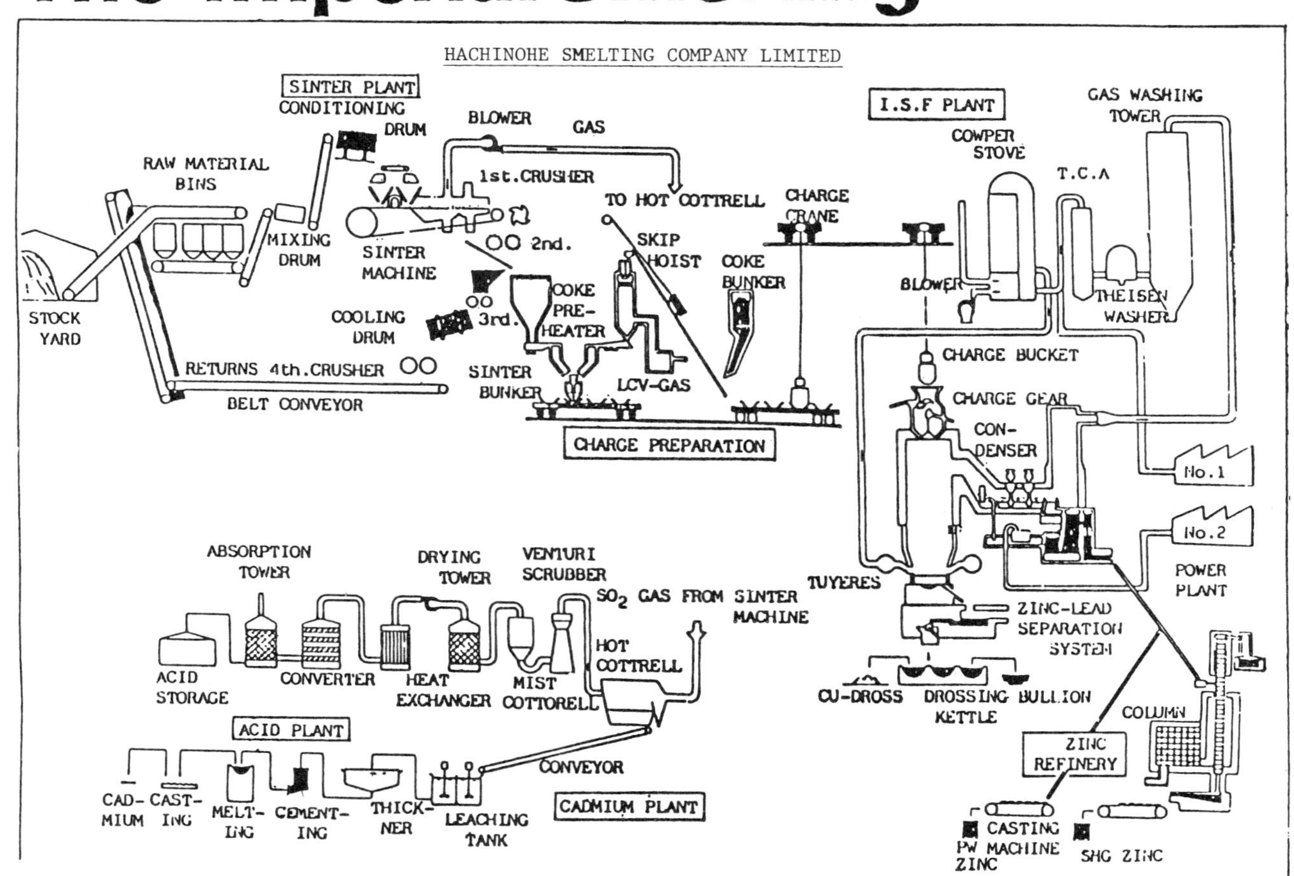

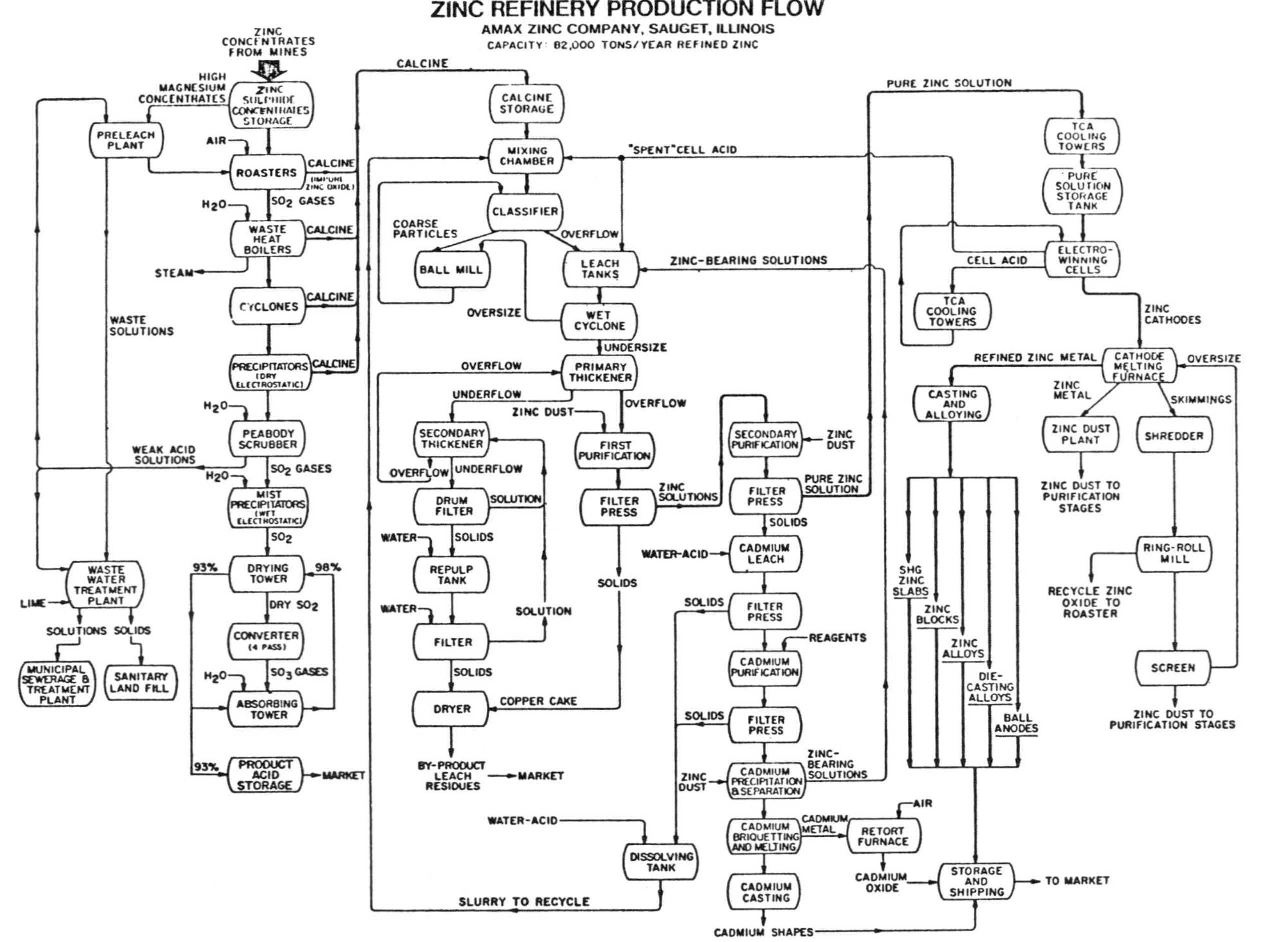
ZINC REFINERY PRODUCTION FLOW
AMAX ZINC COMPANY, SAUGET, ILLINOIS
CAPACITY: 82,000 TONS/YEAR REFINED ZINC
ZINC CONCENTRATES FROM MINES
HIGH MAGNESIUM CONCENTRATES
ZINC SULPHIDE CONCENTRATES STORAGE
PRELEACH PLANT
AIR
ROASTERS
CALCINE
(IMPURE ZINC OXIDE)
SO_2 GASES
H_2O
WASTE HEAT BOILERS
STEAM
CYCLONES
WASTE SOLUTIONS
PRECIPITATORS (DRY ELECTROSTATIC)
PEABODY SCRUBBER
WEAK ACID SOLUTIONS
MIST PRECIPITATORS (WET ELECTROSTATIC)
SO_2
DRYING TOWER
93%
98%
DRY SO_2
CONVERTER (4 PASS)
SO_3 GASES
ABSORBING TOWER
PRODUCT ACID STORAGE
MARKET
WASTE WATER TREATMENT PLANT
LIME
SOLUTIONS
SOLIDS
MUNICIPAL SEWERAGE & TREATMENT PLANT
SANITARY LAND FILL
CALCINE STORAGE
MIXING CHAMBER
"SPENT" CELL ACID
CLASSIFIER
COARSE PARTICLES
BALL MILL
OVERFLOW
LEACH TANKS
ZINC-BEARING SOLUTIONS
OVERSIZE
WET CYCLONE
UNDERSIZE
PRIMARY THICKENER
UNDERFLOW
ZINC DUST
SECONDARY THICKENER
FIRST PURIFICATION
DRUM FILTER
SOLUTION
FILTER PRESS
ZINC SOLUTIONS
WATER
REPULP TANK
FILTER
DRYER
COPPER CAKE
BY-PRODUCT LEACH RESIDUES
WATER-ACID
DISSOLVING TANK
SLURRY TO RECYCLE
SECONDARY PURIFICATION
PURE ZINC SOLUTION
CADMIUM LEACH
REAGENTS
CADMIUM PURIFICATION
CADMIUM PRECIPITATION & SEPARATION
CADMIUM BRIQUETTING AND MELTING
CADMIUM METAL
CADMIUM CASTING
CADMIUM SHAPES
RETORT FURNACE
CADMIUM OXIDE
STORAGE AND SHIPPING
TO MARKET
TCA COOLING TOWERS
PURE SOLUTION STORAGE TANK
ELECTRO-WINNING CELLS
CELL ACID
ZINC CATHODES
CATHODE MELTING FURNACE
REFINED ZINC METAL
CASTING AND ALLOYING
SHG ZINC SLABS
ZINC BLOCKS
ZINC ALLOYS
DIE-CASTING ALLOYS
BALL ANODES
ZINC METAL
ZINC DUST PLANT
ZINC DUST TO PURIFICATION STAGES
SKIMMINGS
SHREDDER
RING-ROLL MILL
RECYCLE ZINC OXIDE TO ROASTER
SCREEN
OVERSIZE
ZINC DUST TO PURIFICATION STAGES

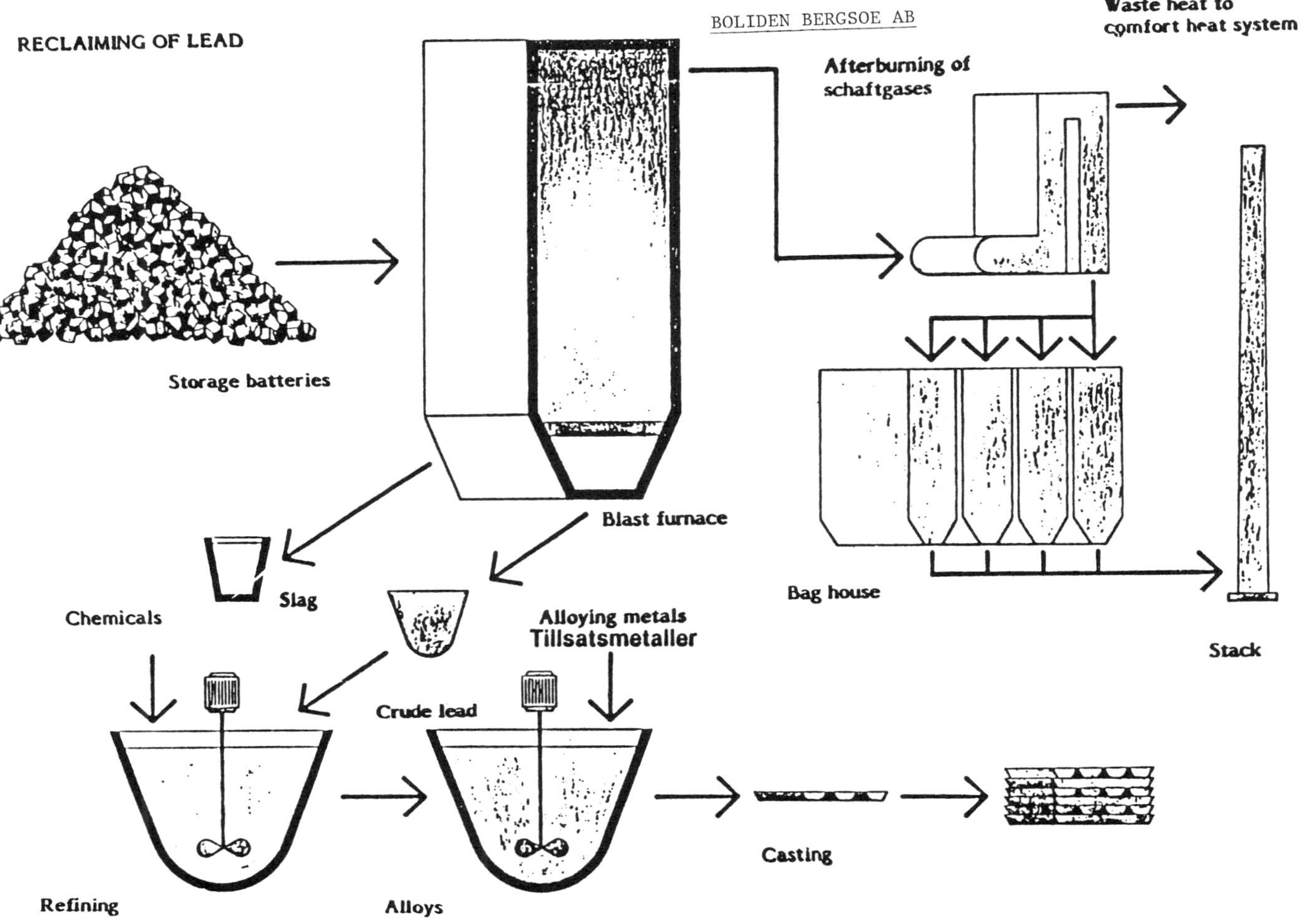
RECLAIMING OF LEAD
BOLIDEN BERGSOE AB
Waste heat to
comfort heat system
Afterburning of
schaftgases
Storage batteries
Blast furnace
Bag house
Stack
Slag
Chemicals
Alloying metals
Tillsatsmetaller
Crude lead
Casting
Refining
Alloys

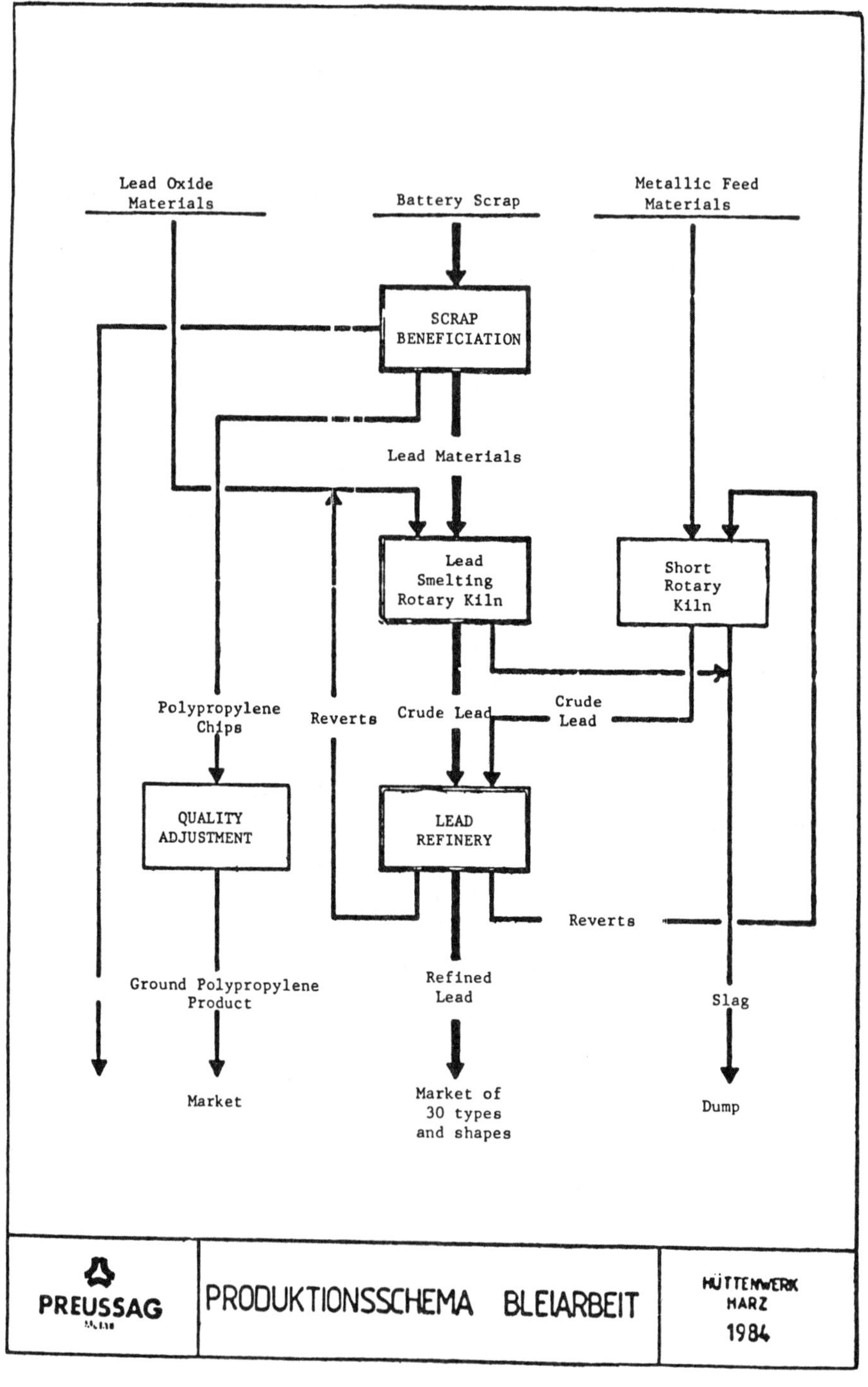
Lead Oxide Materials
Battery Scrap
Metallic Feed Materials
SCRAP BENEFICIATION
Lead Materials
Lead Smelting Rotary Kiln
Short Rotary Kiln
Polypropylene Chips
Reverts
Crude Lead
Crude Lead
QUALITY ADJUSTMENT
LEAD REFINERY
Reverts
Ground Polypropylene Product
Refined Lead
Slag
Market
Market of 30 types and shapes
Dump
PREUSSAG
PRODUKTIONSSCHEMA BLEIARBEIT
HÜTTENWERK HARZ
1984

Flowsheet Tin Smelter
Berzelius Metallhütten GmbH, Duisburg, F.R. Germany

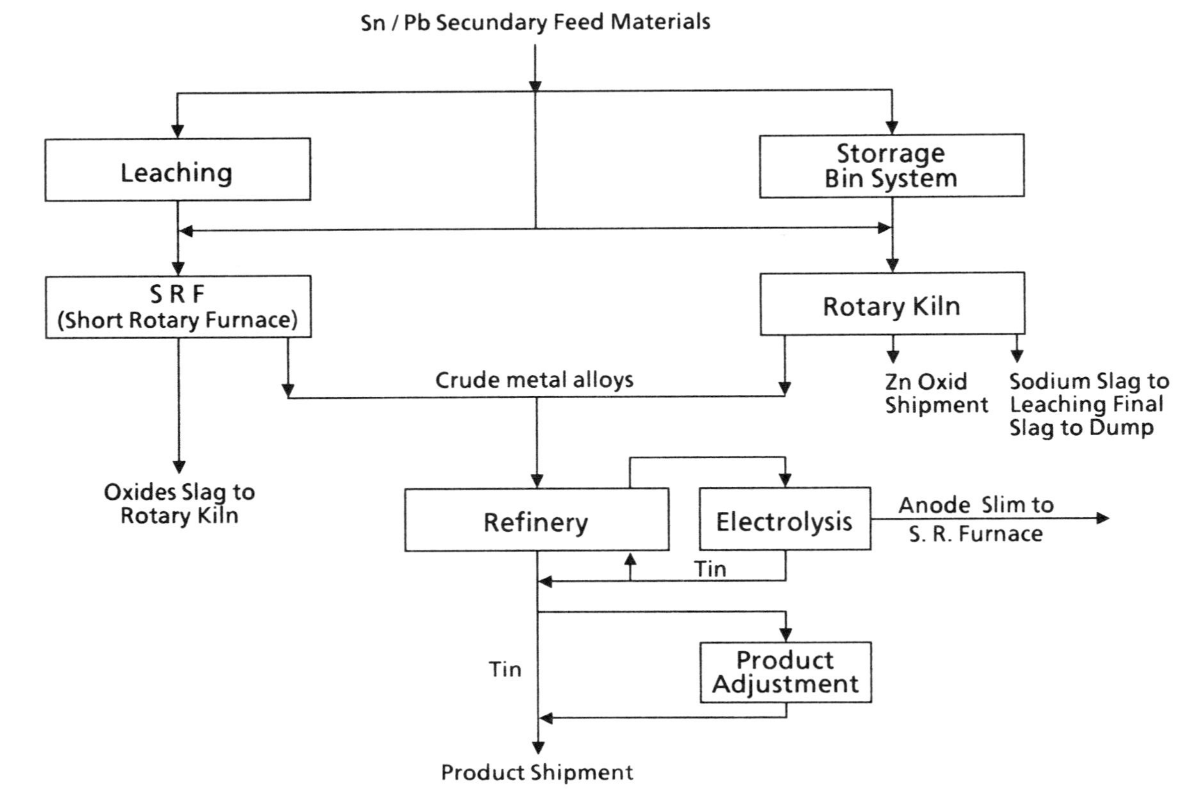

Selected Plant Descriptions

THE PALABORA SMELTER

The following description was supplied in response to the questionnaire.

The Palabora Mining Company Smelter is of conventional design, consisting of a coal fired reverberatory furnace, three converters, two anode furnaces and a casting wheel.

The filter plant and dryer send dry concentrates to the storage shed, where they are blended for charging to the reverberatory furnace.

Sulphur emissions are controlled by a single contact sulphuric acid plant and a "Gas Scrubbing Plant", which also produces a low grade copper concentrate.

Filter Plant and Dryer

Five Eimco disc filters, filter the pulp to give a filter cake containing 12-18% moisture. The filter cake is fed continuously to the rotary dryer, which is 14.65 m long and 2.44 m diameter. Dry concentrates averaging 7-8% moisture are weighed continuously and deposited in the storage shed.

Materials Handling

The dried concentrates, averaging 35% Cu, are taken from the storage shed and dropped onto a conveyor belt by a grab crane. The concentrates going to the furnace are weighed and fluxed with approximately 7% quartz rock, assaying 95% SiO_2, on the runup conveyor. Small quantities of revert materials are also added to the charge as necessary.

Quartz rock is blended with revert material for use as a converter flux, which assays 75% SiO_2.

Revert and quartz crushing is done by a jaw and gyratory crusher.

Reverberatory Furnace

This furnace is a conventional wet side charged furnace measuring 36 x 10.7 meters. Heat is supplied by six coal burners burning 240 tons pulverized coal per day. Secondary air is preheated to 240°C and enriched by 28 t.p.d. oxygen. Additional heat is provided by a roof burner over the settling area burning 15 tons pulverized coal per day.

Charging is continuous by way of shuttle conveyors and charge pipes.

Matte assaying 45-50% Cu is tapped from both sides of the furnace into 7.8 m^3 ladles, on demand.

Slag is also skimmed on both sides into 6.4 m^3 ladles and is dumped. Converter slag is returned via one of two return slag launders at the burner end.

The waste gasses pass through two waste heat boilers each producing about 18,000 kg/h of steam, and then via a balloon flue to an electrostatic precipitator and then to the gas scrubbing plant or the 152 m stack.

The coal is pulverised in three Babcock and Wilcox coal pulverizers, two feeding the furnace and the third being a standby unit and also for producing coal for other heating purposes.

Converters

There are three 3.96 x 10.05 metre Pierce-Smith converters, utilising fifty-two 50 mm tuyeres fitted with Kennecott 4 B 5 punching machines.

Converter hoods are water cooled with stack fluxing and each is fitted with two radiation pyrometers. The converters are serviced by two overhead cranes each with a 44 tons main hoist and two 18 ton auxiliary hoists.

Two or three converters are hot at any time and blown as required.

Air is provided by three 50,000 m^3/hr blowers situated in the power plant; blowing rates vary between 720 (850) kg/min and 830 (1,000) kg/min.

A converter charge produces approximately 120 tons of blister copper from 8-10 ladles of matte. Smelter reverts and copper scrap also being added.

The converter gasses containing 3.5-8% SO_2 are ducted to an electrostatic precipitator and then to the sulphuric acid plant. Surplus gases are scrubbed in the gas scrubbing plant.

Anode Furnaces

There are two 3.96 x 9.14 metre coal fired anode furnaces, which cast onto a 22 mould anode wheel.

Two converter charges make up one anode charge. The charges are aired through a furnace door by lance to remove the last of the sulphur and then poled to the required set by wooden logs.

The preparation takes 6 hours and the casting 8 hours. Anodes weighing 315 kg are cast, quenched, inspected and loaded for despatch to the refinery.

Power Plant

The steam generated in the waste heat boiler is used to preheat the reverberatory furnace combustion air, drive a 10 MW generator which provides emergency power, and drives the converter blowers.

Sulphuric Acid Plant

The acid plant is a conventional single contact plant with a series of scrubbing and mist elimination processes to ensure a clean dry gas is fed to the contact section.

The plant produces 400/450 tons of 98% sulphuric acid per day at 95% conversion.

Gas Scrubbing Plant

Reverberatory furnaces gasses and surplus converter gas are scrubbed by concentrator hydroseparator overflow, which is pumped over the towers, through the flotation cells and back to the concentrator.

This plant serves the double purpose of minimising sulphur emissions to the atmosphere and it recovers copper from the concentrator tailings. This copper as vallerite does not float in the conventional circuit.

Some 90% of the sulphur sent to the G.S.P. is recovered and the gasses leaving the plant assay 0.05% SO_2.

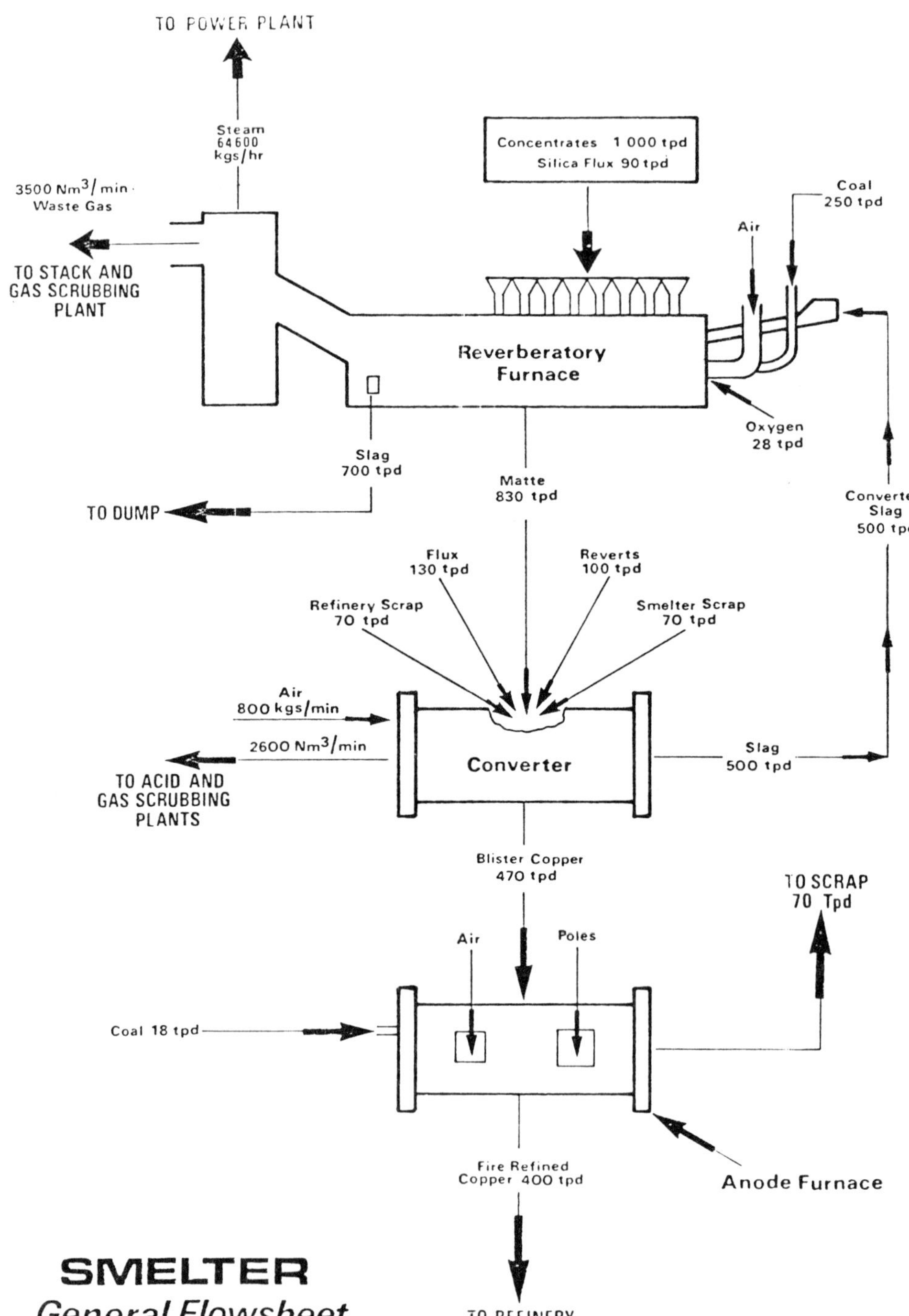

SMELTER
General Flowsheet

SMELTER SULPHUR BALANCE

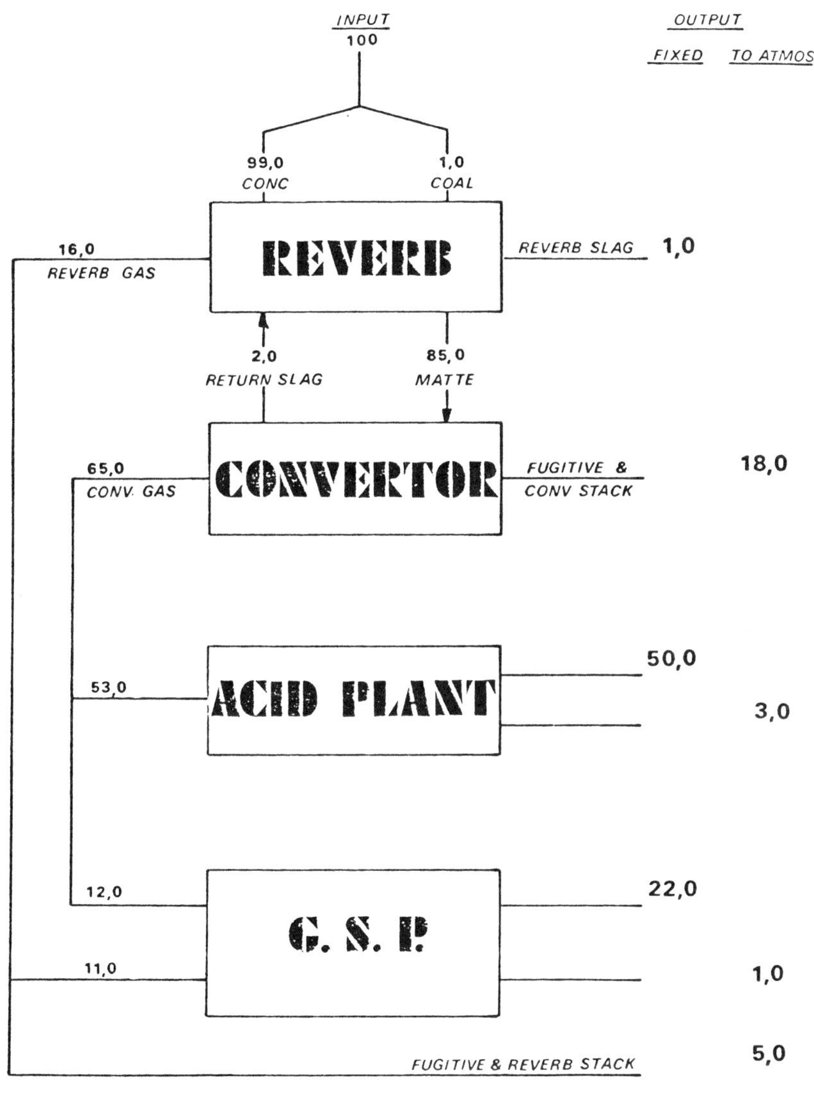

TSUMEB CORPORATION LIMITED

TSUMEB SMELTER

Dr. Robert Haegele
Chief Metallurgist
P.O. Box 40, Tsumeb, South West Africa

Introduction

Tsumeb Corporation Limited operates an integrated copper and lead smelter, treating concentrates from the local mine and its two other mines in South West Africa, as well as all copper and lead concentrates produced in the territory (the latter either on a purchase or toll basis); further concentrates and scrap are purchased elsewhere. The Copper Smelter, Lead Smelter, and the Cadmium and Arsenic Plants (the latter both administered by the Lead Smelter) are described separately in the following. All given data refer to the present state of operations as per January 1986; Si-units are used throughout.

Copper Smelter

General

The Copper Smelter is of conventional design, the operation consisting of charging green feed to a pulverized-coal fired reverberatory furnace and blowing the resulting matte to blister in Peirce-Smith converters. A peculiar feature is the high lead and arsenic content of the charge; because of this, the lead-rich converter slag is treated at the adjacent Lead Smelter, whereas the baghouse dusts are roasted at the Arsenic Plant for the recovery of arsenic trioxide.

Concentrate Dryers

The dryer is rarely used and then only during the rainy season when concentrates tend to become wet while in transit or on stockpile.

Number	one
Type	Rotary
Dimensions	10.7 m long, 1.5 m diameter
Retention Time	5 min
Dryer Fuel	Producer Gas
Calorific Value	6.8 MJ/m^3
Inlet Gas Temperature	1200°C
Concentrate Feed	
Average Rate	25 wet t/h
Average Composition	depends on concentrate being dried
Moisture	before: 15%, after: 5%
Outlet Gas Temperature	600°C
Gas Cleaning Equipment	Baghouse
Number of Units	one
Discharge Gas Temp.	120°C
Discharge Gas Volume	9000 m^3/h NTP

Feed Preparation

Smelting Furnace Flux	Generally, the furnace feed is of a self-fluxing nature; if required, the following fluxes are added:
Silica	Old copper/silver-bearing mill tailings; 33% SiO_2, 4% CaO
Lime	Limerock added to reduce copper/lead losses to slag; 50% CaO, 4% SiO_2
Iron	Coarse aisle reverts are added to the charge as a matter of course, to supply required oxidic iron; 12% SiO_2, 2% CaO, 23% Fe.
Converter Flux	Silica, 94% SiO_2, 2% CaO; sized 2-25 mm

Blending System
Number of Beds — one
Bed size — 250 t
Method of Reclamation — Overhead grab crane

Roasters — none

Sinter Machine — none

Smelting Furnaces

Number	One in operation (to which the following data apply); another (smaller) furnace is presently not used
Type	Reverberatory
Nominal Capacity	21 dry t/h of NMBM
Dimensions	27 m long, 9.7 m wide
Oxygen Enrichment	none
Fuel	Pulverized coal, calorific value 27 GJ/t
Operating Temperature	1150-1200°C (combustion gas at throat)
Campaign Life	5 years
Flux	4.5% silica and lime flux; 20% reverts

Analysis of Typical Furnace Charge, including Coal Ash

29.5% Cu
7.5% Pb
2.5% Zn
2.5% As
15.5% S
15.5% Fe
12.0% SiO_2
4.5% CaO

Furnace Matte Analysis	50.0% Cu 14.5% Fe 22.8% S
Temperature	1100°C when tapped (estimated)
Average Production	325 t/d
Furnace Slag Analysis	0.9% Cu 22.0% Fe 13.0% CaO 38.0% SiO_2
Temperature	1100-1150°C when skimmed (estimated)
Average Production	150 t/d

Furnace Slag Analysis	0.9% Cu 22.0% Fe 13.0% CaO 38.0% SiO_2
Temperature	1100-1150°C when skimmed (estimated)
Average Production	150 t/d

Furnace Off-Gas

Temperature	1150-1200°C, at throat
Volume	45 000 m^3/h NTP
SO_2 Concentration	1.0%
Cooling System	2 waste heat boilers, 1150 to 450°C 2 air pre-heaters, 450 to 280°C 1 trombone cooler, 200 to 155°C Tempering air, baghouse inlet: 130°C
Disposition of Dust	Boiler/air pre-heater dusts are recycled to the furnace, while the baghouse dust is roasted in the Arsenic Plant.

Converters

Number	2 in operation (hot), one blown at any one time ('cascade blowing'); 1 on standby/under maintenance 1 (9 x 4 m) converter available, but presently not used; data given refer to those converters which are in operation.
Dimensions	6 m long, 3 m diameter
Tuyeres	28/converter, 37.5 mm diameter; we are now changing to 50 mm tuyere diameter
Average Blowing Rate	4.14 min/t of matte
Oxygen Enrichment	none
Air Rate	316 m^3/min NTP
Scrap added	8% of matte charged
Flux added	11% of matte charged
Matte Converted	75 t/converter cycle

Blister Copper

Average Production	60 t/d/converter
Analysis	98.5% Cu 1.0% oxygen 0.5% combined Pb, As, Ni, Fe, etc. Because the charges are overblown for the elimination of lead and arsenic, virtually no sulphur is left in the blister.
Converter Slag	Treated in Lead Smelter
Analysis	8% Cu 15% Pb 33% Fe 21% SiO_2

Slag Cleaning

Not applicable. By treating the converter slag in the Lead Smelter, the contained copper is eventually recovered in the form of a matte- and speiss phase while drossing the lead bullion; these 'dross skims' are returned to the Copper Smelter.

Treatment of Smelter Product

Fire Refining	not applicable
Product Casting	
Method	Blister is discharged from converter into 30-t ladles, from which the copper is cast by overhead crane into stationary moulds
Number of Moulds	30
Average Casting Rate	50 t/h
Casting Temperature	1100°C (estimated)
Blister Cakes	
Size	1100 mm long, 1000 mm wide, and 250 mm overall height, tapered
Mass	1650 kg/cake average
Rejects	15% of new blister cast
Annual Production	
Blister Copper	46 000 t/a
Contained Silver	92 000 kg/a
Contained Gold	138 kg/a

Sulphur Fixation Facilities

None - all sulphur dioxide originating from smelting and converting is vented through a 140 m high stack.

Lead Smelter

General

The Lead Smelter is of conventional design, the operation consisting of sintering, blast furnace smelting, separation of copper phases by extensive drossing, and kettle refining; the latter excludes as yet refining for bismuth. A Harris Plant as part of the refinery is presently under construction.

Concentrate Dryers

Number	one
Type	Rotary
Dimensions	18.3 m long, 2.3 m diameter
Retention Time	8 min
Dryer Fuel	Producer Gas
Calorific Value	6.8 MJ/m^3
Inlet Gas Temperature	750°C

Concentrate Feed	
Average Rate	35 wet t/h
Average Composition	Tsumeb Lead Concentrate only, which makes about 33% of the NMBM sinter machine feed: 25% Pb 4% Fe 6% S 17% SiO_2 83% -200 mesh
Moisture	before: 15%, after: 7%
Outlet Gas Temperature	200°C
Gas Cleaning Equipment	Wet scrubbing
Number of Units	one
Discharge Gas Temp.	50°C
Discharge Gas Volume	7000 m^3/h NTP

Feed Preparation

Blending and Proportioning Plant with variable-speed conveyor belts: 8 bunkers of 195 m^3 capacity each, 14 feeder bins with capacities ranging from 20 to 150 m^3.

Composition of NMBM Sinter Machine Feed

	25.5% Pb 3.7% Zn 9.0% S 1.3% As 10.0% Fe 10.5% CaO 11.0% SiO_2
which includes as flux	11.5% limerock (50% CaO; 100% <25 mm)
and as fuel	3.6% fine coke
Moisture	6%

Total sinter machine feed includes 40% NMBM, 5% recycled material (sinter dusts and clean-ups), and 55% return sinter.

Roasters	none
Sinter Machine	
Number	one
Type	Dwight-Lloyd, updraft
Dimensions	4.7 m long, 2.44 m wide; 116 pallets
Nominal Capacity	100 t/h of feed
Auxiliary Fuel	Producer gas (6.8 MJ/m^3) for ignition 1.4% fine coke (25 GJ/t) admixed to feed

Sinter Composition	28 % Pb 4 % Zn 6 % Cu 1.4% As 11 % Fe 1 % S
Exhaust Gas	
Volume	160 000 m^3/h NTP
SO_2 Concentration	2.5%

Smelting Furnaces

Number	One in operation, identical one on standby
Type	Blast or shaft furnace; Lower part water-jacketed, upper part bricked; open top; 2 x 18 tuyeres, 102 mm diameter; continuous tapping into fore-hearth.
Nominal Capacity	23 t/h of sinter
Dimensions	6100 mm long, upper part 2850 mm wide, 1500 mm wide at tuyere level, 7850 mm overall height (crucible to top); tuyeres 508 mm above crucible
Oxygen Enrichment	none
Fuel	11% metallurgical coke admixed to sinter. Calorific value of coke 26 GJ/t
Operating Temperature	>1400°C
Campaign Life	400 days
Flux	none, except for occasionally recycled coarse slag
Furnace Metal Analysis	77.0% Pb 16.5% Cu 3.2% As 0.3% Fe 2.0% S
Temperature	1150°C when tapped (estimated)
Average Production	175 t/d
Furnace Slag Analysis	2% Pb 22% Fe 23% CaO 27% SiO_2
Temperature	1150°C when tapped (estimated)
Average Production	240 t/d
Furnace Off-Gas	
Temperature	120°C when leaving top of furnace
Volume	70 000 m^3/h NTP
SO_2 Concentration	nearly nil

Cooling Method	Air ingressing at top of furnace Water sprays used when required to achieve baghouse inlet temperature of 110°C
Disposition of Dust	Recycled to sinter preparation plant

Drossing Operation (converting not applicable)

Kettle drossing and smelting of dross in a producer gas-fired reverberatory furnace (another identical furnace is on standby);

Dross Reverb Furnaces	
Dimensions	6.1 m long, 2.7 m wide, 2.7 m high
Temperature	1100°C, measured at throat
Production Rate	50 t/d of 'dross skims'
Product	'Dross Skims' = copper matte and-speiss
Analysis	17% Pb 58% Cu 11% As 1% Sb 8% S

Slag Cleaning not applicable

Treatment of Smelter Product

Lead Refining

Kettles	Capacity 225 t; producer gas-fired
Bullion to Refinery	120 t/d
Average Analysis	50 g/t Cu 50-500 g/t As 1000 g/t Sb 40- 75 g/t Bi
Refining	Softening: As and Sb removal with caustic soda and nitrate (there is hardly any tin present); Harris Plant in construction Parkes de-silvering, followed by retorting of the silver crusts after pressing; the resulting retort bullion (silver/lead alloy) is added to the blister copper. Vacuum de-zincing, followed by final refining of the lead before casting.
Refined Lead Analysis	99.994% Pb 2 g/t Ag 1 g/t each Zn, Cu, As, Sb 54 g/t Bi

Product Casting	
Casting Machine	One straight-line Sheppard Ingot Master continuous casting machine
Number of Moulds	300
Average Casting Rate	40 t/h
Casting Temperature	450°C
Refined Lead Pigs	
Size	618 mm long (top), 117 mm wide (top), and 81 mm overall height, tapered
Mass	45 kg/pig
Rejects	1%

On special request by customers, 1000-kg 'jumbos' can be cast in stationary moulds

Annual Production	44 000 t/a of refined lead

Sulphur Fixation Facilities

None - all sulphur dioxide originating from sintering and dross smelting is vented through a 140 m high stack.

Cadmium Plant

The only hydrometallurgical operation (the Harris Plant will be the other one), consisting of leaching water-soluble cadmium compounds from sinter baghouse dust, purifying the leach liquor, cementing cadmium metal with zinc, and finally melting, refining and casting.

1. Sinter baghouse dust is leached in campaigns, once water-soluble cadmium exceeds a concentration of 10% in dust. Leaching medium is water, acidified with sulphuric acid to a pH of not less than 4.

Average Dust Analysis	46.0% Pb
	1.0% Zn
	12.0% Cd
	1.4% Cu
	1.8% As
	24.0% S

2. The leach liquor is purified by the addition of chromate and permanganate, to precipitate solubilized lead, thallium and arsenic as chromates or arsenates; and by the addition of some zinc dust to cement out solubilized copper.

3. Cadmium sponge is cemented with zinc dust at a pH of 1.5-2.0, washed and filtered.

4. The sponge is melted in small producer gas-fired kettles and refined with caustic soda and an ammonium chloride/natrium chloride mixture at temperatures of 475-500°C.

5. The refined metal is cast into 30-kg raw ingots, to be remelted - and if need be to be further purified by retorting - and cast into sticks and balls for sale.

Production	60 - 80 t/a refined cadmium
Product Analysis	150 g/t Cu maximum
	250 g/t Pb "
	20 g/t Tl "
	10 g/t Ni "
	5 g/t Fe "
	0.05 g/t Zn

Arsenic Plant

Copper baghouse dust is mixed with 28% pyrite and 13% coal and roasted to obtain an arsenic trioxide sublimate; the calcine is returned to the Lead Smelter for lead recovery.

1. Roasters	
Number	four
Type	one-hearth Godfrey roasters
Dimension	8 m diameter; 6 rpm
Fuel	Producer Gas, 6.8 MJ/m^3
Temperature	650°C
Oxygen in Off-Gas	4% maximum
2. Dust Analysis	25.0% Pb
	2.5% Cu
	18.5% As
	5.0% S

3. Each roaster is combined with a condensing 'kitchen', which consists of 12 sections of each 63 m^3 volume. The waste gas is cleaned in a small baghouse and discharged to the Lead Smelter stack.

   A roasting campaign lasts for about 40 days, after which the kitchens are emptied.

4. Production	2000 t/a arsenic trioxide
Analysis	99% As_2O_3 minimum
Packing	Sealed steel drums of 190 kg capacity

NEW DEVELOPMENTS AT THE HARJAVALTA SMELTERS IN 1981 - 1987

V. Salmi, P. Tuokkola

Outokumpu Oy, Harjavalta Works, Finland

Abstract

The development work at the Harjavalta smelters has also in the 1980's focused on improving the profitability. Efforts have been made in the first place to lower the unit costs by raising the production capacity. At the same time particular interest has been paid to reducing the energy and labour costs.

With concentrate supplies from the company's own mines diminishing competition about purchased concentrates has become of vital importance for the smelter. The capability to treat impure concentrates facilitates purchases of raw material.

More than before attention must have been paid to improving the working conditions and minimizing emissions to the environment when implementing new projects. This paper discusses the major capital projects undertaken in the period 1981 to date.

Introduction

The Outokumpu flash smelting method has been in industrial scale operation at Harjavalta since 1949. The method was at first used for smelting copper concentrates. Flash smelting of nickel concentrates was started in 1959. The most important improvement to the process has been the introduction of oxygen enrichment in the copper and nickel smelters in 1971 resulting in a considerable rise in the smelting capacity and reduction in the energy costs.

The development work after the introduction of oxygen enrichment in the 1970's and 1980's has aimed at improving the profitability of the smelters. The age, small size and distant location of the smelters are factors that burden their profitability and have to be compensated by other means. To improve the profitability efforts have been made in the first place to lower the unit cost by raising the production capacity. At the same time particular interest has been paid to reducing the energy and labour costs. With concentrate supplies from the company's own mines diminishing competition about purchased concentrates has become of vital importance for the smelter. More than before attention must have been paid to improving the working conditions and minimizing emissions to the environment when implementing new projects.

The planning of all new projects has been based on their implementation in connection with the ordinary summer shut-downs without any considerable production losses. This has been enabled by good planning, thoroughness in preparatory work and accuracy in scheduling.

This paper discusses the major capital projects undertaken in the period 1981 to date under the following general categories:

- Bottleneck removal to allow increased production capacity
- Rationalization of manpower
- Reductions in process downtime through improved equipment reliability
- Increased metal recovery
- Increased energy utilization
- Treatment of impure raw materials

This text includes no details of processes because they can be found elsewhere in the book. The processes with accompanied equipment are shown in the enclosed process flow sheets.

2. Removal of Production Bottlenecks

The capacity of the copper smelter was raised from about 60 000 t to 80 000 t/a by an investment program implemented in 1977 - 1981. The investments were shortly the following:

- Copper drying plant 1977 - 1978
- Expansion of the slag flotation plant 1977 - 1978
- Replacement of the converter aisle cranes 1977 - 1978
- Copper flash furnace electrostatic precipitator 1978 - 1979
- Concentrate storage facilities 1980 - 1982
- Converter revisions 1980 - 1982

The investments started in 1982 aim at raising the capacity of the copper smelter further from 80 000 t to 100 000 t/a. It is expected that upon completion the capital expenditures will total about 50 million US dollars, the largest investments being:

- Modernization of the sulfuric acid plants 1983 - 1985
- New oxygen plant 1983 - 1984
- Second stage of the expansion of the slag flotation plant 1983 - 1984
- Concentrate storage expansion 1985 - 1986
- Enlargement of the settler of copper flash smelting furnace 1987
- Modernization of slag transportation and cooling facilities 1986 - 1987

It is believed that a production level of 100 000 tons in the drying plant, waste heat boiler and converting will be reached without any greater investments. The capacity of the anode casting plant is sufficient.

With the supplies from the new Enonkoski nickel mine the nickel smelter will probably operate the next five years nearly at full capacity. Concentrate smelting is at an annual level of 140 000 - 150 000 t. The bottlenecks are mainly the electric furnace used in slag cleaning and, at times, the converting. In the long term the availability of nickel raw materials is, however, uncertain and therefore investments to raise the capacity of the nickel smelter are not considered possible or necessary.

2.1. Modernization of the Sulfuric Acid Plants

The modernization of the converting section of the smelter was completed in the autumn of 1982. The copper production capacity had at that time been raised to 80 000 t/a. However, already on the completion of the first new copper converter in the autumn of 1981 it was obvious that the gas handling capacity in the sulfuric acid plants was not sufficient to accept the gases of the increased production. Therefore metal production in the copper and nickel smelters had to be restricted to a level determined by the gas handling capacity at the sulfuric acid plants. This situation came to Outokumpu as a surprise that the company was not prepared for.

For historical reasons there has been a unique and different situation at the Harjavalta smelters in that metal production and the processing of SO_2 gas to sulfuric acid have belonged to two different companies. According to the cooperation agreement Outokumpu Oy has sold the sulfur in the gas at a certain price and Kemira Oy has been committed to handle all the gas and to maintain a sufficient gas handling capacity at their plants. The production of sulfuric acid of smelter gases changing widely in the volume requires a considerable overdimensioning of the capacity in the acid plants. This fact had not, however, been paid sufficient attention to by the sulfuric acid company accustomed to the use of steady roasting gases in their other plants. Therefore it is understandable that such a bottleneck had developed in the metal production chain.

The situation was solved in the negotiations between the companies so that a decision was made on building jointly a new sulfuric acid plant. Outokumpu Oy participated in the financing of the new plant by 50 %. In addition Kemira Oy decided to modernize their two existing old sulfuric acid plants and to double the capacity of their liquid SO_2 plant. The modernization of the old plants was carried out in 1983 and 1984. The construction work of the new acid plant was started in 1983 and the plant was commissioned in the beginning of 1985. Then the oldest sulfuric acid plant, built in the 1950's, could be shut down and demolished. With the investments carried out the total nominal capacity of the sulfuric acid plants was doubled being at the moment 2400 t H_2SO_4/d. The production capacity of liquid sulfur dioxide is 150 t SO_2/d.

One of the old sulfuric acid plants is a single contact plant handling mainly converter gases. The other is a double contact plant that can treat both flash furnace and converter gases. The new sulfuric acid plant is naturally also a double contact plant. It is based on Monsanto's licence. The technical data of the new sulfuric acid plant are as follows:

Table 1

Acid plant performance data:

Type of process

Metallurgical double absorption acid plant for treating of flash smelter process gases

Plant design maximum:	
Production rate of 100% H_2SO_4	1000 t/d
- produced acid strength	93 % or 96 %
Design gas flow to converter	69500 Nm^3/h
- process gas SO_2 concentration	8.5 - 13.5 %
Design emissions of SO_2	max 650 ppm
and mist emissions of 100% H_2SO_4	max 50 mg/Nm^3
Conversion guarantee	99.6%

In the design of the plant the reducing effect on the gas volume and the raising effect on the SO_2 content of the new oxygen plant have been taken into consideration. After the modernization of the sulfuric acid plants emissions to the environment have decreased considerably and the total recovery of sulfur has risen to 99 %.

According to the agreement concluded in late 1986 Outokumpu Oy bought the Harjavalta sulfuric acid plants owned by Kemira Oy. So the production of sulfuric acid and liquid sulfur dioxide passed totally into Outokumpu Oy's ownership. Our company finally reached the natural situation which was aimed for several years at and where the entire metal production process is controlled by one company.

2.2. New Oxygen Plant

As known, oxygen enrichment was introduced at the Harjavalta smelters in 1971 when the first oxygen plant was built. The nominal capacity of the plant was 7200 Nm^3/h and the oxygen enrichment of the flash furnaces varied between 30 - 40 %. The increase of residues in the feed mixture and the simultaneous rise in the production led to utilizing the whole capacity of the oxygen plant by the beginning of the 1980s. Oxygen enrichment had then risen to 40 - 50 %. With the further increase of production in the flash furnaces it was necessary to use more and more oil which brought about a considerable rise in the energy costs. Besides the capacity of the waste heat boiler began to become a bottleneck in the production. It was known that an annual copper production of over 80 000 t would not be possible with the existing gas handling equipment without a considerable increase in the oxygen enrichment.

To raise the production capacity and to lower the energy costs it was decided to build a second oxygen plant. Special arrangements were made to implement the project. According to the agreement, Outokumpu Oy concluded with Oy Aga Ab, a Finnish-Swedish gas company, Aga built their new plant in the premises of Outokumpu in the immediate vicinity of the old oxygen plant. Outokumpu's personnel is responsible also for the operation of Aga's plant. The total number of personnel in both oxygen plants is three men in a shift.

The new plant produces 7700 Nm3/h of gaseous oxygen, the production capacity of liquid oxygen is 2600 Nm3/h and the production of liquid argon can raise to 410 Nm3/h. If needed, a corresponding amount of liquid nitrogen can be produced instead of liquid oxygen. Of the production of the plant Outokumpu gets the gaseous oxygen and Aga the liquid oxygen. The production of argon is shared between the two companies. Gaseous oxygen is used in the smelters. Outokumpu's share of liquid argon is delivered mainly to Outokumpu Oy's stainless steel works in Tornio. Aga sells their share of the argon and the liquid oxygen to the Finnish market.

The most important reason for cooperation between these two companies was the economy of the project when building one great air gas plant that produces gaseous oxygen, liquid oxygen and liquid argon. Another alternative would have been that both Outokumpu Oy and Oy Aga Ab would have built their own smaller plants in which case the total capital and operation costs of the two plants would have been considerably higher. The new oxygen plant was commissioned in the autumn of 1984.

The copper and nickel smelters now have 15 - 16 000 Nm3/h oxygen available. The oxygen enrichment of the flash furnaces ranges between 70 and 95 %. The consumption of oil has decreased by about 10 000 t/a. The gas volumes have decreased considerably and the contents of sulfur dioxide in the gas have increased correspondingly. The SO2 contents of the flash furnace gases before the acid plants are 20 - 24 % and therefore they can be very well used also in the production of liquid sulfur dioxide.

After the increase of the oxygen enrichment the height of the reaction shaft of the copper flash furnace could be shortened by 1.5 m, whereby heat losses in the reaction shaft decreased by about 20 %. At the same time jacket water cooling system reached the whole reaction shaft area.

2.3. Second Stage of Flotation Plant Expansion

After the first expansion stage carried out in 1978 the capacity of the slag concentrator corresponded to a copper production level of 65 000 t/a. The following targets were set for the second stage expansion project:

- The capacity of the existing equipment will be raised to a copper production level of 80000 t/a with minimum possible changes.
- The capacity of new equipment will be designed for a copper production level of 100 000 t/a.
- The most recent equipment and process development of Outokumpu Oy will be utilized and applied for slag flotation.

In the implementation of the project all the flotation cells were replaced. The new main cells are of OK 16 U type. The number of shafts was reduced from 30 to 13. At the same time, however, the cell volume increased from 90 m3 to 110 m3. The classification of the lump mill and the pebble mills was improved.

The automation degree of the slag concentrator was raised considerably. The process automation was designed and implemented by Outokumpu Electronics Division. The whole plant is operated by a Proscon process computer. E.g. the feed volumes of flotation air and chemicals are controlled by the computer.

In the analyzing of slurries the Minexan analyzer was taken into operation. The Cu, Ni, Fe, Pb, and Zn contents are analyzed continuously. The slurry density is also measured. The information from the Minexan analyzer is transferred to the process computer for control parameters.

Slag concentrate is transferred to the copper drying plant by pumping. Due to environmental protection a closed circulating system was installed for the waters of the drying plant, slag concentrator and waste area.

The expansion of the concentrator was commissioned in December 1984. The target set for the capacity increase was very well reached. It is believed that the capacity of the slag concentrator is sufficient also at the copper production level of 100 000 t/a.

2.4. Concentrate Storage Expansion

Nowadays more than 70 percents of the copper concentrates are imported. Due to difficult winter circumstances at sea the concentrates to be used in winter must be shipped to Harjavalta already in autumn. Thus the amount of concentrates to be stored is highest in autumn. The capacity of the earlier concentrate storage was inadequate. The concentrates cannot be stored in the open air because of dust and solutive losses as well as environmental reasons. So the capacity of the concentrate storage was doubled, being today about 60 000 tonnes.

This expansion project also included three new day bins in addition to the seven old ones. The capacity of the new bins, 700 tonnes each, is the same as the capacity of the old ones. As a result of a greater number of bins, it is possible to feed more and more different raw materials to the flash furnace, and still have a control over the composition of the feed mixture. The new bins were taken in use in the autumn 1986. The capacity of the belt conveyors will be doubled in 1987. The increase in the capacity of bins and belt conveyors will reduce the number of labour by 3 - 4 persons in the concentrate unloading section.

2.5. Enlargement of the Settler of Copper Flash Smelting Furnace 1987

To increase the campaign life of the flash furnace and to secure the production capacity the furnace will be renovated during the summer shut-down in 1987. The brickwork of the furnace as well as the steel constructions will be entirely renewed. The arch of the settler will be changed to a suspended flat roof. We believe that in this way the life time of the roof will increase and the repairs during the campaign will decrease and become easier.

To secure the sufficient capacity of the flash furnace the volume of the furnace will be increased by widening the settler with one meter. The settler volume will increase with appr. 20 %. In addition the residence time of slag and matte will increase correspondigly. Thus the separation of slag and matte will be intensified resulting in a better operation of the furnace and a lower copper content of slag.

2.6. Modernization of Slag Transportation and Slag Cooling Facilities

When aiming at the annual copper production of 100 000 tonnes, the worst bottleneck in the copper smelter (probably at the moment) is the crane capacity in the converter aisle. In the smelter there are three cranes, only two of which can be simultaneously in use while one is standing by. The two cranes in use must handle the following transportations:

- matte from both flash furnaces to the converters
- slag from the copper flash furnace and the copper converter out to the cooling area
- slag from the nickel converter to the electric furnace
- matte from the electric furnace to the converter
- high grade nickel matte from the converter out to the granulation plant
- blister copper from the converter to the anode furnace
- cold materials to the converters.

With the present production the cranes can handle these transportations, but when the production increases there will be difficulties. The capacity of the cranes is mostly tied by slag transportation from the copper flash furnace to the cooling area outside. This is why the slag transportation will be totally rearranged.

In this new system the slag will be tapped into large 40 t ladles which will be carried by a special truck from the flash furnace to the cooling area. The slag will not be cooled in open pits any longer, but in the same transportation ladles. Altogether 30 - 40 ladles will be needed. The slow cooling of the slag can be better controlled in the ladles. As a result the copper content of the slag will be reduced and the copper recovery improved.

The cooled slag will be removed from the ladles by tipping the ladles over by the transportation truck. The slag will be crushed by a hydraulic hammer and fed to the slag concentrator bins by belt conveyors. The new cooling system will have a closed water circulation. The project is being implemented just now, and the new slag transportation and cooling system will be started up in the autumn 1987. Then it will also be possible to reduce the amount of labour in the smelter by 2 - 3 persons per shift.

2.7. Other Investments

The dryer is not likely to become a bottleneck at the copper production rate of 100 000 tonnes. As the pretreatment plant for complex concentrates, described later in this paper, produces dry concentrate direct to the dry charge bin of the flash furnace, the existing dryer is not needed for that purpose. In addition, the disc filters now filtering the slag concentrate will be replaced by Larox pressure filter which is much more efficient. This means that the moisture content of the slag concentrate to be fed in the drum dryer will be reduced to a half. Thus the amount of water to be evaporated in the drum dryer will be reduced by 15 - 20 per cent.

At the present level of oxygen enrichment, the amount of exhaust gases from the flash furnace is so small that the capacity of the waste heat boiler will be sufficient even for a greater production. However, as the waste heat boiler became older, repeating leakages caused problems. For this reason there were plenty of production losses in the copper smelter in 1985. The wall panels of the boiler will be changed stepwise during the summer shut-downs. The work was started in 1985 and it will be finished in 1987.

With the revision of the waste heat boiler the convection sections II and III will also be renewed.

One investment to improve the on-line availability of the copper and nickel smelters was the overlifting device of the cranes in the converter aisle. As earlier stated, there are three cranes in the converter aisle, two of which are working while one is standing by. If the crane in the middle went broken, the third crane could not be used. The crane operation with only one crane limited the production considerably.

There were two alternatives to improve the situation: either to buy a fourth crane, so that if one crane went broken there would always be two cranes available, or to construct a so called overlifting device which could change the location of the cranes. The latter alternative was chosen at the Harjavalta smelters because of lower capital costs. The overlifting device is used when the crane in the middle is being maintained or repaired. The device lifts the broken crane, and the crane behind is driven under the broken one. Thus two cranes are available, and the third one can be repaired without hurry. The change of cranes takes at the shortest only 15 minutes.

3. Pretreatment Plant for Complex Concentrates

As the company's own copper mines are being depleted, more and more concentrates must be imported each year. Finland's remote situation weakens our competitiveness in the raw material business, particularly when competing for pure concentrates. It is easier to buy impure concentrates which can also be profitably treated because of higher treatment charges and penalty clauses for impurities. For this reason Outokumpu Oy decided to build a pretreatment plant at Harjavalta for copper concentrates containing arsenic, antimony and bismuth. The method has been developed within Outokumpu Oy and it is patented /1/, /2/.

The main features of the process are as follows:
The moist concentrate is dried by heavy fuel oil in a drum dryer. The exhaust gases are cleaned in an electric filter. Then the dry concentrate is fed to a rotary kiln. Also sulfur and possibly fuel as well as oxygen are fed there. The reaction temperature in the kiln is 700 - 800°C and impurities of the concentrate like arsenic, antimon and bismuth will be evaporated.

The clean product concentrate leaving the rotary kiln is cooled and transported by a pneumatic conveyor to the dry charge bin of the copper smelter, from where it is fed together with other concentrates into the flash smelting furnace.

The exhaust gas of the rotary kiln containing 20 - 30 % of sulfur dioxide is cooled and cleaned in a scrubber and then led to the sulfuric acid plant. The precipitate in which the impurities are concentrated is filtered and further refined for sales products.

Special attention during the design of the plant has been paid to the working conditions and hygiene in the working facilities as well as the eliminating of pollution. The plant is highly automated, and thus the operating personnel can have been minimized. 20 - 25 per cent of the annual concentrate consumption of the copper smelter can be covered by pretreated concentrates. The start-up of the plant began at the end of 1986. A refining plant for arsenic and antimon precipitates is under construction and it will be started towards the end of the year 1987.

4. New Sampling Department

Due to greater quantities of imported raw materials, their sampling and analysing has become a very important task involving great economical risks.

The equipment and facilities of the old sampling plant did not meet modern requirements. Thus a decision was made to build a totally new sampling plant. The new plant was commissioned in 1984. It has the following sections for the treatment of different kind of raw material samples:

- raw materials containing precious metals
- concentrates
- precipitates and catalysts
- scraps

A baling press was installed in the sampling plant to make the scrap handling and smelting in the converter easier. The samples are taken in connection with the baling.

The ventilation of the plant and the gas cleaning system of the smelting furnaces were designed and built with special care.

There are 14 employees at the sampling plant; one foreman and 13 workers taking samples.

Last year over 7 000 samples were prepared to define the moisture content of concentrates. Further, 2 500 samples were prepared for commercial raw material analyses. In addition, numerous samples are continuously being prepared for the process control and research purposes. All samples are analysed at the laboratory of the Harjavalta works.

5. Other Actions

In order to improve the metals recovery, a belt conveyor with auxiliary equipment was built for the transportation of high grade nickel matte from the smelter to the nickel refinery. The equipment removes the matte from the granulation plant directly to the bin of the nickel leaching plant. Earlier the matte was loaded in a bucket and carried by a fork lift truck to an intermediate storage in the storage area of the nickel refinery.

For the same reason, a pipe line was built from the leaching plant of the nickel refinery to the copper drying section. Along this pipe line the copper precipitate produced in the nickel refinery is pumped by a high-pressure pump on a belt conveyor which feeds concentrates to the copper dryer. Earlier the precipitate was carried in a bucket by a fork lift truck to the copper concentrate storage, where it was mixed with copper concentrate and then fed through day bins to the dryer.

As a result of the reorganization of the whole company, the Harjavalta works were divided in 1983 into two profit units with the official names 'Copper and Nickel Smelters' and 'Nickel Plant'. The smelters are responsible for their operation to the Vice President & General Manager of the Metallurgical Division. The purpose of this change was to clarify the operating principles and particularly to motivate the personnel to strive for better profitability.

6. Future Plans

A continuous automatic analyser for the feed mixture of the copper flash furnace is under development at the moment. Connected with the earlier constructed temperature measuring system it will enable the computer control of the process. The target is to minimize changes of the furnace temperature and copper content of matte. Thus the operating cost of the smelter will decrease.

During the next few years the entire control system of the smelter will be renewed to be based on a computer system. At the moment computer controlled process is used by the flotation plant, power plant, oxygen plant, the new sulfuric acid plant and the pretreatment plant of impure concentrates.

The long distance plans of the copper smelter do not include an increase of the production capacity over 100 000 tonnes per year. The capacity increase would imply great investments which would be difficult to realize in the old and cramped smelter building. Even the modifications made so far have been very difficult for the same reason.

To make the raw material supplies easier and to maintain profitability, Outokumpu will concentrate more than ever on the treatment of concentrates containing different impurities. This will involve investments in the development of the process.

More attention will also be paid to the human resources. Working conditions must continuously be improved so that qualified workers would like to work in the smelter in future, too. Nowadays many young men do not much appreciate smelter work which is considered dirty and heavy. Work motivation of the personnel must be improved by affecting attitudes by training and by developing the wage and rewarding system.

Appendix 1

FLOWSHEET OF THE HARJAVALTA COPPER SMELTER
UNLOADING OF CONCENTRATES
CONCENTRATE STORAGE
FEED PREPARATION
DRYING
FLASH SMELTING FURNACE
WASTE HEAT BOILER
ELECTROSTATIC PRECIPITATOR
GASES TO ACID AND SO_2 PLANTS
MATTE
SLAG
CONVERTERS (3 PCS)
WASTE HEAT BOILER FOR ALL Cu- AND Ni-CONVERTERS
GASES TO ACID PLANT
ANODE FURNACE (2 PCS)
ANODE CASTING
TO REFINERY
SLOW COOLING OF SLAG
GRINDING
FLOTATION
TAILINGS
SLAG CONCENTRATE

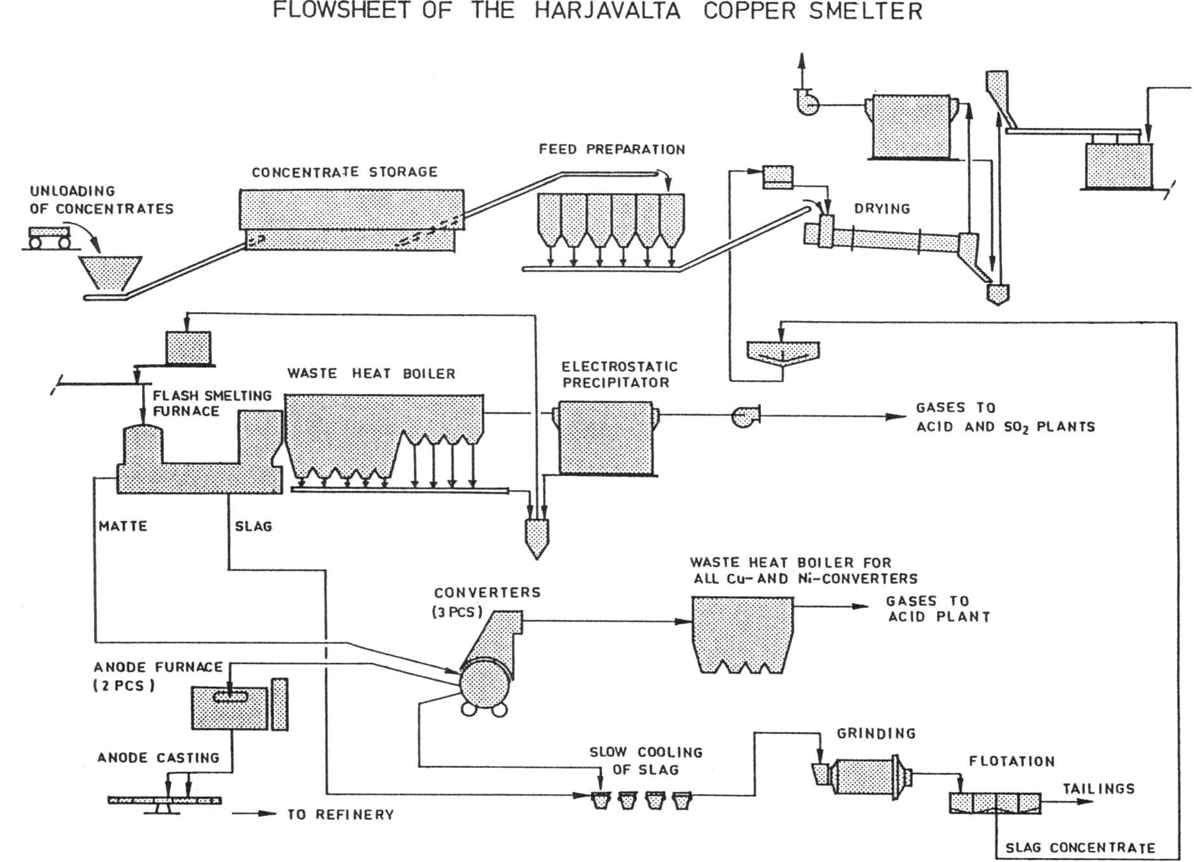

FLOWSHEET OF THE HARJAVALTA NICKEL SMELTER

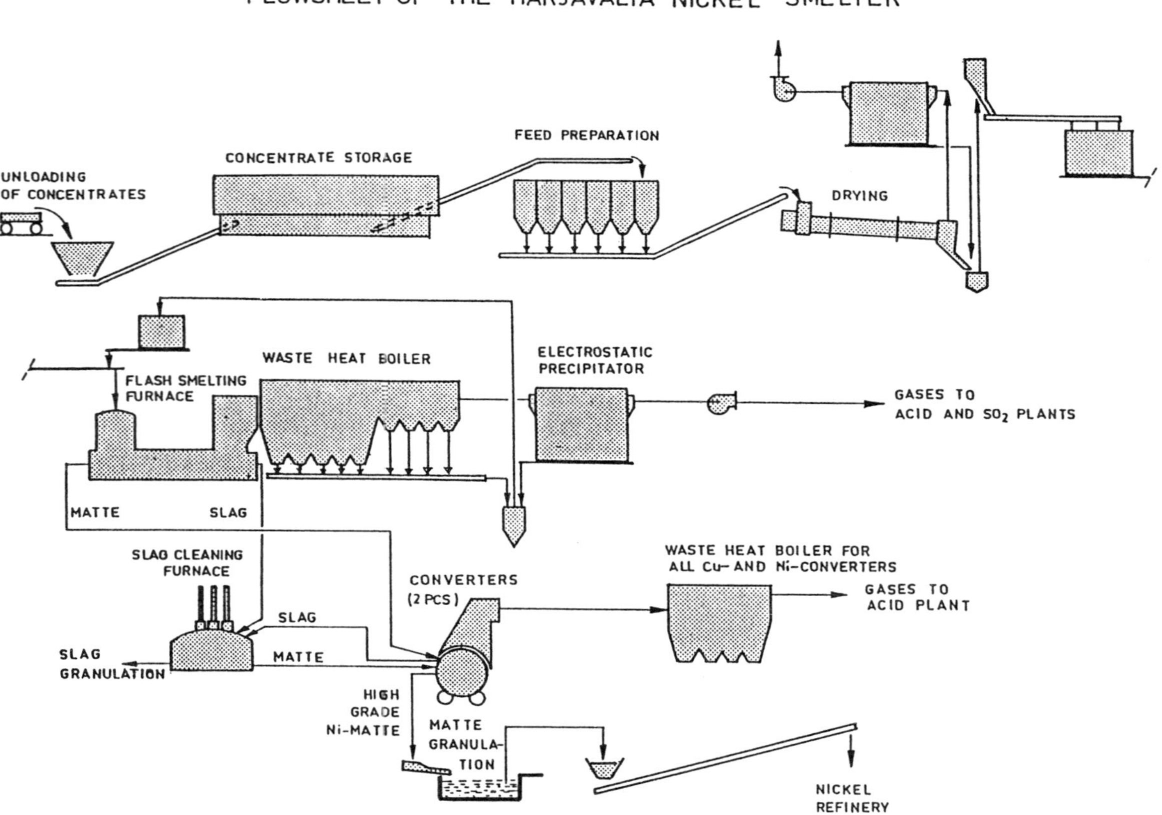

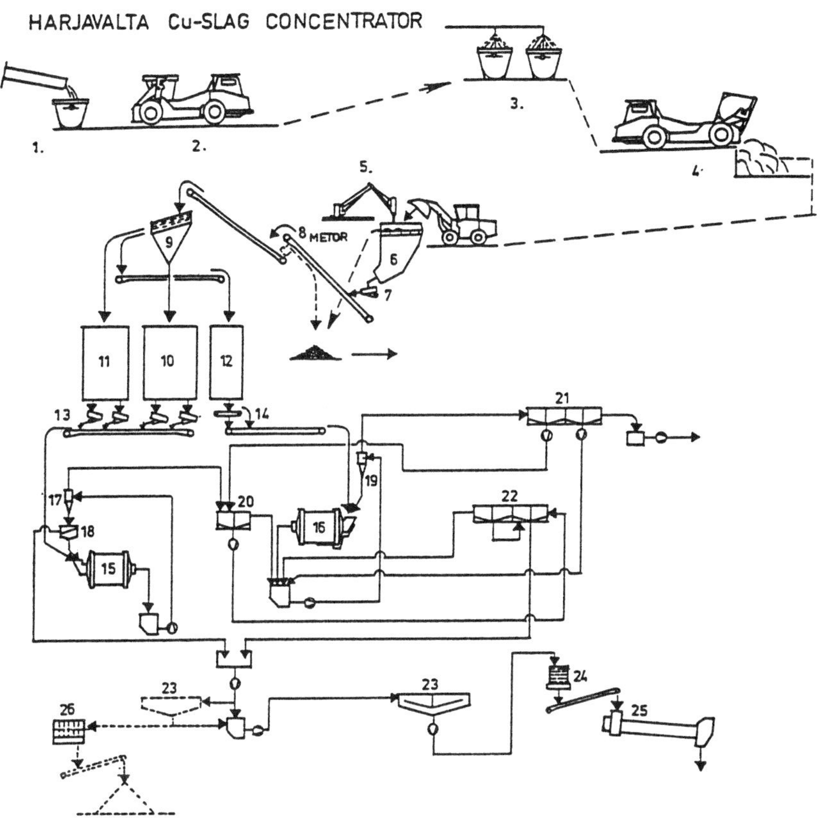

1 SLAG TAPPING
2 TRUCK
3 SLAG COOLING
4 SLAG TRANSPORTATION
5 HYDRAULIC HAMMER, RAMMER
6 HOPPER 100 t
7 ELECTROMAGNETIC VIBRATING
8 METOR METAL SORTER
9 DOUBLE-DECK VIBRATING SCREEN -1000 MM × 2500 MM
10 CRUSHED SLAG BIN 600 t (+400 t).
11 LUMP BIN, 480 t
12 PEBBLE BIN, 240 t
13 4 ELECTROMAGNETIC VIBRATING FEEDERS
14 BELT FEEDER
15 LUMP MILL Ø 4500 MM × 5000 MM
16 2 PEBBLE MILLS, Ø 3200 MM × 4500 MM
17 CYCLONE, Ø 350 MM
18 SKIM-AIR FLOTATION MACHINE OK 2,2-1C, CELL VOLUME 2 M^3
19 CYCLONE, Ø 260 MM
20 TWO-CELL FLOTATION MACHINE-(OK 16-U) CELL VOLUME 16 M^3
21 FOUR -"- -"- -"- -"- -"- -"-
22 FOUR-CELL FLOTATION MACHINE(OK3), CELL VOLUME 3 M^3
23 2 THICKENERS, Ø 8 M AND 12 M
24 PRESSURE FILTER
25 ROTARY DRYER (AT THE SMELTER)
26 DISC FILTER, Ø 1800 MM/4

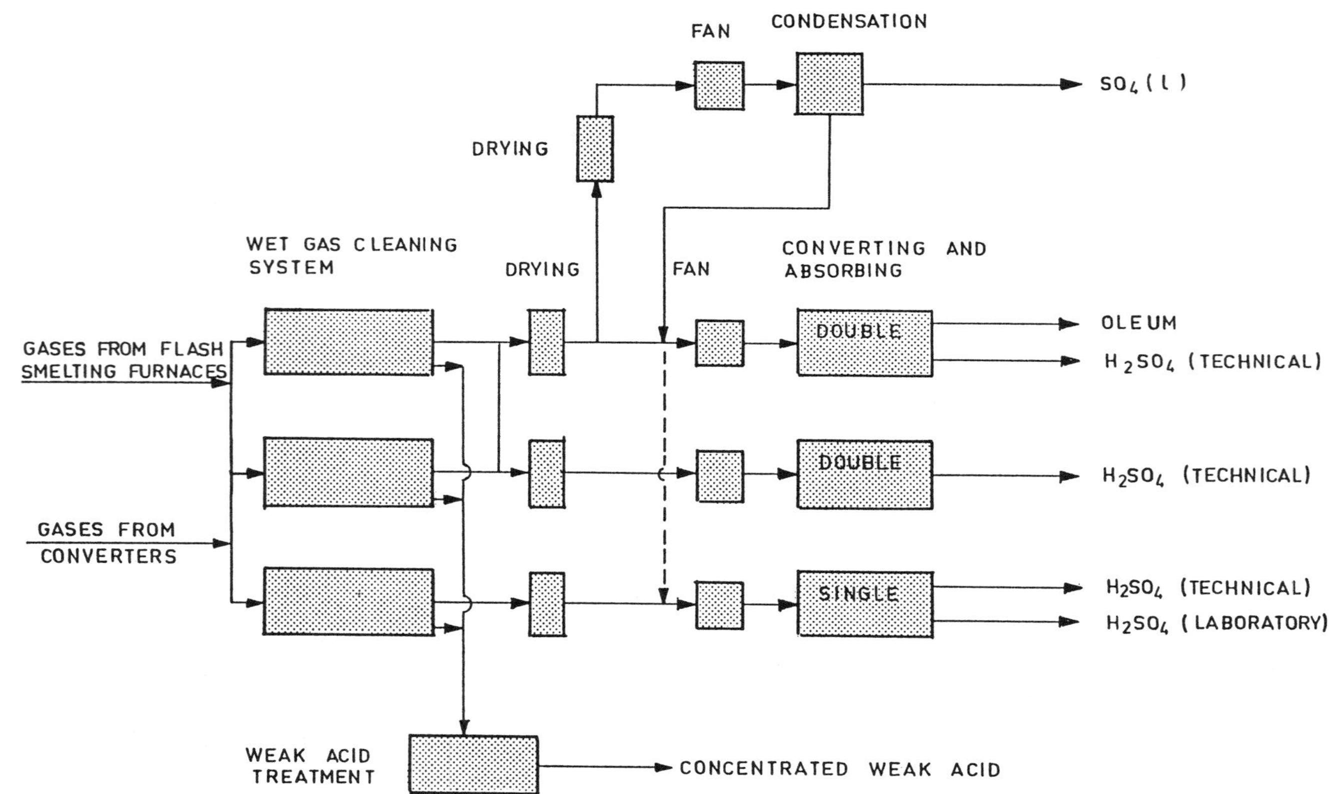
FLOWSHEET OF THE HARJAVALTA SULPHUR DIOXIDE PROCESSING
FAN
CONDENSATION
SO_4 (l)
DRYING
WET GAS CLEANING SYSTEM
DRYING
FAN
CONVERTING AND ABSORBING
GASES FROM FLASH SMELTING FURNACES
DOUBLE
OLEUM
H_2SO_4 (TECHNICAL)
DOUBLE
H_2SO_4 (TECHNICAL)
GASES FROM CONVERTERS
SINGLE
H_2SO_4 (TECHNICAL)
H_2SO_4 (LABORATORY)
WEAK ACID TREATMENT
CONCENTRATED WEAK ACID

RECENT INNOVATIONS AT THE HUELVA SMELTER

By:

D. de la Villa - Metallurgical Manager
F. Cevallos - Maintenance Superintendent
P. Barrios - Superintendent of Metallurgical Department

ABSTRACT

Since 1975 RIO TINTO MINERA is operating with an Outokumpu designed Flash Furnace.

An Environmental Compliance Programme was developed to renew and improve the existing process gas collecting and handling systems for the converters and anodes furnaces, by installing new evaporative spray cooling systems. Reduction of the secondary emissions was achieved by computerised converter movements and the design of a new system to feed the scraps while the converter is at the blowing position.

Due to the increasing cost of energy, in 1981 oxygen enriched operation was adopted, by using 2000 Nm^3/h pure oxygen. During last 1984 a 300 tonnes per day Oxygen Plant was erected by SEO which supplies oxygen to RTM over the fence. This supply permits an oxygen enrichment of air of 35-40 %. A Restructuring Plan was started, Planning and Programming procedures were adopted to achieve a higher productivity with lower operational costs.

Recent innovations including the shortening of the furnace reaction shaft, installation of a central jet type concentrate burner, waste gas boiler recirculation and various energy savings, are also described.

INTRODUCTION

Since 1970, in Huelva, on the South West coast of Spain, Río Tinto Minera is operating a metallurgical complex that assumes a millenarian tradition in the mining and smelting of copper, gold and silver ores from the Río Tinto mines. (1)

At first, the installation consisted of two Momoda type blast furnaces, with a capacity of 250 tpd each, Pierce Smith converters, refining furnaces and a 40.000 tpy Electrolytic Refinery. A Sulphuric Acid Plant, with a maximum production rate of 1000 tpd treated the off gases from the converters and furnaces. (2)

Having in mind the availability of concentrates in the market, Río Tinto Minera decided to increase its treatment capacity and in 1975 a 1000 tpd Outokumpu Flash Furnace was erected, to replace the two Momoda Furnaces. Previously, the capacity of the Electrolytic Refinery had been increased to 105.000 tpy and a double absorption acid plant with a capacity of 600 tpd, was installed to operate simultaneously with the former one. (3)

Great effort has been done in the last two years to reduce:

- fugitive emissions.
- unit energy consumption.
- maintenance costs.

Detailled descriptions of the more recent innovations are given below.

ENVIRONMENTAL COMPLIANCE PROGRAMME

The Smelter is placed in an industrial zone situated at about 4 Km from the town of Huelva. The neighbouring works are dedicated to the production of sulphuric acid and fertilizers. In 1979, the Government prescribed an Urgent Plan directed to the improvement of the environmental conditions of the city and, at the same time, to the elimination of the causes of such conditions.

In 1981 RTM finished an Environmental Compliance Programme with the following objectives:

- Improve the existing process gas collecting and handling the systems for Converters and Refining Furnaces.

- Reduce the secondary emissions of process gases to the converter aisle and surrounding areas.

Reduction of the converters emissions

To decrease the converter emissions to the aisle, the following objectives were followed:

- Substitution of the balloon flues.
- Reduction of the emissions when the converters come in or go out of the stack.

The arrangements of the converter hoods did not permit the dilution of off gases by the action of the entering secondary air and so the temperatures in the baloon flues rose above 700°C originating great deformations and cracks in the shell. Likewise, the sulfates accretions on the shell surface progressively reduced heat radiation in the baloon flues, and therefore their cooling capacity as the converter campaign advanced.

The cracks in the baloon flue permitted infiltrations of air into the chamber which caused a reduction in the draught at the hood and the subsequent escape of gases at the mouths of the converters. The high temperatures of the insufficiently cooled gases surpassed the design temperatures of the precipitators, forcing the interruption of the blowing operations, specially during the copper blows and at the end of the converter campaigns . The maintenance costs of each campaign were extremely high since it was necessary to have the balloon flues fully reconstructed.

Regarding the emissions caused during the time the converter goes in and out of the stack these were mainly due to the following:

Due to the small value of the blowing angle of the converters, -5°, and to the design of the hood, conditioned by the height of the converters aisle, the mouths of the converters remains uncovered by the hood when the converters starts going out of the stack.

The high consumption of scrap and blister copper in the copper blow made it necessary for the converter to go in and out of the stack very often.

Evaporative cooling chamber

Río Tinto Minera asked SELTRUST ENGINEERING to study and evaluate the possible ways to solve the related problems. The option chosen was to substitute the ballon flues by evaporative cooling chamber using SONIC type nozzles. (SELTRUST and RTM have in preparation a paper in which the selection of the options, the design parameters and the experience of operation are described extensively).

The water is injected in extremely fine particles and its volume is regulated in proportion to outlet gas temperature, which permits an even temperature in the electrostatic precipitators.

Modifications to the gases collecting hoods

In spite of the limitations caused by the position of the crane rail in the converter aisle, the hood was modified, improving its design and providing it with a new hydraulical closure at the rear.

In stack scrap charging machine

With the purpose of minimizing the number of turnouts of the converter, when adding metallic copper during the copper blow, a hydraulic driven charging system was developed.

The metallic copper is added to the converters using a tray, which penetrates into the hood, driven by a hydraulic system.

With this system we have obtained a complete copper blow with no interruption, a controlled temperature and a reduction of gas emissions.

Automatic air flow control

Río Tinto Minera has developed its own system of automatic air flow control by the application of a microprocessor in the converter operation.

At the time in which the converter is coming into stack, the amount of air flow is fixed to a necessary minimum value which prevents the tuyere filling with molten material the air flow is then increased step by step simultaneously as it turns into stack and is located under the hood, where it reaches its set point value.

The angle at which air is admitted to the tuyeres is variable and depends on the relation between the position of these and the depth of the bath. As the level of the bath is reduced during the blow, the starting angle of the blow becomes bigger, i.e., the mouth of the converter is nearer to the hood every time.

When the converter leaves the hood, the flow is decreasing progressively till a complete shut off is produced when reaching a point which depends on the original level of the bath.

This system allows the converters to be turned in and out of stack having their emissions to the aisle reduced to a minimum, it also avoids the installation of secondary hood or simplifies the design of these in the event we decide to put them in in the future.

Reduction of gas emission from refining furnaces

There are two 13' x 30' refining furnaces which are normally in operation. The reduction is performed through injection of propane gas into the bath through a tuyere. The combustion of propane is not fully accomplished and only 30 % of the gas is burnt inside the furnace. Gases were then passed on to a postcombustion chamber from which they were exhausted through a stack, producing an emission of soot derived from the defficient combustion.

The objective pursued was to prevent this emission from going to the atmosphere, by designing a new system to handle and treat the gases, making good use of the existent installation.

The system chosen was the one traditionally used in the steel industry for scrubbing the gases and automatic pressure control in the furnaces. It consists of a scrubber which allows the control of pressure inside the furnace and therefore the inner atmosphere.

The washing process is performed through the following stages:

Cooling and scrubbing off gases

The gas, which is drawn trough a pipe shaped duct connected to the furnace, passes to the cooling and saturation chamber of the scrubber where special nozzles are set to reduce the temperature and saturate the gases with water.

Subsequently by submitting the saturated gas again to the action of a series of nozzles, the bigger particles are removed.

Washing the gas and controlling the pressure inside the furnace

This washer has a ventury shaped section in which a cone shaped control element which moves vertically reducign or increasing the pressure drop through the scrubber which permits a very close control of the anode furnace draft.

Evacuation of the residual gases

The washed gases are passed through a separator and then exhausted to a stack provided with another drop separator.

Recirculation of the washing water and recuperation of solids

The water used is recirculated through a "lamella" thickener and is reused. Solids collected with a high copper content are recycled to the Smelter.

FURTHER IMPROVEMENTS AND ENERGY SAVINGS

Several improvements and modifications have been carried out so far aiming at more efficient and economical operation.

Higher oxygen enrichment

As a consequence of the higher cost of energy, oxygen enrichment operation was adopted in 1981.

As a preliminary test, the oxygen amount was increased up to 2.000 Nm^3/h.

During 1984 a 300 tpd Oxygen Plant was erected. The Plant started to operate in January 1985 and oxygen enrichment was raised step by step to 40 % just before the general shut-down in June 1985, to refurbish the furnace after the completion of its second campaign.

The Flash Furnace, with a shortened shaft, is currently on its third campaign and tonnage oxygen operation has been adopted.

Table 1 shows typical operating data and heat balance in the reaction shaft before and after oxygen enrichment.

Advantages of using oxygen enriched air in the flash furnace are the following:

- Smelting capacity increases.
- Dust production decreases.
- Fuel-oil consumption decreases.

The increment in smelting capacity for the last decade is shown in Fig. 1.

Fig. 2 shows the change of fuel-oil consumption as 100 % for the level in 1976. It has decreased greatly with the expansion of the smelting capacity by the oxygen enriched operation.

Table 1

Typical Mass and Energy Balance

		Atmospheric air	Enriched air (After shortening Reaction Shaft)
Mass balance			
INPUT			
- Concentrate	tpd	1.000	1.300
- Flux		127	194
Total			
OUTPUT			
- Matte	tpd	472	544
- Slag	tpd	532	780
Matte grade	%	50	56
Process air	Nm^3/h	57.900	21.700
Oxygen enrichment	%	21,00	50,1
Gas flow	Nm^3/h	64.300	25.400
Anodes production	tpd	330	427
Acid production	tpd	856	1.113

Heat balance			
INPUT			
- Chemical reaction	Mcal/h	22.300	32.300
- Fuel oil combustion	Mcal/h	22.300	4.500
- Preheated air	Mcal/h	6.200	1.100
- Others	Mcal/h		
Total		51.800	37.900
OUTPUT			
Heat in matte	Mcal/h	4.200	4.800
Heat in slag	Mcal/h	8.700	12.800
Heat in dust	Mcal/h	1.500	2.000
Heat in gases	Mcal/h	32.400	14.000
Heat losses	Mcal/h	5.000	4.300
TOTAL	Mcal/h	51.800	37.900
Steam production	t/h	53	21
Fuel-oil consumption (*)	Kg/h	3.200	480

(*) Consumption in air preheater included.

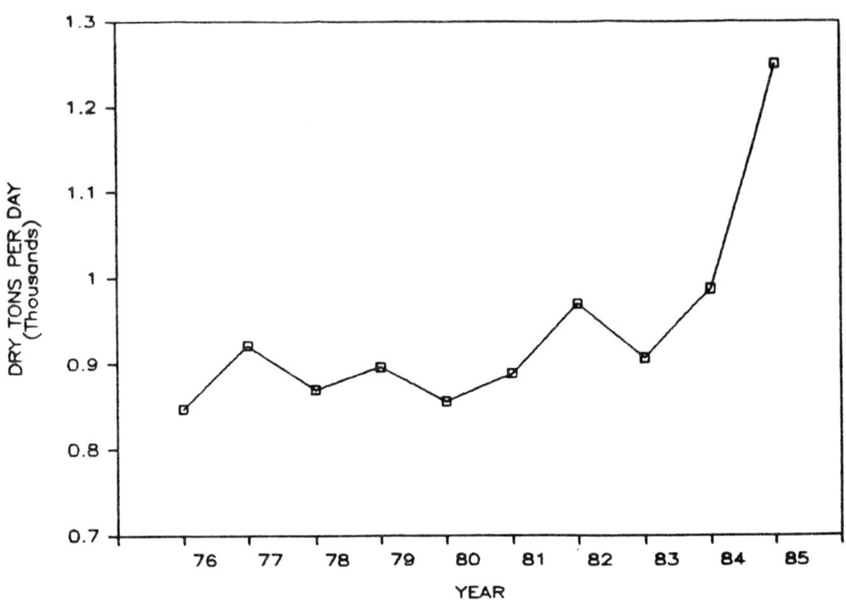

Fig. 1 Progress of Smelting Capacity
(1985 figure corresponds to last quarter estimation)

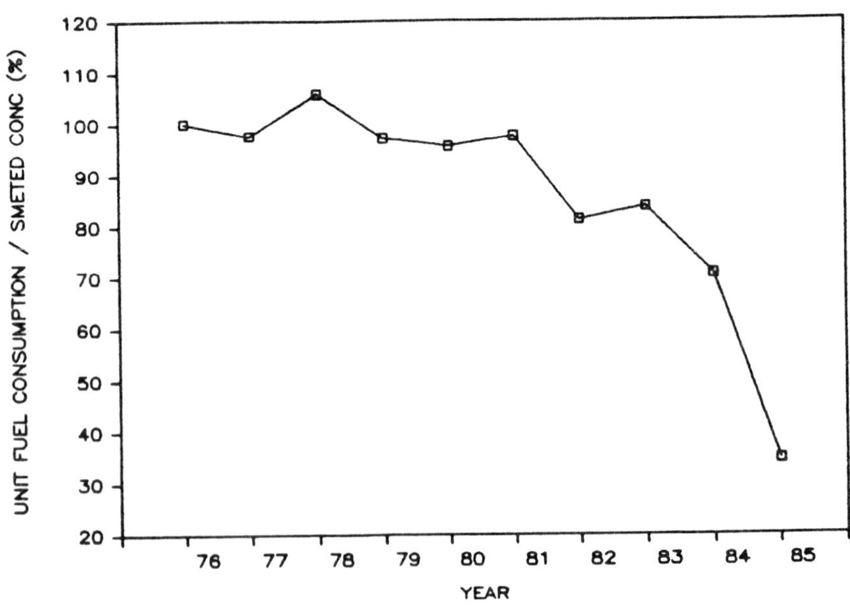

Fig. 2 Reduction of Unit Fuel-oil Consumption
(1976 is 100%)

Shortening of the reaction shaft

The use of tonnage oxygen in the shaft modifies the conditions of the combustion and the kinetics of the chemical reactions taking place inside the shaft. The focus of the flame is situated nearer the burner and therefore, the "effective height" of the shaft changes. The molten particles fall towards the bath from a higher level and the radiation losses can be reduced easily, by shortening the reaction shaft in a length proportional to the rise of the focus.

Reaction shaft has been shortened by 3 meters although its diameter has not been modified. The present measures are 6.8 m x 6.5 m Ø.

Improvement of the concentrate burners system

According to the earlier Outokumpu design, four concentrate burners were installed in the reaction shaft.

As a consequence of the use of tonnage oxygen, the flow of process air decreased in a great extent and, therefore, the velocity of the air in the burners decreased also, giving way to a higher dust production and consenquently to a higher consumption of energy.

A single Outokumpu central jet type concentrate burner has been installed, with a treatment capacity of 70 t/h.

The reason why this burner was chosen were as follows:

- As atomizing air is available to regulate the width of the flame, the operation turns more flexible since the air-concentrate mixture is not dependent on the volume of air.

- A single central burner, having the focus of the flame in the center of the chamber, decreases wear of the refractory linning.

- The number of auxiliary equipment for the burners is reduced.

Off gas recirculation

The gas handling system has been modified in such a way that the gases can be recirculated from the outlet of the electroprecipitator back to the radiation section of the waste heat boiler.

This system results in a more flexible operation since it is possible to have a control of the temperature of the gases inside the boiler.

Exhaust gases heat recovery

Oxygen enrichment operation has eliminated the need to heat process air. Formerly the combustion gases from the air preheater were used in the concentrate dryer. At present, the exhaust gases of the steam superheater are used and so the energy of a waste gas is recovered and the heat balance of the dryer is improved.

This modification will carry out a saving of 1.042 tpy of fuel-oil.

Summary of energy savings

Energy requirements has decreased drastically in the last years as shown in fig. 3.

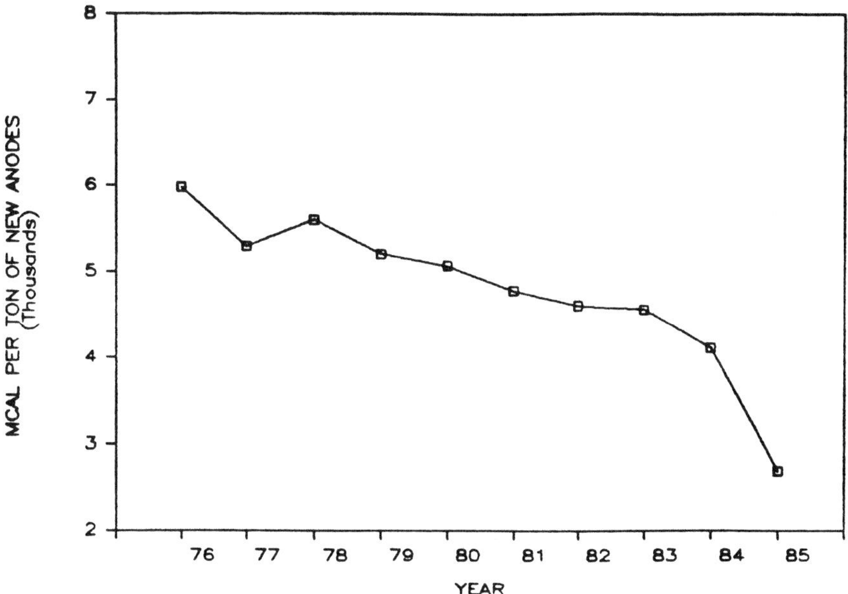

Fig. 3 Energy requirements per new copper
(Includes overall process in the smelter except acid plants)

RESTRUCTURATION PLAN

In mid 1982, Río Tinto Minera started a Restructuration Plan, which envolved all the Company, and was directed to:

- Modernization of the operational systems and procedures.

- Restructuration of the man power at all levels.

- Financial restructuration.

- Involvement of supervision in attaining objectives.

With respect to the organization at the Huelva Smelter, the following has been implanted:

- Increase of the Metallurgical Department, a Staff acting as a support to the operation supervision and studying possible innovations of the plant.

- Planning and programming in the maintenance services.

Río Tinto Minera has decided on the application of a computerized maintenance programme in which predictive and preventive maintenance has been enfasized, warehouse procedures have been modified decreasing the stocks of spares to more reasonable levels. Redistribution of maintenance personnel, organization of the electric and instrumentation department have been done to reduce contractors.

To attain these objectives the Supervision in Maintenance has been impelled to increase the planning and programming functions, establishing new lines of the activity, which are directed to the consecution of the following purposes:

- Reduce equipment failure to a minimum.

- Reduce down time of equipments to a minimum.

- Increase training.

- Reduce maintenance costs.

- Reduce accidents to a minimum.

The necessary tools for the attainment of these objectives have been defined and they are:

- Improvement of the work order systems.

- Time cards.

- Control of pending work orders.

- Backlog in manhours.

- Weekly control of contractors.

- Repetitive work order programs.

- Planning of work orders.

- Scheduled preventive maintenance.

- History file.

- Availability.

- Damage analisys to equipment.

- General information.

- Etc.

The software used in this program is basically the Palabora Mining Company maintenance program which has been duely adapted to RTM.

Fig. 4 shows the change of the maintenance unit cost as 100 % for the level since 1976.

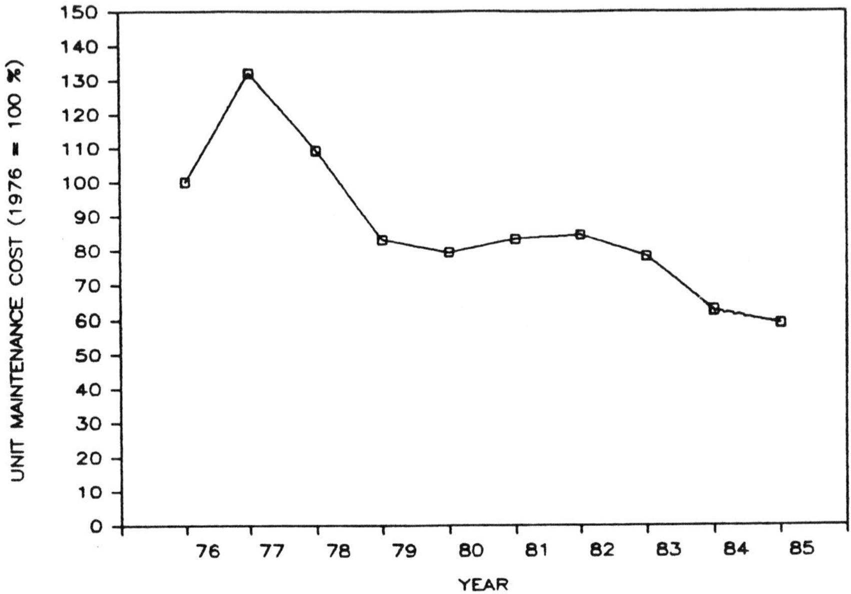

Fig. 4 Unit Maintenance Cost
(1976 is 100%)

CONCLUSION

During the last years, Rio Tinto Minera has accomplished several modifications and many improvements in order to reduce the secondary emissions, to increase the production capacity and to reduce costs by increasing productivity.

The results of the performed innovations can be summarized as follows:

- Reduction of secondary emissions.

- Increase of the capacity of production by 30 %.

- Great reduction of the energy consumption.

- Reduction of the maintenance costs.

ACKNOWLEDGMENT

The authors wish to express their acknowledgment to Rio Tinto Minera for having permitted the publication of this paper. Likewise, they thank the whole staff in the Huelva Plant for their valuable contribution.

REFERENCES

1. Avery, D., Not on Queen Victoria Birthday (The story of the Río Tinto Mines). Collins, 21-31, 1974.

2. Luzón, F., Una Nueva Planta Integrada de Cobre Electrolítico y de Acido Sulfúrico, Revista de Metalurgia, Vol. 5 n° 6, pag. 697-705, 1969.

3. Palacios, C., Huelva Flash Smelter. The Third International Flash Smelting Congress. Helsinky-Hamburg-Huelva, May 16-20, 1977.

COPPER RECYCLING PLANT AT BRIXLEGG (AUSTRIA)

H. Woerz and J. Wallner

Austria Metall AG
Montanwerke Brixlegg
P.O.Box 19
A-6230 Brixlegg/Austria

Summary

The Montanwerke Brixlegg is a copper recycling plant. The raw material, productions plants, envoronmental problems and some aspects of the future technical development are described.

The copper plant in Brixlegg was founded in 1463, that was more than 500 years ago. The plant in these days was surrounded by several mines, which in the course of time were all shut down. The last mine was closed in 1969. So the basis for the plant is not anymore ore, but scrap and copper containing residues, which are the main part of the raw material input. The rest is blister copper, which is imported from overseas.

In the historical development of the plant it may be of interest, that recycling was started in the early 1920 and that electrolytic copper is being produced since 1886. Brixlegg is one of the first refineries in the world which has used a dynamo machine for producing electrolytic copper.

Today the electrolytic copper production is 33 000 tpy. The Montanwerke Brixlegg is a plant of the Austria Metall AG (formerly Vereinigte Metallwerke Ranshofen-Berndorf AG) and employs a staff of approx. 400.

Raw Materials

The types of pyrometallurgical furnaces now in use are: shaft furnace, converters and anode furnaces. Depending on the furnace type the raw material input varies.

The shaft furnace of a recycling plant is charged mainly with oxidized materials, and in order to improve the energy consumption, to support the reduction and to achieve a low viscosity slag, low grad copper scrap containing iron is being added.

The shaft furnace material mainly consists of slags, drosses, ashes, grinding dust, ball mill fines, hydroxide-, galvanic - and metal slimes as well as scarp from shredders and electrical and electronical scarp, containing precious metals. The average content of the raw material input of the shaft furnace is 30 % copper and 20 % iron.

It becomes a common practice to charge so called converter materials into the shaft furnace in order to increase the zinc content in the filter dust on one hand and on the other hand to increase the liquid input into the attached converter. An other advantage of charging converter material into the shaft furnace is that the zinc oxide content of the converter flue dust decreases, causing an increase of the specific tin content and thus easing the further processing of converter filter dust.

In the converter, being located right after the shaft furnace an oxidizing process is under way and metallic alloy scrap is being charged, mainly consisting of mixed brass, bronzes, gunmetal, alpaka, car-radiators and so on. The average copper content of the scrap input into the converter is approx. 43 % Cu.

The next step in the process is the anode furnace. The input consists of copper scraps, starting with a copper content of approx. 85 %. These are mainly mixed industrial- and demolition scraps and isolated cables which were treated to burned-off cables and wire granules.

Description of the Production Plants

Shaft Furnace

The charge of a shaft furnace is being reduced and molten down by means of coke. The aim of this process is to achieve

- the metallic phase whereby all valuable metals to a great extend are being dissolved in the liquid black copper

- the slag phase with the lowest possible content of copper (below 1 %) and other valuable metals which makes further treatment not necessary and

- a flue dust with the highest possible content of ZnO and PbO.

The cross section of the shaft furnace is rectangular and equiped with 12 tuyeres, the tuyere sectional area being 3,2 m^2. The shaft rises another 4,6 m above the tuyeres. The amount of process air enriched with oxygen is 6500 m^3/hour. The coke-content is 8 % related to the total input. The temperature in the throat of the blast furnace is between 650 and 950° C.

A combustion chamber serving as an after burner (which is in the planning phase) is going to be arranged after the throat in order to reduce the content of organic carbon below 200 mg/Nm^3.

The waste heat vessel has a heating surface of 300 m^2 and a performance of 1,3 t steam per hour.

The flue gas is cooled down to 100° C by a forced flue gas cooler. The filter equipment is a bag filter with a capacity of 20 000 Nm^3/h and a net surface of 400 m^2. In order to suck off the flue dust in the area of the taphole and the throat there is another bag filter with 25 000 Nm^3/h and a net-surface of 420 m^2.

The throughput of the furnace is 175 t/day of copper bearing materials.

The daily production runs at

50 tons black copper,	containing approx. 80 % Cu, 4 % Ni, 3,5 % Sn, 5 % Fe, 2 % Pb, 4 % Zn
80 tons slag,	containing approx. 1 % Cu, 0,3 % Ni, 6 % Zn, 0,3 % Sn, 0,3 % Pb, 33 % FeO, 28 % SiO_2, 6 % CaO, 8 % Al_2O_3
6 tons flue dust,	containing approx. 50 % Zn, 0,1 % Ni, 12 % Pb, 3 - 4 % Cu, 0,8 % Sn, 3 % Cl

The slag is being quenched in a watersump and sold for sand blasting.

Converter

In the case the charge of the converter is not yet liquified this will be achieved by means of an oil-burner. Into the molten metal bath oxygen of the free air is injected. During this process the metal components will react accordingly to their affinity to oxygen and creat mixed metaloxides which are being expelled by means of the flue gas or in turn will go into the liquid phase and oxidize to slag.

The converter has an outside diameter of 2,1 m and is 3,8 m long. The process air in the amount of 4 000 m^3/h is directed into the furnace by means of 10 nozzles, each of them having an inside diameter of 1 inch.

A heavy oil-burner with a capacity of 2000 l/h is mounted at one of the frontsides of the converter and serves to melt down the solid input. The converter flue gas direction is horizontal, that means: the flue-gas is discharged along the furnace axis, passing a combined deposition and combustion chamber, and subsequently is cooled down to 180° C by a 640 m^2 tube type flue gas cooler.

The flue dust is collected in a bag filter with a capacity of 35 000 Nm^3/h and a net-surface of 6440 m^2. Finally the flue gas is washed by means of a new developped scrubber using $Mg(OH)_2$ as reagent to reduce the SO_2-peaks down to 200 mg SO_2/m^3 or lower.

The primary slag that forms in the converter during the oxidation phase is rich in iron but the copper and tin contents are poor. Therefore the slag has to be removed from the converter in order to avoid, as the oxidation goes on, that the SnO2 now being produced extensively will form an iron and tin slag which is hard to reduce.

After the primary slag has been drained, the metal bath is charged with coke, as the oxidation once again is started, the now forming SnO_2 is being reduced to Sn and SnO.

As in the course of the process Sn and SnO evaporate, the flue dust in consequence will be enriched with SnO and thus its value is improved.

The secondary slag, which forms as the oxidation goes on, consists of SnO_2 which has not been reduced and is rich in copper. It goes back to the shaft furnace just the same way as the primary slag does.

Running on a two-shift per day operation 3 - 4 charges are being produced. Each charge rending about

15 tons	raw copper	with 97 % Cu, 1,3 % Ni, 0,3 % Sn, 0,3 % Pb, 0,1 % Sb
2,4 tons	flue dust	with 56 % Zn, 10 % Sn, 14 % Pb, 0,9 % Cu
5-6 tons	slag	with 25 % Cu, 20 % Fe, 5 % Ni, 5 % Sn, 3 % Zn, 5 % Pb

Anode Furnace

The final pyrometallurgical process takes place in the anode furnace. The aim is to produce the purest and most dens anodecopper possible.

Due to the higher affinity of impurities towards oxygen selective oxidation is achieved by means of injection of air or air enriched with oxygen whereby the impurities will turn into slag as oxides.

After the oxidation and after removal of the slag the metal bath is being reduced resp. poled by dipping alder- and beech poles.

The bath surface of the anode furnace is 29 m^2. The charging weight is 160 tons. The cycletime for one charge is 24 hours. Heavy fuel oil is being used whereby the air can be enriched with oxygen if necessary. In relation to 1 ton of anode the oil consumption amounts to about 100 kg, that of fire resistant materials to about 11 kg.

A radiation- and convection boiler connected to a superheater laid out for a steam pressure of 25 bar is installed after the funace. This is followed by a bag filter. The capacity of the filter is laid out for 25 000 Nm^3/h of process gases and an additional 100 000 Nm^3/h of cooling air which cooles it down to 95° C. The cooling air is taken from the working area thus improving environmental and working conditions of the job site.

A casting wheel with 14 moulds and an automatic weighing system taps anodes weighing 220 kg each. The hourly production rate amounts to about 28 tons. The daily production amounts to

145 t anode with	99,3 % Cu	0,04 % Sn	600 g/t Ag
	0,4 % Ni	0,03 % Zn	5 g/t Au
	0,3 % Pb	0,05 % As	
	0,3 % O_2	0,03 % Sb	

12 - 15 t slag with 50 - 60 % Cu, 5 % Ni, 1 % Sn

0,5 - 0,8 t flue dust with
12 % Cu, 11 % Zn, 9 % Pb, 2 % Sn
3 % S (as sulphate), 13 % SiO_2

Tankhouse

The tankhouse has a capacity of 33 000 tpy and is run by a process which was developed in Brixlegg and is known as the PCR-process which stands for "periodic current revers".

The current density in the production cells is 330 A/m^2, in the starter sheet-cells because of the lesser number of electrodes 420 A/m^2. There are 25 anodes and cathodes in a production cell.

The anode spacing is 96 mm. For the contact between anodes and cathodes the Baltimore contact is used.

Out of one anode, three cathodes are produced with a weight of 65 kg each. The anode-journey is 16 days, the cathode journey 5 - 6 days. The anode rest figure is 16 %.

The electric power consumption is 400 kWh/t, the current efficency 93 %. The average steam consumption over the year amounts to 0,7 tons steam per ton of cathods.

Composition of the electrolyte: 42 g Cu/l, 10-15 g Ni/l, 170 g H_2SO_4/l; Inhibitors: 120 g glue/t and 50 g thiourea/t;

Flow rate of the electrolyte: 15 - 20 l/min, flowing from the bottom of one side to the top of the other front side.

2 % of the total amount of electrolyte in use is being regenerated daily passing through 6 liborator cells, followed by the nickelsulphate plant.

For decopperisation unsoluble lead anodes are used, reducing the copper content of the electrolyte below 0,5 g Cu/l. In the following nickelsulphate plant a very pure $NiSO_4 \cdot 6 H_2O$ is produced by precipitation with limestone and H_2S. This nickelsulphate is used in the galvanic industry. The yearly production is approx. 400 t.

The anode slime being developed is treated by a purely hydrometallurgical process, which was also developed by the Montanwerke Brixlegg, was industrially introduced to our plant and has proved to be very successful for more than a year.

The anode slime amount per year is 380 - 400 tons, respectively 10 kg/t anodes with a precious metal content of 60 000 g Ag/t and 500 g Au/t.

Precious Metal Recovery Plant

In the first process step the acid soluble copper and nickel is dissolved in a sulphuric acid solution and recycled into the tankhouse. The more or less copper- and nickel free anode slime undergoes first refining stages, whereby gold and platinium metals are being dissolved which go to the separating plant. The silver now left in the residue is treated in a further refining process reaching a purity of 99,99 % achieved purely by a hydrometallurgical treatment. There is no electrolysis involved.

Gold and the platinum metals are recovered by a selective cementation. Se, (Te) and the rest of copper remains in the residue. Gold is obtained in highest purity as a metal (again without electrolysis), the platinium metals as enriched salts. The Selenium and traces of Tellurium are also quantitativly separated in a further process stage.

All waste waters are centrally collected, neutralized and separated from heavy metals by precipitation, so that no waste water problems occure concerning the environment.

The remaining slime is fed back into the shaft furnace thus creating a closed material circle with the lowest loss of precious metals.

Casting Shop

In the casting shop there are 3 semi-continuous casting machines with a capacity of 30 000 tpy. Due to a special arrangement of the channel induction furnaces, billets are produced in the quality "grade 1", other oxygen-free qualities with approx. 10 ppm O_2, P-desoxidized copper, Ag-alloyed copper and Sn-alloyed copper. The electric power consumption is 360 kWh/t.

Environment Protection

Air:

The problems of a recycling plant are encountered in the SO_2- and heavy metal emissions.

SO_2 emits not only because of the sulfur in the heavy fuel oil, but also because of the sulfur which is reduced into the black copper during the shaft furnace process. During the oxidizing process in the converter short SO_2-peaks arise, which have to be reduced by a scrubber. As mentioned this scrubber is a new development especially for rapidly arising SO_2-peaks (up to 12 000 mg/Nm^3) and a high O_2-content (approx. 20 %) in the flue gas, using $Mg(OH)_2$ as a very reactive reagent and producing $MgSO_4$ which can be utilized in the wood chip board industry.

Another SO_2-emission caused by the heavy fuel oil is also encountered because of the operation of the anode furnace. These SO_2-quanities may however be reduced by means of dry absorption, which is achieved by adding limestone in the flue gas stream.

Following the anode furnace a bag filter separates the flue dust. As the amount of dust is neglegible it is recycled to the shaft furnace.

Behind the shaft furnace and the converter there are also high efficency textile bag filters which reduce the dust content in the pure gas down to 10 mg/Nm^3 and lower.

Although the gas of all furnaces is filtered heavy metal immissions are found in the surrounding area of the plant. This pollution is partly due to the contamination which happened in the past and is partly caused by wind drifts. The reduction of these immissions will be extraordinary difficult to achieve.

In order to reduce the heavy metal immissions to the lowest possible level and also to avoid metal losses nearly all storage areas are paved and regularly swept clean. Some storage areas are covered by roofs.

Water:

The waste water consists mainly of cooling water and to some extent of water coming from the neutralization section of our chemical plants, but this does not constitute a problem.

Conclusion

Copper recycling plants become more and more important as an increasing copper production in the past is nowadays with a time lag reappearing as scrap. This effect is emplyfied by the fact, that copper containing products today have a shorter lifetime than in the past.

Recycling also generally reduces environmental problems as otherwise dumping grounds would be additionally loaded. Since dumping costs show an upward trend it is of increasing interest also to accomodate waste materials with low metal contents. The higher treatment charges will be offset by avoiding dumping costs.

Besides the further development of existing technologies in the future following actual problems should be brought to a sulution:

- o the development of processes to treat raw materials with a low metal content and side products which occur in the own plant,

- o the further development of separating plants, filters and cleaning equipments for environment protection. Such reagents should be preferred, which are converted into a product that does not create dumping problems but can be significantly used,

- o the development of high performance reactors by using oxigen-, vacuum- and plasmatechnologies

- o the development of systems to increase the efficency of energy input and

- o the improvement of comunication and automatisation.

a large field which has to be solved by the universities, research centers and by the engineers in the recycling plants.

DESCRIPTION OF THE BOR SMELTER

BOR, YUGOSLAVIA

The following was supplied to supplement the questionnaire.

The end of the nineteenth century marked the beginnings of organized prospecting of copper ore in Bor, and very significant copper content was found in the ore which initiated the mine development. Construction of the first smelting capacities started, and the new Bor Smelter, comprising two shaft furnaces with capacity of 100 tons of rich ore per day and four converters with capacity of 3 tons of blister copper, was put into operation in 1905.

The satisfactory start-up resulted in an increase in melting capacities and another two shaft furnaces were constructed in 1908 with capacity of 160 tons of ore per day.

One smelter, modern for that time, was completed in the period between the two wars and in 1934 it consisted of seven shaft furnaces and six converters with capacity of 26 tons per operation.

During these years, the richest ore was exploited and the output of 43,000 tons of copper was achieved in 1940.

A decrease in copper content of the ore resulted in construction of the first flotation plant in Bor in 1937, and one year later, a tankhouse of 12,000 t/y capacity was put into operation. Its capacity was increased after World War II with the construction of the new tankhouse and the total capacity was 38,000 tons of copper per year.

In 1961 a new copper smelter was put into operation - consisting of a bedding system for charge preparation, roaster plant comprising five hearth furnaces, one shaft furnace, three converters and two anode furnaces. All this was followed by the construction of other plants: copper mine and copper flotation plant in Majdanpek, fertilizer plant in Prahovo, sulphuric acid plants in Bor, etc.

The latest development at the Bor Smelter in the 1970s gave it its present shape and form - concentrate storage, two fluo-solid partial roasting reactors, one reverberatory furnace, converter and anode furnace as well as three new sulphuric acid plants and a technical oxygen plant.

The Bor Smelter now has a production of up to 165,000 tons of anode copper per year. The basis is so-called standard procedure of pyrometallurgical copper production and this includes charge preparation in bedding system, roasting of concentrate in fluo-solid reactors, melting of matte in shaft furnaces, matte converting and anode refining of blister copper.

Concentrates from flotation plants at Bor, Majdanpek, Veliki Krivelj and Bucim are treated in the smelter, as well as imported concentrates. Their storage and preparation is carried out in bedding and storage facilities at the smelter. Charge is formed in bedding on two beds and by means of a conveyor belt system is transported to the bins located above the roasting reactors.

Partial roasting is carried out in two fluo-solid reactors. Sulphide oxidation from charge is carried out by means of oxygen from air used for fluidization at temperatures of 620 to 650°C. The greater part of the calcine (80%) is discharged from reactor with roaster gases, while the smaller part (20%) is discharged at the lower part of the reactor through a so-called fluo-seal.

Reactor gases are dedusted in primary and secondary cyclones (90% of dust is eliminated), cooled by means of supersonic spray in water cooling tower and at temperature of 350°C they proceed to electrostatic precipitators for fine dedusting, and after that to sulphuric acid production. Calcine from fluo-solid reactors is collected in two bins, from where it is charged into reverberatory furnaces by telescopic devices.

Melting of calcine is carried out in two reverberatory furnaces.

The furnace is lined with chrome- magnesite bricks, has a suspended roof which is serviced by crane for quick repairs. At the slag level, water cooled jackets are incorporated. Charging is carried out on the melted surface of the bath, that is, "mirror". Discharge is performed through 6 holes for matte and two for slag. Offgases enter the waste heat boilers (Furnace No. 1 has one boiler and Furnace No. 2 has two boilers), and after dedusting in cyclones and electrostatic precipitators, are discharged to atmosphere through a stack of 110 m height. Reverberatory Furnace No. 1 is fired by coal dust and Furnace No. 2 by fuel oil. Each furnace has 4 burners and combustion air is enriched by oxygen. The oxygen plant is of 100 t/d capacity. Slag from the reverberatory furnace is taken to the dumping area and is not used further.

Copper matte is transferred by ladles to the converter section.

There are 4 Peirce-Smith converters lined with magnesite and chrome-magnesite bricks of 457 mm thickness at the tuyere line and side wall, and 330 mm on the cylinder wall. Campaign life of converters between two repairs is 140-160 operations. For melt blowing, air at pressure of 1.1 atm is used, and sometimes it is enriched by oxygen. Gases are dedusted in chambers and electrostatic precipitators before mixing with roaster gases and further treatment in sulphuric acid plants. Converter slag is returned in liquid form into the reverberatory furnace and blister copper is taken to the anode refining section for the purpose of further processing.

Fire refining of copper is carried out in three anode furnaces. Casting is performed on two casting machines. Oxidation is carried out with air and reduction using beech trunks. Anode copper product is cast into anodes of 220 kg and are sent to the electrolytic refinery.

Metallurgical gases from the smelter are mostly used for sulphuric acid production. The first sulphuric acid plants were constructed in the 1970s at Bor. They were of Petersen tower type.

Up to now, three sulphuric acid plants have been constructed, based on a single contact catalysis procedure. Their operation is autothermal at a gas strength of 4% SO_2. By the 1980s the tower plants were out of use.

Metallurgical gases from all sources: roasting, converting and reverberatory furnace melting, are previously cooled (in boilers for the purpose of energy utilization) and dedusted in electrostatic precipitators before they are taken for treatment in sulphuric acid plants or taken out through stacks and discharged to atmosphere. Gases from roasting process are completely used and converter gases up to 50%. The rest of converter gases and gases from reverberatory furnace smelting are discharged into the atmosphere. Sulphur recovery from metallurgical gases reached 60%.

The Bor Smelter deals intensively with the problem of how to use the rest of the low strength gases for sulphuric acid production. Acid is used in great quantities in highly developed chemical industry of fertilizers, located on the Danube River in Prahovo, near Bor.

This problem and further increase of sulphuric acid production and reduction of air pollution should be resolved in the following ways:

- Additional roasting of pyrite concentrates in reactor of fluosolid type, production of gases rich in sulphur dioxide and mixing with poor smelter gases for purpose of their common treatment.

- Improvement of smelter gas quality.

- Production process intensification in metallurgical plants (technical oxygen introduction) or finally, replacement of reverberatory furnace melting process by continuous processes (INCO, Noranda, melting in liquid bath, flash smelting, etc.).

INCO FLASH SMELTING OF COPPER CONCENTRATE *

T. N. Antonioni et al

INCO Limited

* Extract from the British Columbia Seminar on Copper Smelting and Refining Technologies - Vancouver, B. C., 1980

Copper Cliff Operation

Introduction

The Ontario Division of the Inco Metals Company typically mines about 15 million MT of ore per year and produces about 150,000 MT of nickel and 140,000 MT of copper, almost as much as the nickel. Inco also recovers cobalt, Se & Te, gold, silver and other precious metals. In addition, 80,000 MT of liquid SO_2 and 800,000 MT of sulfuric acid are recovered from the process off-gases at Copper Cliff, Ontario.

Copper Cliff Smelter Operations

Nickel and copper concentrates produced at Inco's mills in the Sudbury district are supplied to the Copper Cliff smelter. The smelter circuits are quite complex because of the copper in the nickel concentrate and nickel in the copper concentrate (1-4). A simplified version of these two parallel smelting operations is shown in Figure 1.

The fluxed nickel concentrate is treated in multi-hearth roasters and reverberatory furnaces. The slag produced is discarded, while the matte is delivered to Pierce-Smith converters. The converter product, called Bessemer matte, and containing about 50% Ni and 25% Cu, is partially cooled in designated converters prior to delivery molten to the matte casting, cooling and separation processes (5) which yield three products, a nickel sulfide concentrate, a copper sulfide (chalcocite) concentrate and a metallics fraction. The nickel converter slag is recycled molten to the reverberatory furnaces.

The fluxed copper concentrate, containing about 30% Cu and 1% Ni, is fluid bed dried prior to injection into the flash furnace using a stream of oxygen. The resulting off-gas containing 70-80% SO_2, is delivered to liquefaction and the matte, typically containing 40-50% Cu, is delivered to Pierce-Smith converters. The furnace slag, containing 0.6-0.7% Cu, is discarded. In the converters, the furnace matte is blown with flux and the chalcocite concentrate (6), produced in matte separation, to yield a blister

Figure 1 COPPER CLIFF SMELTER FLOWSHEET

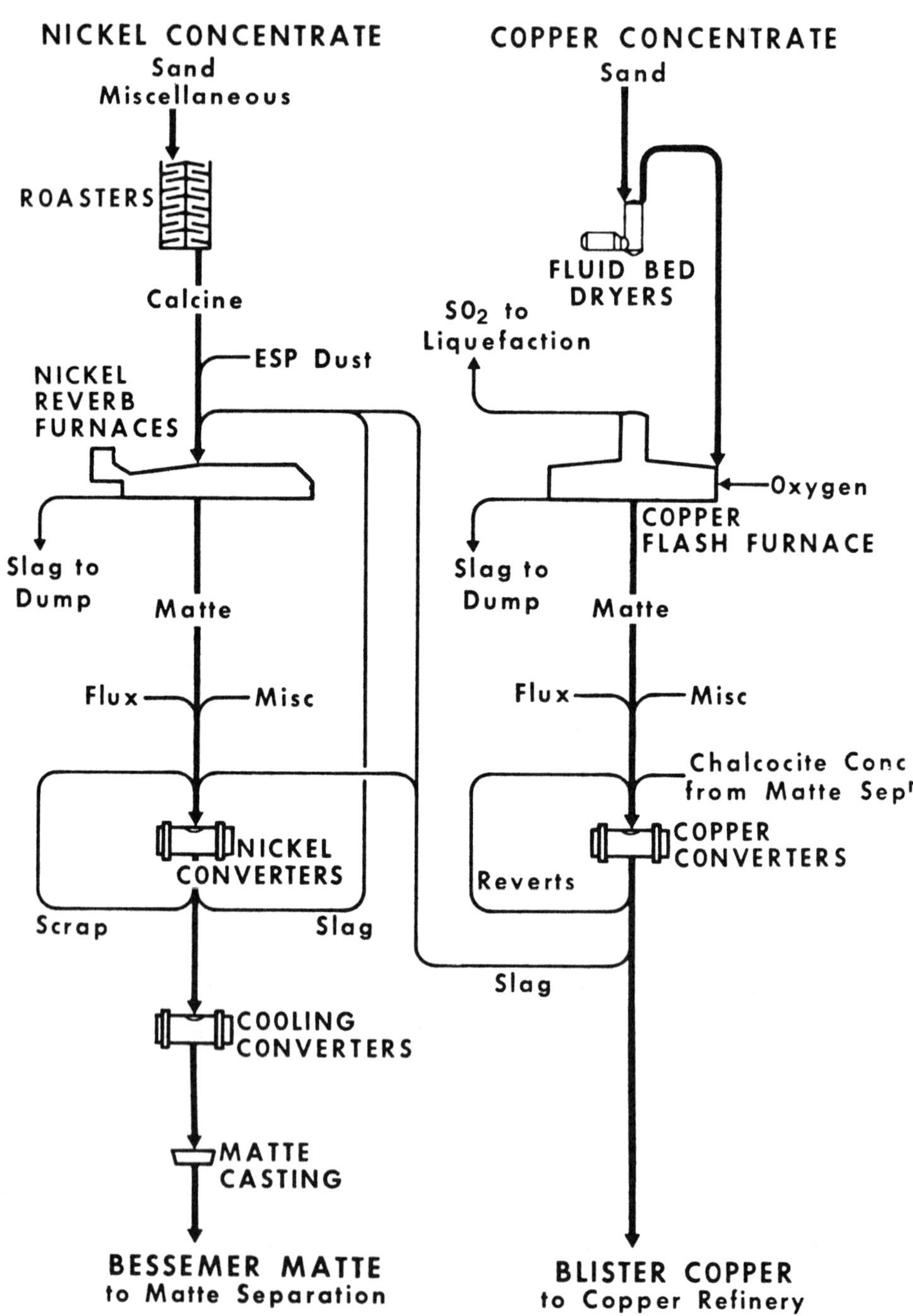

copper, which is delivered molten to the copper refinery for anode furnacing and casting. A portion of the copper converter slags are reverted to the flash furnace molten while the remainder is reverted to the nickel circuit.

Until quite recently none of the converter slag was reverted to the flash furnace as this stream was used to bleed nickel from the copper circuit. However, with improved separation in the mill, and as a result of slag reversion testwork, about 50% of the converter slag is now being reverted to the flash furnace.

Flash Smelting Process Development

Laboratory and pilot plant testing of the Inco Oxygen Flash Smelting Process was initiated at Copper Cliff in 1946 (7,8). Based on the data obtained, the first commercial furnace was commissioned in early 1952. This unit was rated at 450 MTPD of concentrate and had the dimensions shown in Figure 2. The second furnace with double the capacity and slightly larger in size was commissioned in late 1953. This unit operated until 1968 when its capacity was increased by extending the shorter end of the furnace to make it symmetrical about the center line. This unit was rated at 1360 MTPD.

The development and present operational practice of Inco's Oxygen Flash Smelting Process has been extensively covered in the technical literature (9-18). Its application to the smelting of nickel concentrates has also been reported (17-19).

Current Flash Furnace Characteristics

The current operating furnace is 24.4 m (80') long, 7.3 m (24') wide and 5.7 m (18') high at the ends, has a volume of 600 m^3 (21,200 cu ft) and a hearth area of 140 m^2 (1,500 sq ft). The roof is a sprung arch with the off-gases leaving through the center uptake. About 20% of the sidewall is cooled with water jackets located in the mid furnace area. The furnace is totally enclosed in a mild steel shell except for the slag line area which is externally cooled by air jets. Slag is skimmed through a single hole at one end of the furnace and matte is tapped through one of two holes located on one side of the furnace also near the mid point.

The furnace refractories consist of a magnesite bottom and chrome magnesite in the walls and roof.

Fluid Bed Drying of Copper Concentrates

At Copper Cliff, the concentrates, at about 8% moisture, are delivered by conveyor along with about 10% sand flux to the top of the fluid bed dryer. These units are fired by natural gas, though light oil has also been used. The dried product, at about 0.1% moisture, is conveyed by the fluidizing air to the baghouses where the furnace feed is collected for delivery to the furnace feed bins. The flowsheet is shown in Figure 3.

There are two dryers each nominally 2.4 m (8') dia. at the hearth with fluidizing shafts about 4.9 m (16') high. They are each capable of drying about 900 MTPD of concentrate. The windbox temperature is typically 315°C (600°F), the bed temperature 120°C (250°F) and the off-gas 105°C (220°F). The fluidizing blowers have a capacity of 37,600 Nm^3/hr (25,000 scfm), and the combustion chamber rating is 5.2×10^6 kCal/hr (20.5×10^6 BTU/hr).

The dust loading at the baghouse inlet is typically about 900 g/Nm^3 (370 Gr/scf) while the outlet is about 0.5 g/Nm^3 (0.19 Gr/scf) for a dust

Figure 2
STAGES OF FLASH FURNACE DEVELOPMENT

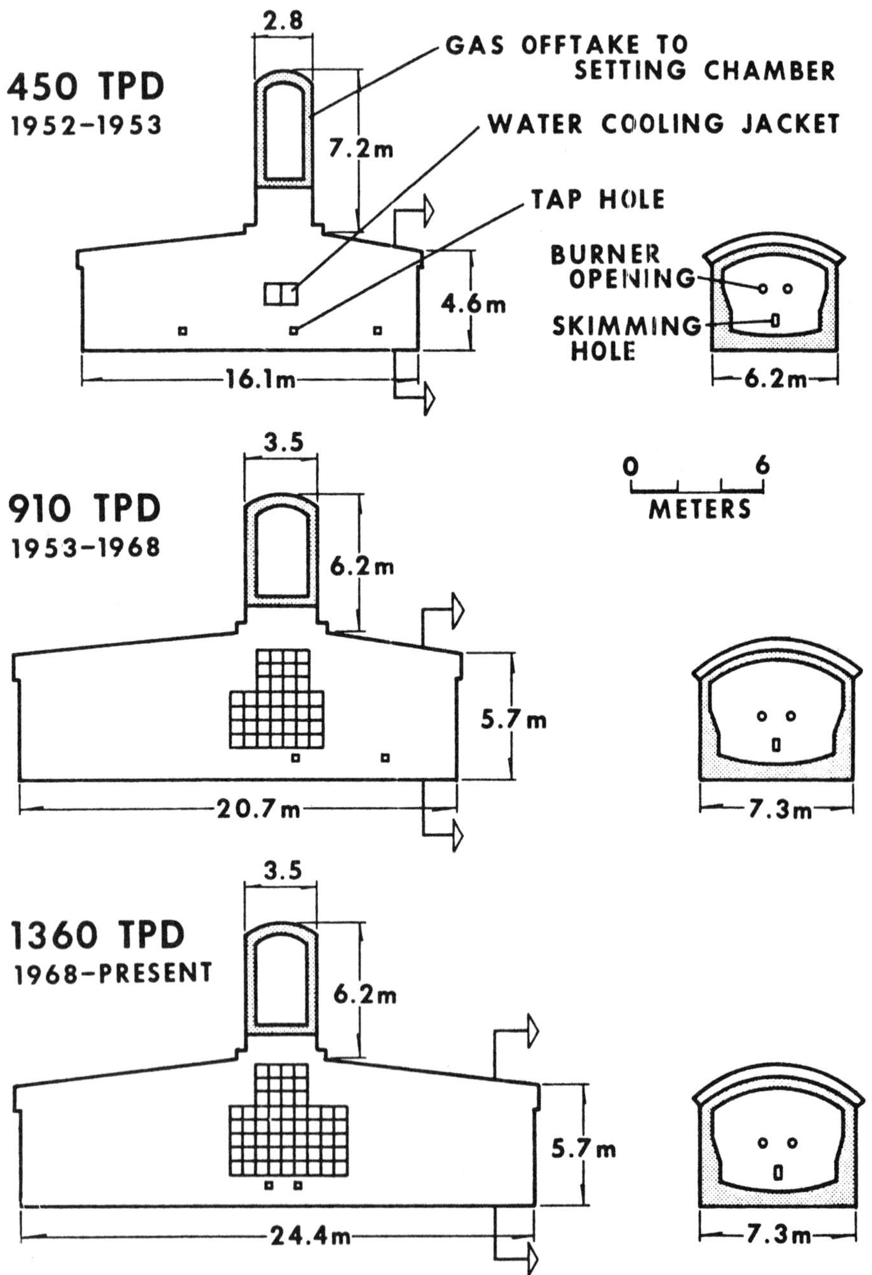

Figure 3
FLASH FURNACE GENERAL ARRANGEMENT

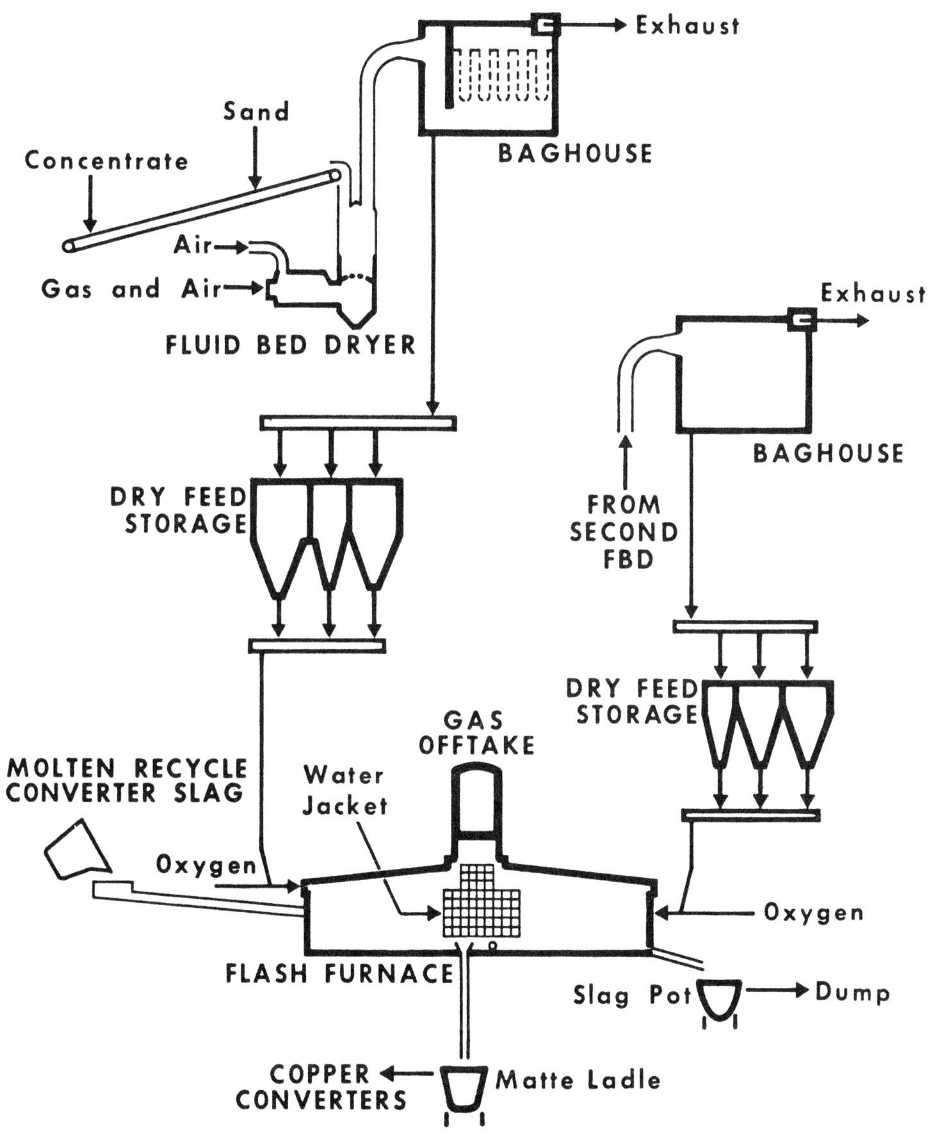

collection efficiency of about 99.95%. The unit is designed to handle about 49,000 Nm^3/hr (32,500 scfm) with typical inlet and outlet temperatures of 105°C (220°F) and 90°C (195°F) respectively. The bag differential varies between 0.5-2.0 kPa (2-9" WC) depending on the condition of the bags, which are typically changed once per year.

Flash Smelting of Copper Concentrates

The flash smelting flowsheet is also depicted in Figure 3. The dried feed, from the baghouse hoppers, is delivered by screw conveyors to feed bins located above and at either end of the furnace. In addition to the feed bins, recycle dust and touch-up flux bins are also located in this area. The dust collected in the furnace off-gas settling chamber and in the copper converter cottrell are both recycled to the furnace. The touch-up flux is added as required to make minor adjustments to the silica content of the furnace slag but is now also used to raise the silica level of the molten converter slag recycle, which enters the furnace at about 20% SiO_2 and is discharged at about 32% SiO_2. The furnace feed rate is controlled by variable flow control valves associated with weightometers located under these bins. The feed materials are then transported by screw conveyors to drop pipes which discharge directly into the furnace burners. In these burners the falling feed is injected into the furnace by a horizontal flowing stream of oxygen. The oxygen flow is ratioed to the burner feed rate and at Copper Cliff is currently 20-21 weight % of the feed.

A cut-away view of the flash furnace is shown in Figure 4. The feed is delivered as just described to 4 concentrate/oxygen burners, two located at each end of the furnace. These burners are water jacketted. Spontaneous ignition occurs with the oxygen reacting with part of the sulfur and iron in the concentrate to form SO_2 and iron oxides. Silica, contained in the flux and concentrate, combines with these oxides to form slag. Copper, except for the minor amount lost in the slag, and the remaining iron and sulfur collect in the matte. The converter slag is reverted down a launder into the end of the furnace opposite to the slag skimming hole. The off-gases leave through the furnace uptake and pass to the adjacent settling chamber.

The oxygen/concentrate reactions supply all the heat required by the process so no extraneous fuel is required - or in other words, the smelting operation in the Inco furnace is totally autogenous.

Our current experience has given us 3 years life on the furnace refractory though with recent modifications we are expecting four or possibly five years life. The furnace bottom has never shown any evidence of magnetite build-up and is the original bottom installed in 1953 except for the portion added during the 1968 extension. It has been in use for almost 30 years and shows no sign of failure.

The slag is skimmed from one end of the furnace. This operation is only carried out with a full hole to preclude the possibility of the escape of furnace gases. The hole is fitted with a vent hood as is the launder and hood over the slag pot. This launder is water cooled. Both the slag skimming and the matte tapping holes are of refractory block and are opened and closed manually.

There are two matte holes and two matte launders, with one in operation and the other on stand-by. The matte flows down the launders into ladles located in a tunnel below the matte tapping platform. The matte holes are vented at the furnace as are the launders and tunnel.

Figure 4
INCO FLASH FURNACE

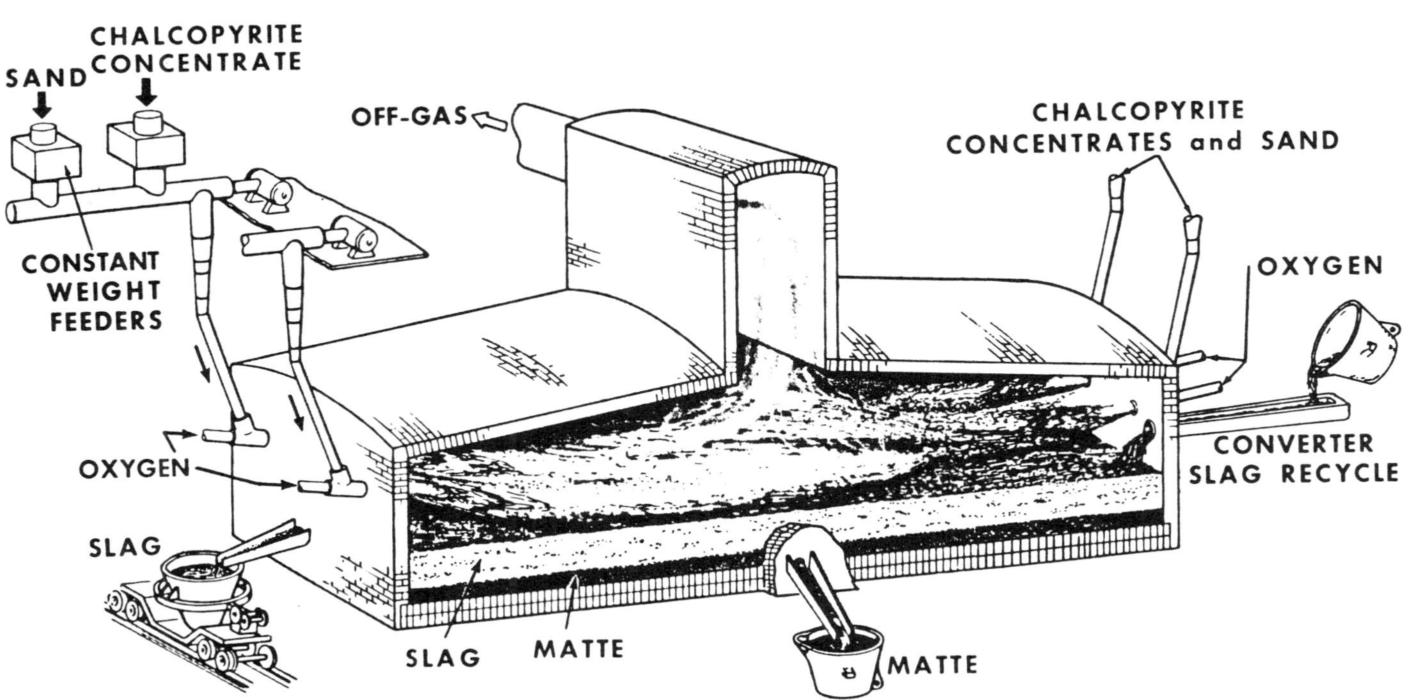

A total of 250-350 MTPD of converter slag is being reverted to the furnace at present. This represents about 50% of the slag generated. The slag is poured from ladles down a 13 m (43') launder and enters the furnace through an opening in the end of the furnace opposite to that from which the discard furnace slag is skimmed.

This return hole is only opened when slag is about to be returned and is closed immediately after the operation is completed. A door is used to cover the opening and is packed with clay as a seal. There is a venting system over the hole. The furnace, which typically operates at about neutral pressure, is put under slight draft during this slag reversion operation.

Gas Cleaning System

The furnace off-gases leaving through the uptake enter a settling chamber which is fitted with cooling fins and air cooling pipes. The furnace gas enters at about 1200°C (2200°F) and leaves at about 750°C (1350°F). In this unit approximately half the dust generated by the furnace is collected, and is removed, by screws located in the hoppered bottom for return, through the recycle dust bins, to the furnace.

Prior to delivery to an SO_2 liquefaction plant the flash furnace gases are cleaned in a wet scrubbing system which follows the settling chamber and is shown in Figure 5.

The system consists of a splash tower, three venturi scrubbers and a wet mist precipitator in series. The scrubber water passes to a settling cone with the overflow being recycled through heat exchangers to the scrubbers. The cone overflow, containing the collected solids, passes to aeration for SO_2 removal followed by neutralization with lime and return to the Copper Cliff concentrator for recycling (20). The cleaned furnace off-gas is delivered to a liquefaction plant operated by Canadian Industries Limited (21).

If the furnace gases were to be used for sulfuric acid production the recommended gas cleaning system would consist of a water spray cooler to lower the gas temperature to about 320°C (600°F) followed by an electrostatic precipitator.

Flash Furnace Operational Data

The furnace feed rate has been varied from about 700 to 1500 MTPD as production requirements dictate. These changes can be easily achieved by shutting off either one, two or even three of the burners. Rates of over 1800 MTPD have been attained with the ultimate capacity of the unit still not established because of limitations external to the furnace such as concentrate and oxygen availability, material handling constrictions, converter capacity, anode casting capability, etc. A turn down of almost 3 to 1 is achievable very readily and quickly.

At a rate of 1200 MTPD the specific smelting rates are 200 MTPD/100 m^3 (62 STPD/1000 cu ft) of furnace volume or 8.6 MTPD/m^2 (0.9 STPD/sq ft) of hearth area.

Typically the furnace holds 100-200 MT of matte and 100-150 MT of slag. The matte is tapped at about 1170°C (2140°F) and the slag is skimmed at about 1230°C (2240°F) while the off-gases also leave at about 1230°C (2240°F).

Because commercial oxygen, at about 97% purity, is used in the furnace, off-gases from the furnace are very low, averaging only 140-170 m^3/MT of

FIG. 5 FLASH FURNACE GAS CLEANING SYSTEM

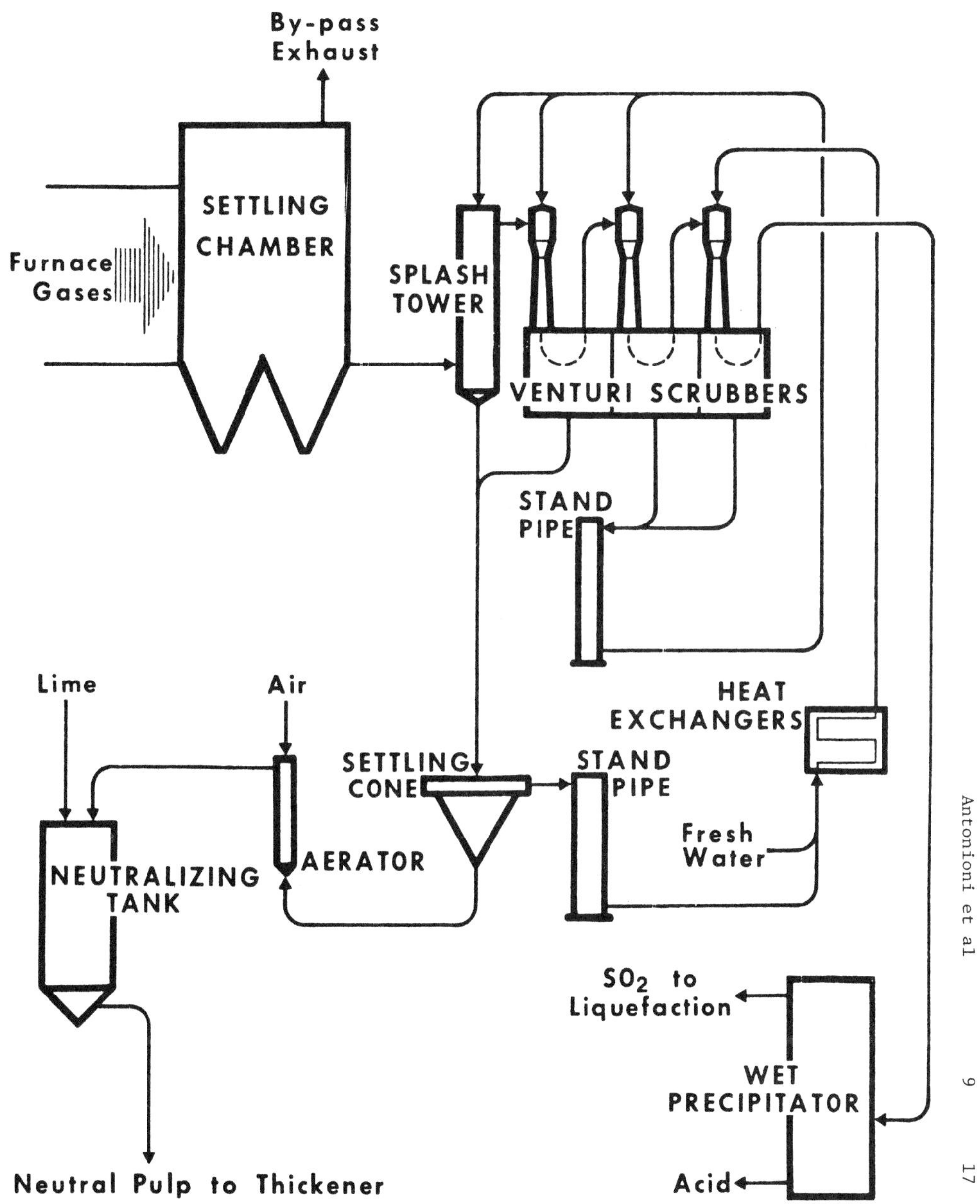

concentrate (4500-5500 scfm/ST). As a result of the low volumes being produced by the furnace, the furnace uptake velocities are only 1 M/S (3-4 ft/sec) so the dusting rate from the furnace is also very low, at only 2-3% of the concentrate rate. The gas handling and cleaning system is also small. In our opinion the small off-gas volume and therefore its small heat content do not justify the installation of a waste heat recovery system.

The refractory utilization during our most recent campaign amounted to 1.1 kg/MT feed (2.0 lbs/ST). This includes all the refractory installed in the furnace and its associated equipment at the time of the major rebuild prior to the campaign and the minor amounts installed during the campaign itself.

Process Metallurgy

Introduction

As already discussed Inco oxygen flash smelting is an autogenous process. The finely divided sulfide concentrate, carried by the oxygen, ignites and burns as soon as it enters the furnace. The oxygen-concentrate reactions are extremely fast.

Flash Smelting Flame Characteristics

Observation of the oxygen-copper concentrate flame at Inco's Research Stations during a pilot plant flash smelting campaign conducted in a top blown rotary converter shows visible ignition of the concentrate occurs at a very short distance from the tip of the burner. Sampling this flame as a function of distance from the burner tip, to investigate how fast the reactions are occurring, has been carried out.

It was found that at about one meter from the burner the SO_2 content of the gas was 75%, proving that most of the oxygen had already reacted. At two meters only slag was collected in the sampling steel spoon, indicating that the matte was liquid. The slag sample had the same composition as the slag in the converter, aside from matte prills. This evidence showed that the reactions in the oxygen flame are extremely rapid and occurred in under one tenth of a second (22).

Flash Smelting Chemical Reactions

In Table I are listed the main chemical reactions which take place in flash smelting. Also shown in this table are the corresponding heats of reaction. The heat generated in the flash smelting flame corresponds mainly to the combustion of the labile sulfur and FeS. A sufficient amount of oxygen per unit weight of concentrate must be supplied to satisfy the heat balance of the operation. If the furnace temperature drops the oxygen to concentrate ratio is increased. Consequently, more iron sulfide is combusted and the furnace thermal equilibrium is re-established. And, vice versa.

It follows, then, that the furnace heat balance determines the matte grade to be obtained from a given concentrate. However, as will be discussed later, there are methods available to permit operation of the process to obtain a matte product with a desired composition.

Table I. Inco Oxygen Flash Smelting Chemical Reactions

			$\Delta H_{25°C}$ kCal/MOL
$CuFeS_2 + \frac{1}{2}O_2$ (Chalcopyrite)	=	$\frac{1}{2}Cu_2S + FeS + \frac{1}{2}SO_2$	- 23.48
$Cu_5FeS_4 + \frac{1}{2}O_2$ (Bornite)	=	$5/2\ Cu_2S + FeS + \frac{1}{2}SO_2$	- 16.88
$CuS + \frac{1}{2}O_2$ (Covelite)	=	$Cu_2S + \frac{1}{2}SO_2$	- 32.57
$FeS_2 + O_2$ (Pyrite)	=	$FeS + SO_2$	- 54.07
$FeS + 3/2\ O_2$	=	$FeO + SO_2$	-112.07
$Cu_2S + 3/2\ O_2$	=	$Cu_2O + SO_2$	- 92.34
$FeO + 1/6\ O_2$	=	$1/3\ Fe_3O_4$	- 24.28
$FeO + \frac{1}{2}SiO_2$	=	$\frac{1}{2}Fe_2SiO_4$	- 4.33

Flash Furnace Heat Balance

Oxygen flash smelting of Inco's chalcopyrite copper concentrate at Copper Cliff at the normal throughput of about 1300-1400 MTPD (1400-1500 STPD) requires 20-21 weight percent tonnage oxygen. About 43,000 kCal/MT of concentrate (1.55×10^6 BTU/ST) are released by the oxygen-concentrate reactions. About 60% of this heat satisfies the heat requirements of the matte and the slag. The heat content of the off-gas, represents only 20% of the heat generated, because of its small volume. Typical heat losses of the Copper Cliff flash furnace are about 72,000 kCal/MT of concentrate (0.26×10^6 BTU/ST). Because most of the heat generated is used in heating up the products, the flash flame temperature is only slightly above 1450°C (2640°F). The foregoing data are given in Table II.

Table II. Inco Oxygen Flash Smelting Heat Balance

Throughput = 1,100 MTPD of concentrate at 28% Cu, 29% Fe, and 32.6% S

Matte Grade = 50% Cu

	kCal/MT Concentrate	BTU/ST Concentrate	Distribution %
Heat Generated			
By O_2/Feed Reactions	430,000	1,548,000	100
Heat Requirements			
Heat In: Matte	182,000	654,000	42
Slag	77,000	279,000	18
Off-Gas	88,000	318,000	20
Dust	11,000	39,000	3
Furnace Heat Losses	72,000	258,000	17
TOTAL	430,000	1,548,000	100

Flash Furnace Mass Balance

The matte grades produced at Copper Cliff vary between 40 and 50% Cu, depending on furnace throughputs and the amount of secondaries reverted to the furnace. Typical mass balance data for the operation of Inco's flash furnace are given in Table III. The information was derived from average shift data for three months of steady state operation, just prior to the converter slag recycle experimental campaign. Within this period the furnace throughput was 1,360 MTPD of concentrate. The dust recovered from the furnace settling chamber, about 10-12 MTPD, is an internal recirculation and it is not shown in this table. The oxygen requirement for smelting the Copper Cliff concentrate, at these throughputs, is 20.6% by weight of the concentrate to yield a 40.5% Cu matte. The copper content of the slag, 0.57%, is typical of this operation. The data collected during many years of operation show that the slag losses in the flash furnace are directly proportional to matte grades, the best correlation being

$$\frac{\%\ Cu_m}{\%\ Cu_s} = 70 \pm 10$$

This copper partition coefficient is valid for matte grades up to about 55% Cu.

TABLE III. Inco Oxygen Flash Smelting Mass Balance for Operations without Converter Slag Recycle (August to October 1977)

	Rate DMTPD	Composition - Wt. %			
		Cu	Fe	S	SiO_2
INPUTS					
Concentrate	1,360	30	31	33.5	2
Recycle Dust	32	42	8	19	1.7
Flux	95	-	2.2	-	78
Tonnage Oxygen	280 (20.6%)	-	-	-	-
OUTPUTS					
Matte	1,030	40.5	28.8	24.5	-
Slag	295	0.57	43	1.3	33
Sludge	20	27	16	18	2.6
Off-Gas	410	-	-	-	-

PARTITION COEFFICIENT: $\frac{\%\ Cu\ IN\ MATTE}{\%\ Cu\ in\ SLAG} = 70$

Composition of Furnace Off-Gas

The use of commercial oxygen in flash smelting results in a small volume of off-gas with a high SO_2 content, as shown in Table IV. Oxygen efficiency in the flash furnace is 100%. Because there is some air in-leakage into the furnace, the composition of the gas in the furnace uptake is typically 83% SO_2. After cooling and cleaning, the gas analyzes 75 to 80% SO_2, because of the additional air dilution which normally occurs during gas treatment.

TABLE IV. Flash Furnace Off-Gas Composition (in volume percentage

	SO_2	A	N_2	CO_2	O_2
At Furnace Uptake					
Without dilution	94.5	2.8	1.0	1.7	0
With dilution	83.0	2.5	13.0	1.5	0.02
After Gas Cleaning	77	2	20	1	1

The high strength gas is used at Copper Cliff for producing liquid SO_2. In most cases, this gas will provide an excellent continuous base load for a sulfuric acid plant as can be noted from the following data.

The sulfur distribution in a hypothetical Inco flash smelting operation is shown in Figure 6. The smelter feed is a chalcopyrite concentrate analyzing 28% Cu, 29% Fe and 32.6% S which yields a 50% Cu matte at a throughput of 1,100 MTPD.

In the flash smelting operation, 57% of the sulfur is eliminated from the concentrate as a strong, low volume gas stream. Except for the minor amount in the slag, the remainder of the sulfur reports to the matte and is eliminated during conversion.

Molten Converter Slag Recycle

As previously explained, the presence of Ni in the Cu concentrate and of Cu in the Ni concentrate results in substantial intercircuit transfer streams at the Copper Cliff smelter. Copper converter slags are an important bleed for nickel from the copper circuit. Until very recently, 100% of these slags were reverted to the nickel reverb furnaces. However, this practice greatly complicated the traffic along the converter aisle. It was thought, then, that at least partial reversion of the copper converter slags to the flash furnace could alleviate this problem. At the same time, the question of the capability of the Inco flash smelting furnace for cleaning converter slags has been raised. An experimental campaign was then conducted at Copper Cliff early in 1978 to answer these questions.

For this purpose one of the furnace burners was removed from the wall closest to the converter aisle to provide a converter slag recycle port. The slag was poured into the furnace via a 13 m (43') long launder. During the experimental campaign, which lasted for almost two months, the furnace operated on only three burners at an average throughput of 1,300 MTPD with a maximum of 1,630 MTPD.

The mass balance data shown in Table V reflects a seven-day operating period during which the average converter slag recycle to the furnace was 214 MTPD, corresponding to about 43% of the total converter slag generated. Feed rate during this period averaged 1,345 MTPD of concentrate.

Production of flash furnace slag was about 600 MTPD, which is double that produced in the normal operation. In spite of this, the copper partition was similar to that usually observed without slag recycle. It should be noted that the same is true for the nickel and the cobalt partitions, a clear indication that the showering of matte through most of the slag layer

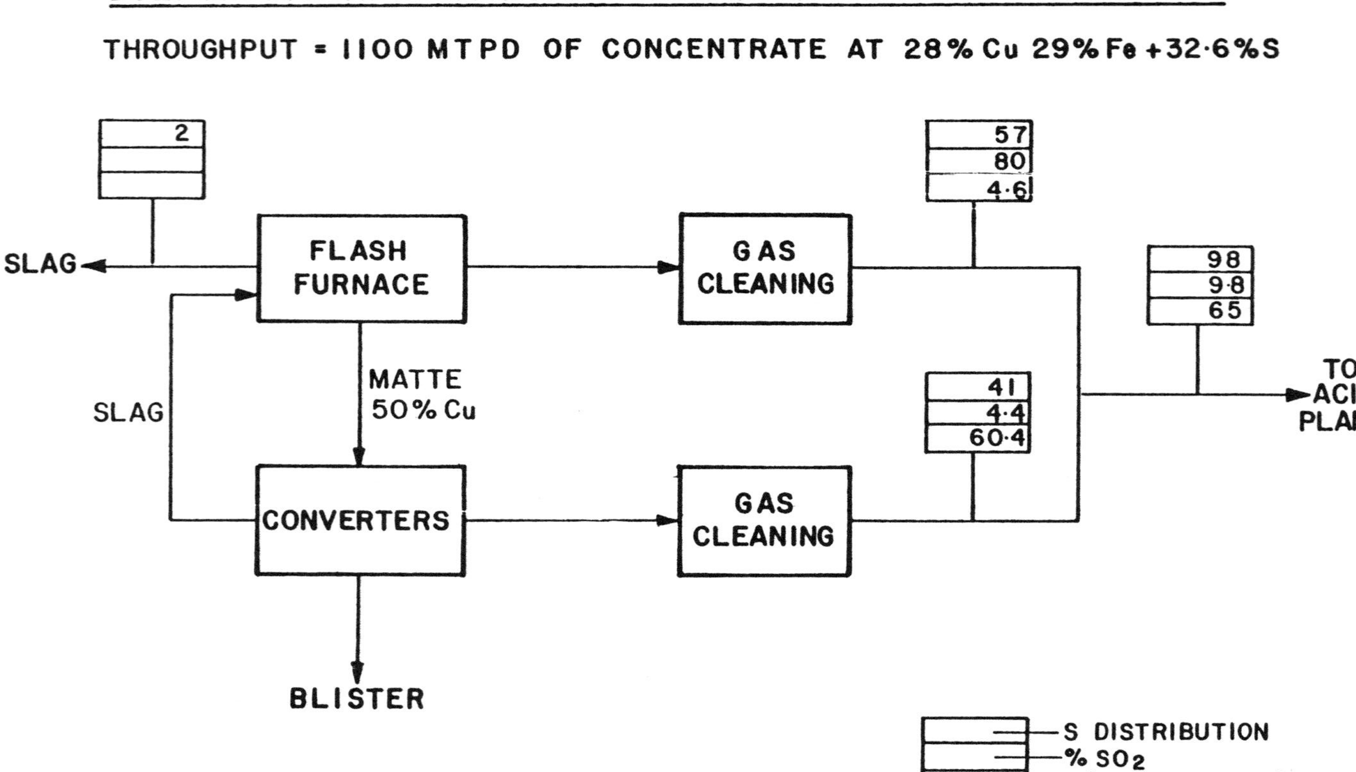
FIGURE 6
SULFUR DISTRIBUTION IN INCO OXYGEN FLASH SMELTING
THROUGHPUT = 1100 MTPD OF CONCENTRATE AT 28% Cu 29% Fe + 32·6% S
2
SLAG
FLASH FURNACE
SLAG
MATTE 50% Cu
CONVERTERS
BLISTER
GAS CLEANING
GAS CLEANING
57
80
4·6
41
4·4
60·4
98
9·8
65
TO ACID PLANT
S DISTRIBUTION
% SO2
VOLUME, 1000 SCFM

in the Inco flash furnace provides excellent conditions for matte/slag chemical interaction.

TABLE V. Inco Oxygen Flash Smelting Mass Balance for Operation with 43% Converter Slag Recycle (January 1978)

	Rate DMTPD	Composition - Wt. %			
		Cu	Fe	S	SiO_2
INPUTS					
Concentrate	1,345	30.1	30.1	33.3	2.0
Recycle Dust	32	41.8	7.9	197	1.7
Converter Slag	214	6.9	47.9	0.9	18.9
Flux	156	-	2.2	-	77.8
Tonnage Oxygen	286 (21.3%)	-	-	-	-
OUTPUTS					
Matte	940	45.0	25.2	24.1	0.3
Slag	598	0.68	43.5	1.4	31.1
Sludge	22	24.0	16.8	18.0	2.2
Off-Gas	434	-	-	-	-

PARTITION COEFFICIENT: $\frac{\text{\% Cu IN MATTE}}{\text{\% Cu IN SLAG}} = 68$

Converter slag recycle results in an increase in the matte grade, 45% versus 40.5% for the normal operation at the same throughput. This is mainly a consequence of the extra heat required to melt the additional flux fed to the furnace to compensate for the low SiO_2 in the converter slag.

The possibility of bottom build-up when returning converter slag to the flash furnace was a particular concern during the campaign. Periodic magnetite and chromia balances indicated that no bottom build-up was taking place. In fact, these balances showed that substantial reduction of Fe_3O_4 was taking place in the furnace.

In addition, the flash furnace was drained before and after the experimental campaign. There was no visual evidence of any bottom build-up.

References

1. "Smelting: I. Copper Cliff Smelter", Canadian Mining Journal, pp 431-45, May 1946.

2. "Metallurgical Improvements in the Treatment of Copper-Nickel Ores", The Staff, International Nickel Company of Canada (Presented by P. Queneau), CIM Transactions, Vol. 51, pp 187-98, 1948.

3. "Extractive Metallurgy at International Nickel - A Half Century of Progress", L. S. Renzoni, Can. Journal of Chem. Eng., February 1969.

4. "The Sudbury Operations of Inco Metals Co.", W. Schabas, Can. Mining Journal, May, 1977.

5. "Treatment of Nickel-Copper Matte", K. Sproule, G.A. Harcourt and L. S. Renzoni, J. of Metals, Vol. 12, No. 3, pp 214-19, March 1960.

6. "Converter Operating Practice at the Copper Cliff Smelter of the Inco Metals Company", R. J. Neal, R. A. Reyburn, Inco Metals Company, presented at 108th AIME Annual Meeting, New Orleans, LA, February 1979.

7. "Flash Smelting of Inco Ores Creates New Industry", Northern Mining, July 8, 1948.

8. "Autogenous Smelting of Sulfides", J. R. Gordon, G.H.C. Norman, P. E. Queneau, W. K. Sproule and C. E. Young, U.S. Patent 2,668,107, filed June 25, 1949, issued February 2, 1954.

9. "The Oxygen Flash Smelting Process of the International Nickel Company", The Staff, CIM Transactions, Vol. 58, pp 158-66, 1955.

10. "Oxygen Flash Smelting Swings into Commercial Operation", The Staff, International Nickel Company of Canada, Journal of Metals, Vol. 7, pp 742-50, June 1955.

11. "Recent Developments in the Inco Oxygen Flash Smelting Process", S. Merla, C.E. Young and J.W. Matousek, 101st AIME Annual Meeting, San Francisco, 1972.

12. "Copper Smelting by the International Nickel Company of Canada", The Staff, Copper Cliff Smelter, International Symposium on Copper Extraction and Refining, Las Vegas, 1976, in Extractive Metallurgy of Copper, Vol. 1, ed. by J.C. Yannopoulos and J.C. Agarwal, TMS-AIME, pp 218-33, 1976.

13. "Flash Smelting", J.W. Matousek, Seminar on Extractive Metallurgy of Copper, McGill University, Montreal, Canada, January 1977.

14. "The Inco Oxygen Flash Smelting Process", C. Diaz, Inco Metals Company, Process Development, Seminar on Copper Metallurgy, Bogota, Columbia, February 1978.

15. "Tonnage Oxygen for Nickel and Copper Smelting at Copper Cliff", R. Saddington, W. Curlook and P. Queneau, J. of Metals, April, 1966, Vol. 18, No. 4, pp 440-452.

16. "Operation of the Inco Flash Smelting Furnace with Recycle of Converter Slags", T. N. Antonioni, A. D. Church, C. Landolt and E. Partelpoeg, 18th Annual Conf. of Metallurgists, CIM Sudbury, August 1979.

17. "Inco's Oxygen Flash Smelting Process for Copper and Nickel Concentrates", C. Diaz, H. C. Garven, M. Y. Solar, Inco Metals Company, Process Research, November 1978.

18. "Inco's Oxygen Flash Smelting Process for Copper and Nickel Concentrates; Off-Gas Handling and Impurity Distributions", M.Y. Solar, A.D. Church, T.N. Antonioni, Inco Metals Company, EPA Symposium on Control of Particulate Emissions in the Primary Non-Ferrous Metals Industries, Monterey, California, March 1979.

19. "Smelting Nickel Concentrates in Inco's Oxygen Flash Furnace", M.Y. Solar, R.J. Neal, T.N. Antonioni, M.C. Bell, Inco Metals Company, Journal of Metals, Vol. 31, No. 1, pp 26-32, January 1979.

20. "Handling Acid Slurries at the Flash Smelting Plant of the International Nickel Co. of Canada", J.N. Lilley, R.W. Chambers and C.E. Young, CIM Annual Meeting, Toronto, April 1960.

21. "Sulphuric Acid and Liquid Sulphur Dioxide Manufactured from Smelter Gases at Copper Cliff, Ontario", Trans. CIM Vol. LV, 1952, pp 123-125.

22. "Oxygen Flash Smelting in a Converter", The Staff, Inco Metals Company, Process Research, presented at 106th Annual Meeting, Atlanta, Georgia, March 1977.

SMELTING PROCESS UPDATE AT

FALCONBRIDGE LIMITED - SUDBURY OPERATIONS

R.R. Hoffman and G.H. Kaiura

Originally Presented at TMS-AIME Annual Meeting
New York, February 1985

Introduction

Falconbridge Limited started up its nickel smelting operation in the Sudbury District of Ontario, Canada in 1930. As a result of several expansions and revisions, the initial annual production of nickel of three million lbs. was increased to 100 million lbs. by 1971.

Prior to 1978 the smelting operation consisted of rotary dryers, down draft sinter machines and blast furnaces for the treatment of partially dried and slurried nickel-copper concentrate. Molten matte and slag were separated in settlers following the blast furnaces with the matte further processed in Peirce-Smith converters to a high grade, nickel-copper matte for shipment to the refinery.

To meet environmental improvement requirements, a new smelting process was commissioned in mid-1978 which replaced the drying, sintering and blast furnace operations. The new complex consists of two, surry-fed, fluid bed roasters for partial roasting of concentrate and two electric furnaces for melting of the calcine produced from the roasters. Off-gases from the roasters are treated for sulphur recovery as sulphuric acid in a contact acid plant. Products from the electric furnaces are a slag that is hauled to waste, and a matte that is processed in the converter aisle to produce the final matte for the Norwegian refinery.

The objectives of the new process were to reduce the emission of sulphur dioxide and particulates to atmosphere as well as improve the working environment while maintaining production levels and metal recoveries.

This paper provides the smelting process update on the developments carried out since the initial operation. These developments have contributed to significant improvements in the environmental control, metal recovery and reliability of the operation. A brief discussion of the focus for future process development is also included.

The smelting process update will be presented under the following headings:

- Process Description
- Major Process Change - Increased Degree of Roast
- Process Modifications and Performance
- Environmental Improvements
- Process Development Focus.

Process Description

The overall process flowsheet is shown in Figure 1.

Concentrate Slurry Handling

Feed for the smelter is a nickel-copper sulphide concentrate produced at the Strathcona and Falconbridge concentrators as shown in Table I.

TABLE I

TYPICAL CONCENTRATE ANALYSIS - 1983

	Ni	Cu	Co	Fe	S
Falconbridge	7.0	5.2	.36	40.8	30.6
Strathcona	6.5	6.2	.26	42.6	32.5

Concentrate from the Strathcona concentrator, 65 km away, is shipped as a 75% solids slurry in 90-tonne capacity rail cars. These are emptied into an agitated receiving sump at the smelter.

Slurry is pumped from the receiving station into two agitated storage tanks. Concentrate slurry from the Falconbridge concentrator, located adjacent to the smelter, is pumped directly into these tanks on a batch basis. Three settled storage tanks provide off-line capacity for 5400 tonnes concentrate as slurry. These tanks have outlets at 75 cm intervals for manual decanting. Slurry from the settled storage tanks and the clarifier, used for cleaning decant water, is transferred to the agitated storage tanks for processing.

A portion of the slurry from the agitated storage tanks is filtered, using vacuum disc filters, and then repulped with slurry to a final prepared feed density of 72% solids. The flow of slurry entering the repulp tank is controlled to a preset level by density measurement on the discharge of the repulper. Final slurry density is set by operating requirements for an autogenous heat balance in the fluid bed roaster. Tramp material is removed from the slurry at the repulper discharge by a vibrating screen prior to entry into two agitated roaster prepared feed tanks.

Roaster Operation

Prepared concentrate slurry is roasted to the desired partial oxidation level of iron and sulphur in two fluid bed roasters. These have a 5.6 meters bed diameter and an 8 meter diameter freeboard.

Concentrate slurry is fed via variable speed positive displacement pumps which discharge slurry directly into the roaster through a centrally located feed gun. All flux, required in the electric furnace, is added to the roasters for preheating and to provide roaster bed material. Flux addition is controlled by weightometers on conveyors and is fed into the roaster through double flap valves.

FIGURE 1

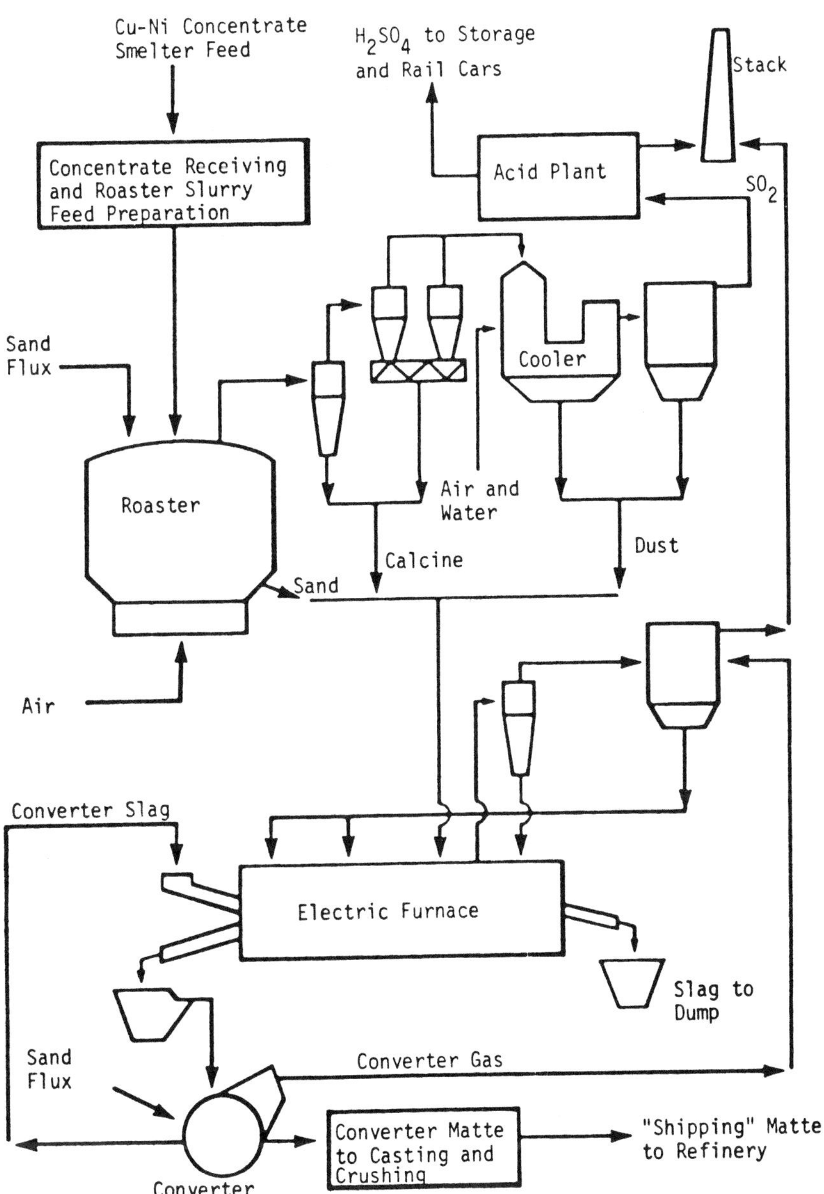

FALCONBRIDGE SMELTER FLOWSHEET

Currently about 60% of the sulphur in the concentrate is oxidized to SO_2 in the roaster. The air to feed ratio is set to control the degree of roast and the excess heat is controlled by the feed density to produce a bed temperature of approximately 680°C for autogenous operation.

The roaster bed depth is controlled at approximately 1.4 meters by cone valves on the bottom discharge from the roasters which are modulated automatically to maintain a constant windbox pressure of 25 to 30 kPa. The roaster freeboard pressure is automatically controlled at -0.1 kPa by modulating the roaster off-gas fan inlet vanes.

The first roaster was started up in June 1978 and the second in January 1980. In 1979, the first roaster treated 1,141 tonnes per operating day.

Roaster Gas Cleaning

Approximately 85% of the roasted calcine leaves as calcine dust via the off-gas system through two off-gas ducts, two primary and four secondary cyclones. The temperature is reduced to 330°C in a dry gas cooler prior to entering two hot electrostatic precipitators in parallel followed by the off-gas fan. The flues, from the roasters to the cyclones, are refractory-lined and fitted with wafer-type expansion joints. Gas cleaning equipment and flues, through the electrostatic precipitator, are unlined.

Over 99% of the metal-bearing calcines are removed in the off-gas cleaning system. Clean gases, containing SO_2, are routed to the acid plant.

Acid Plant

Clean roaster off-gases, at approximately 10% SO_2, enter the acid plant for the recovery of the SO_2 as sulphuric acid. The acid plant flowsheet is shown in Figure 2.

The gases enter the scrubber-cooler for the removal of moisture and particulate. Weak acid, circulating through the scrubber, is cooled and a continuous bleed-off of weak acid is neutralized with mill tailings and pumped to the disposal area. Six mist precipitators are provided, four primary units in parallel, followed by two secondary units, to remove the balance of the moisture and dust.

The remaining moisture in the gas, after adjustment to 7.0% SO_2 by ambient air addition, is removed by sulphuric acid in the drying tower. Dry, cold gas passes through the SO_2 blower and then through a series of heat exchangers utilizing heat from the conversion operation. The gas is heated to 426°C for entry to the converter. Conversion of the SO_2 to SO_3 is carried out in a single pass, through four vanadium pentoxide catalyst beds in the converter. The gas, at 166°C, enters the absorption tower where the SO_3 is removed as sulphuric acid (98.5%) after contact with lower strength acid.

Tail gas, containing less than 0.3% SO_2, is heated to 175°C prior to exit via the smelter stack.

A balance of sulphuric acid is maintained between the absorption and drying tower with product acid, at 93.5%, drawn off through a stripping tower to reduce the SO_2 content below 50 ppm. Hydrogen peroxide is injected into the product acid to remove colour. The acid is pumped 2 km to two storage tanks, with a capacity of 33,000 tonnes.

FIGURE 2

ACID PLANT FLOWSHEET

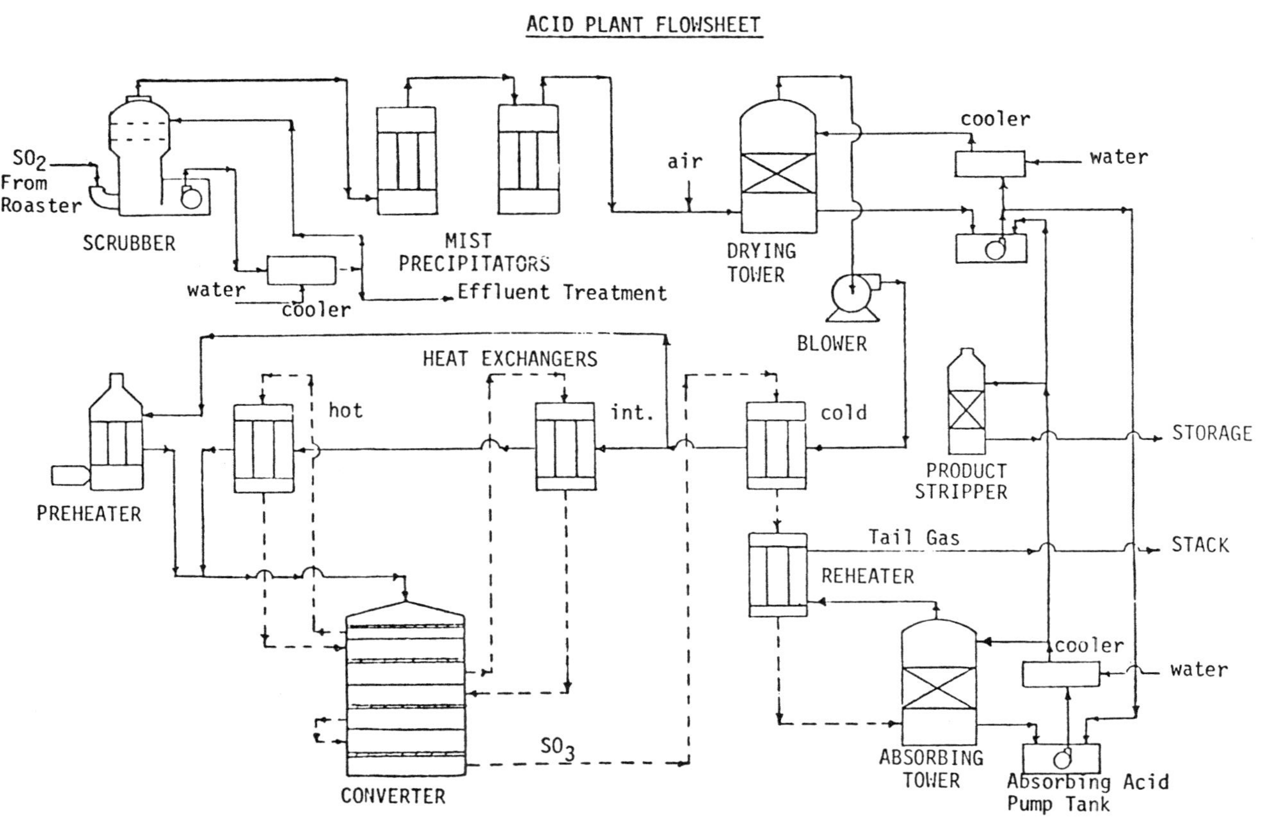

The acid plant was designed to produce 1,180 tonnes of sulphuric acid per operating day with the current production level around 900 tonnes per operating day.

Electric Furnace

Calcines, from the roasting system, are delivered by enclosed drag conveyors along both sides of each electric furnace. For each roaster, the cyclones and precipitators are arranged in two lines so that calcine collected by each line feeds one side of the furnace. Calcine is distributed to each side of the two electric furnaces through fettling pipes located in the furnace arch. The calcine is melted with the silica flux in the electric furnaces to produce a nickel-copper sulphide matte and an iron silicate base slag.

Four additive bins, for each furnace, supply petroleum coke reductant and miscellaneous feeds.

The two electric furnaces have inside dimensions of 9m x 30m x 2.7m. Power for each electric furnace is supplied through six Soderberg self-baking electrodes 1.4 m in diameter. Each furnace has three 12 MVA transformers located in adjacent vaults.

The electric furnaces are operated with the surface of the slag bath covered by 40-60 cm of unsmelted calcine to provide a "semi-black top" condition. Power levels are set to match roaster feed rates with matte and slag temperatures regulated by electrode position, slag depth and applied voltage. The slag composition is maintained at an iron to silica ratio of between 0.85 and .90 by adjusting the flux to concentrate ratio to the roaster. Petroleum coke reductant, at approximately 4% on concentrate, is added to the fettling drag conveyors to reduce metal losses in the furnace. Slag temperature is maintained at 1230°C ± 20°C.

The slag from each furnace is tapped 1.9 meters above the hearth through water-cooled copper blocks fitted with a stainless steel insert. Slag flows via water-cooled launders into slag pots for disposal by rail. Hydraulically-operated clay guns are used to close the slag hole. Approximately 1.25 tonnes of slag are made per tonne of concentrate treated.

Matte is tapped from the furnace through two 4.4 cm tap holes located in one end wall. These holes are located 57 cm from the furnace hearth and 2.13 meters from the centre line of the furnace. The matte end wall is 1.22 meters thick and the chrome-magnesite tap blocks are 23 cm x 23 cm x 11.5 cm.

Matte flows by refractory-lined steel fabricated launders to a single ladle. Ladles are moved in and out of the fume containment enclosure by a self-propelled car. The matte flow is stopped in a similar manner as that of slag. Gases from the electric furnace are drawn from each end of the furnace and passed through cyclones for dust removal. All dusts are returned directly to the furnace freeboard. The gases are quenched with air to 300°C prior to the fans used for transfer through high velocity flues to an electrostatic precipitator for final particulate removal. Clean gases exit via a 90 meter stack to atmosphere.

Typical electrical furnace operating conditions are shown in Table II.

TABLE II

TYPICAL ELECTRIC FURNACE OPERATING CONDITIONS

Calcine Depth	50 cm
Slag Depth	150 cm
Matte Depth	70 cm
Bottom Buildup	15 cm
Slag Temperature	1230°C
Matte Temperature	1130°C
Electrode Immersion	25 cm
Freeboard Pressure	-.025 kPa
Off-gas Volume	28,300 Nm^3/hr
Off-gas Temperature	700°C
Power - Voltage	400
- Current	15,000 Amps/Phase
- Factor	.98
Power Consumption	338 kWh/Tonne Dry Solid Charge
Paste Consumption	1.2 kg/Tonne Dry Solid Charge
Coke Consumption	2.45% of Dry Solid Charge

Process Control

A control room supervisor has complete process control over the roaster and electric furnace operating conditions.

The control supervisor has computer access to select any instrument loop variable, set the measurement span and alarm limits. All operating information is stored in a dual PDP8 computer and can be recalled via a teleprinter or television monitor. Interlock sequences and programs were developed to provide "first out" alarm information which simplified trouble-shooting and identification of process upsets. Sufficient information is available on the control panel to allow plant operation in the event of computer failure.

In addition, the computer performs area power level monitoring and trend analysis to provide real time system control of operating variables within set limits.

Converting and Matte Handling

The existing converter, crane, matte casting and crushing operations were incorporated into the new smelting process with start-up in 1978. The converters process the electric furnace matte to remove the remainder of the iron as a slag and produce a finished product matte for shipment to our Norwegian refinery. All converter slag is returned to the electric furnace for cleaning prior to disposal. The converting operation consists of four, 4m x 9m Peirce-Smith converters that are serviced by two 54-tonne overhead cranes handling 5.8 m^3 ladles. Air is supplied to the converters at about 500 Nm^3/min. under 2 kPa pressure through forty-eight 5 cm diameter tuyeres inclined 2 degrees downward from the horizontal. Tuyeres are punched automatically with punchers located on the converter shell. Flux and other additives are delivered to the converter via conveyors and Garr gun. Each converter is equipped with a radiation pyrometer for continuous bath temperature measurement. A spectrophotometer is used for rapid determination of converter matte iron content.

Two converters are normally on-line with the two roaster/furnace operation with a third "off-line" for refractory repairs.

Converter "off-gas" is transferred, via a common balloon flue, to an electrostatic precipitator. Dust collected is returned to the electric furnace with clean gas exiting via a 90 meter high stack.

Finished product matte is cast in 10 tonne molds and broken by pneumatic hammer and jaw crusher.

The 1983 overall smelter balance is shown in Table III.

TABLE III

TYPICAL SMELTER BALANCE - 1983

		Analysis						% Distribution					
Input	% Wt.	Ni	Cu	Co	Fe	S	SiO2	Ni	Cu	Co	Fe	S	SiO2
Concentrate	62.6	6.7	5.6	.3	42.4	31.2	7.1	95.5	96.5	90.9	94.0	99.6	13.8
Flux	36.1				3.5		75.7				4.5		85.5
Miscellaneous	1.3	15.3	9.8	1.3	33.8	5.6	19.0	4.5	3.5	9.1	1.5	0.4	0.7
TOTAL	100.0							100.0	100.0	100.0	100.0	100.0	100.0
Output													
Dump Slag	74.4	.11	.21	.10	35.6	0.8	40.7	2.0	4.5	43.9	99.2	3.2	100.0
Product Matte	10.0	40.7	32.6	1.0	2.3	22.9		98.0	95.5	57.1	0.8	12.3	
Other	15.6											84.5	
TOTAL	100.0							100.0	100.0	100.0	100.0	100.0	100.0

Major Process Change - Increased Degree of Roast

The new smelter was designed to operate at a roaster sulphur elimination of 50 percent and started up in 1978 at this level. Efforts to further reduce sulphur dioxide emissions from the smelting complex resulted in the development of process parameters and the implementation of process changes to provide for 60% sulphur elimination in the roaster. This is a major change in the smelting operation and is described in detail.

Benefits

Because acid is produced from roaster off-gas, more sulphur can be captured by roasting to higher degrees of sulphur elimination. The relationship between degree of roast and sulphur captured in the smelter is shown in Figure 3. Increasing the roaster sulphur elimination from 50 up to 60% reduces emissions by 28%. Also, extrapolation of the curve shows that with degrees of roast in excess of 80%, sulphur emissions would be less than 10%. However, Figure 3 overlooks a number of limitations to the process as the degree of roast is increased. These factors are detailed later in this paper.

A second benefit of increasing the degree of roast is that the electric furnace matte grade increases and the tonnage decreases as shown in Figure 4. This results in less converting, a major process cost and source of metal losses in smelting. An increase in the roaster sulphur elimination from 50 to 60% reduces the furnace matte quantity and thus converting by 18%, from 50 to 41% of the concentrate weight.

Thirdly, with less converting there is a reduction in converter slag tonnage. Highly oxidized converter slag, when recycled to the electric furnace, adversely affects the slag cleaning process which occurs during smelting. The converter slag flows to the matte-slag interface and is not readily stripped of its metal values. This stripping mechanism improves by reducing the quantity of recycled converter slag to be cleaned in the electric furnace.

Finally, a fourth benefit occurs during periods of reduced feed supply which decreases sulphuric acid production. Increasing the degree of roast increases acid yield per unit weight of concentrate.

Limitations

Serious limitations to increasing degree of roast were overcome. Each factor is discussed in conjunction with laboratory and plant development.

Sulphate Formation During Roasting. Semi-pilot scale roasting tests showed that a kinetic barrier existed in sulphur removal as the air/feed ratio was increased. Oxygen utilization decreased and shifted from a sulphide-oxide reaction to a sulphide-sulphate reaction. The conversion of sulphides to sulphates does not contribute to sulphur elimination and these sulphates cause serious accretion formation in the off-gas cleaning system, particularly in electrostatic precipitators.

However, the semi-pilot plant tests also showed that a temperature increase would overcome the kinetic barrier. A bed temperature increase from 620°C up to 680°C was required to increase the degree of roast from 50 up to 60% sulphur elimination while maintaining the same off-gas oxygen level of less than 0.5%.

Roaster "Off-Gas" Handling Capacity. As the degree of sulphur elimination during roasting is increased, more air is required per unit of feed. The air/feed ratios for 50 and 60% sulphur elimination are 0.87 and 1.10 Nm^3 per kilogram of concentrate, respectively. Furthermore, the higher air/feed ratio results in a greater generation of heat per unit weight of concentrate and more water is required in the form of lower density feed slurry. Thus, the feed slurry densities for a 50 and 60% degree of roast are 75.5 and 72% solids, respectively. The two effects combine to increase the off-gas volume by approximately 25% at the same concentrate treatment rate. Currently, because the roaster input is at 72% of design, additional capacity is available in the off-gas system to roast to 60% sulphur elimination.

Electric Furnace Slag Losses. One of the most important impacts of increasing the sulphur elimination is the potential increase in slag losses in nickel smelters. A comparison of slag value is made with copper smelting in Table IV. As a result of lower concentrate grades and higher metal prices, particularly for cobalt, the value of slag losses for nickel smelting is up to 25 times greater than that of copper smelting, per unit weight of metal produced.

FIGURE 3

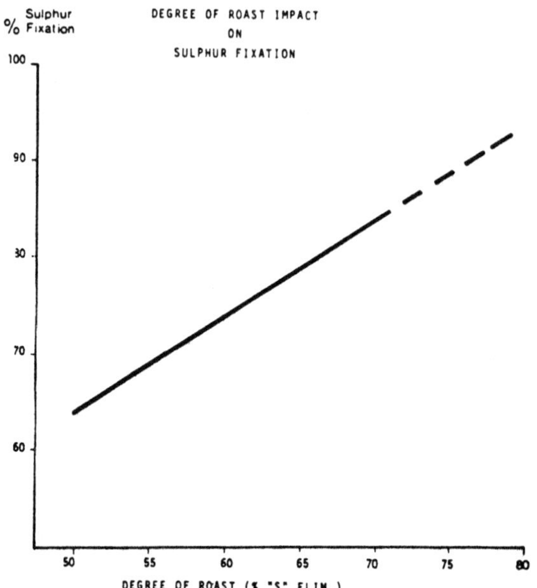

FIGURE 4

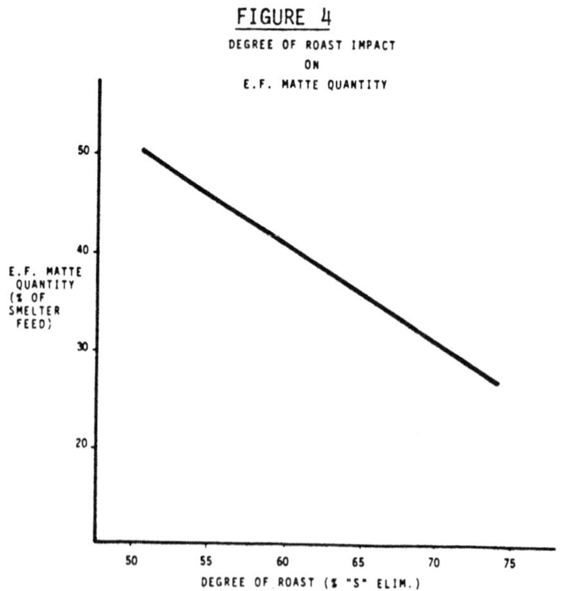

It was previously stated that increasing the degree of roast decreases the converter slag tonnage returned to the electric furnace and that this was beneficial to slag cleaning. However, a detrimental effect also exists. When the degree of roast is increased, the oxidation burner is shifted from converting to roasting. In roasting, iron is oxidized to the higher oxide states, magnetite and hematite, while in converting, only wustite and magnetite are formed. Therefore, the total oxygen input into the electric furnace as roaster calcine and recycled converter slag increases. Based on calcine chemistries determined in the pilot scale tests, it was estimated that the "reducible" oxygen (in excess of wustite) fed to the electric furnace increased by 33% when the degree of roast was raised from 50 to 60% sulphur elimination. As a result, more reductant is required in the furnace.

TABLE IV

COMPARATIVE SLAG VALUES
NICKEL VS. COPPER SMELTING

		SMELTING	
		COPPER	NICKEL
Concentrate Composition (Wt. %)	- Cu	25	5
	- Ni		7
	- Co		0.3
	- Fe	30	44
Pay Metal Value ($US)/Ton		350	450
Iron To Metal Ratio		1.2	3.6
Slag Make Per Ton Of Metal (Tons)		3.3	10.0
Metal In Discard Slag (Wt. %)	- Cu	0.5	0.2
	- Ni		0.2
	- Co		0.2
Values In Slag $US/Ton		7	59

A further difficulty in controlling slag losses is the increase in matte grade as more iron is oxidized in the roasting process. Therefore, to maintain the same slag losses, higher distribution coefficients, (% metal in matte/% metal in slag) must be attained in the electric furnace.

Figure 5 presents the distribution coefficient required to maintain the same slag losses for nickel between 50 and 75% sulphur elimination roasting.

Laboratory studies determined that the distribution coefficients for nickel and cobalt could be increased by the addition of more reductant during smelting. These results are presented in Figure 6 which shows that as the ratio, iron in matte/magnetite in slag, increases, the distribution coefficients for nickel and cobalt increase. The x-axis in Figure 6, iron to magnetite ratio, increases with reductant quantity as more iron oxide in slag is reduced to iron which reports to the matte phase. These results demonstrated that losses specifically of nickel and cobalt in slag can be controlled by raising the degree of reduction.

The level of copper in the slag did not decrease with stronger reducing conditions. As the copper grade in matte increased, its level in slag rose proportionately. These observations are consistent with Nagamori (Ref. 3) who proposed that copper was not dissolved in slag as an oxidic species under smelting conditions.

Increasing reductant addition to control slag losses of nickel and cobalt is limited by two factors, the furnace freeboard temperature and matte saturation with metallics.

The first limitation is caused by the combustion of some of the reductant or the related carbon monoxide in the furnace freeboard which raises the temperature in that location. Because the draft air cooling capacity is limited, the reductant feed rate to the furnace is also restricted in order to maintain acceptable freeboard temperatures. Thus, higher reductant to feed ratios are possible only at decreased throughputs with current operating equipment restraints.

The second limitation is the existence of a metallic saturation boundary in the iron-nickel-copper-sulphur system. The addition of reductant to the smelting process reduces iron oxides and causes a transfer of iron to the matte phase which decreases the sulphur content. It has been determined that when the sulphur content of the electric furnace matte falls below 25% an iron-nickel rich alloy precipitates at 1050°C to cause a metallic buildup on the furnace hearth. Therefore, reductant additions to the electric furnace are limited by the physical chemistry of matte.

Plant scale development confirmed the laboratory findings. As the degree of roast was increased from 50 to 60% sulphur elimination, the matte grade (% Ni + % Cu + % Co) rose from 25 up to 32 wt. %. The nickel and cobalt contents in the electric furnace slag were maintained or improved by increasing the coke feed rate from 3.0 up to 4.1% of the dry concentrate weight - however, the copper assay in slag increased. These results were obtained at the expense of a rise in furnace off-gas temperature of 20 to 60°C. The furnace operation shifted away from a "black top" condition. The higher freeboard temperatures led to steeper calcine banks with a greater area of open slag surface. This compounded the difficulty of maintaining lower freeboard temperatures. Additionally, coke feed rates in excess of 4.1% of the concentrate weight were attempted and resulted in a buildup on the furnace hearth.

Operation at 60% Sulphur Elimination

During the recent downturn in the nickel market, capacity became available in the smelter to increase the degree of sulphur elimination during roasting and to raise the coke to feed ratio to the electric furnaces. In order to capitalize on this opportunity and based on the laboratory, pilot scale and plant testwork, as described previously in this paper, it was decided to operate the smelter at a degree of roast of 60% sulphur elimination. The smelter has now been operated in this mode since early 1983.

FIGURE 5

K_{Ni} REQUIRED TO MAINTAIN METAL RECOVERY AS ROAST IS INCREASED

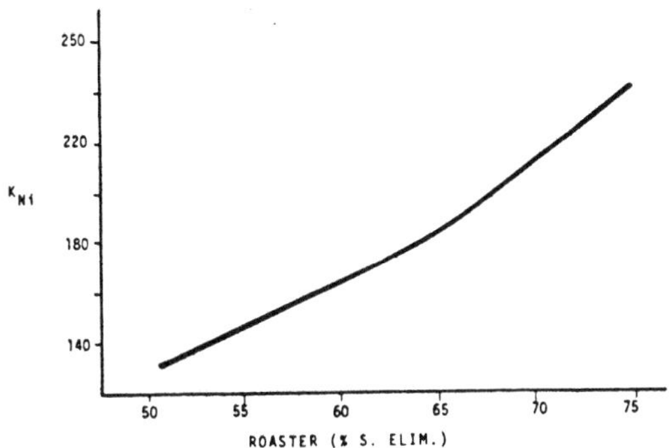

FIGURE 6

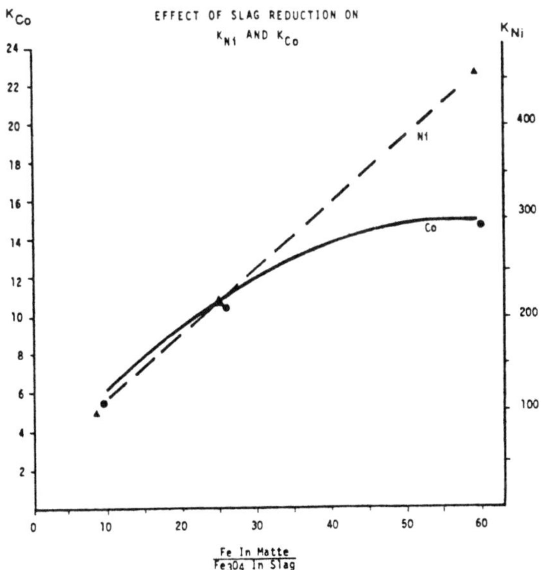

Roasting. The roasting parameters at 60% sulphur elimination are summarized in Table V and compared with previous operations at a degree of roast of 50%. The concentrate feed rate of 770 tonnes per operating day is substantially below the design value of 1070 tonnes per operating day which is based on 50% sulphur elimination. The sand flux feed rate has increased because the iron removal burden which requires fluxing has shifted towards the roaster-electric furnace stage of the process. Previously, at the lower degree of roast, the additional flux was added to the converting process.

The air to feed ratio and bed temperature were increased simultaneously to achieve the higher sulphur elimination while controlling sulphate formation and off-gas oxygen concentrations. The roaster off-gas oxygen contents in Table V show that oxygen utilization has been maintained at the higher degree of roast. Although sulphate formation data are not available for comparison, it can be stated that accretion formation, particularly in the electrostatic precipitators, has remained the same at 60% sulphur elimination.

TABLE V

COMPARISON OF ROASTING PARAMETERS
AT 50 AND 60 PERCENT SULPHUR ELIMINATION

Nominal Degree of Roast (% of Feed Sulphur Eliminated)		50		60
Concentrate Feed (tonne/day)		932		770
Sand Flux (% on feed)		33.5		44.2
Air to Feed Ratio (Nm^3/kg)		.87		1.10
Slurry Density (wt. % solids)		75.5		72.0
Bed Temperature (°C)		620		680
Roaster Bed Screen Analysis (% cum)				
	+ 3 Mesh		5.4	
	+ 10 Mesh		25.0	
	+ 28 Mesh		49.3	
	+ 48 Mesh		89.5	
	- 48 Mesh		100.0	
Roaster Off-Gas Oxygen Content (vol. %, dry gas)		.5		.4
Acid Production (tonne/tonne feed)		.45		.56

The roaster start-up procedure was revised. Previously, the bed temperature was raised to 600°C before feeding commenced and several hours were necessary to bring the temperature up to 620°C. The starting bed temperature has been increased to 640°C to minimize the opportunity for sulphation.

Finally, the acid production per unit weight of concentrate has increased to reflect the greater proportion of sulphur removal in the roasting stage. Thus, the first benefit of reducing sulphur dioxide emissions to the atmosphere was realized.

Electric Furnace Smelting. The electric furnace operating parameters in Table VI for 50 and 60% sulphur elimination roasting. Many of the conditions have remained unchanged, and were presented earlier. A small increase in bottom buildup has occurred as a result of testwork with higher reductant rates.

One parameter, which has been controlled with the higher degree of roast and reductant rate, is the off-gas temperature. The upper limit of 700°C has been maintained. The semi-black top condition in the furnace, caused by the reduced flow characteristics of the calcine, has also been controlled by closely monitoring the feed distribution.

The electrical energy consumption per unit weight of dry solid charge increased 5.7% with 60% sulphur elimination roasting as more of the smelting process was shifted from the autogenous converters to the electric furnaces. The increased energy was required to melt the higher flux addition and to supply the endothermic carbon reduction reaction.

The reductant feed rate was raised in order to control slag losses at the higher degree of roast. The stronger reducing conditions are reflected in the drop in slag magnetite assays. More importantly, in spite of the higher matte grades with more roasting, the nickel and cobalt levels in the discard slag have been lowered. However, as predicted, the copper level in slag increased. At current metal prices the net result is a reduction of value of metals in slag.

The converter slag tonnage has decreased per unit of feed by approximately 14% as a result of the higher electric furnace matte grade. Thus, the third benefit outlined previously has been achieved. However, because of the simultaneous increase in reductant feed rates, it is not possible to isolate the effect of the lower converter slag tonnage on the cleanliness of electric furnace slag.

Finally, at the current matte sulphur content of 26%, precipitation of metallics on the hearth is not a problem.

TABLE VI

TYPICAL ELECTRIC FURNACE OPERATING CONDITIONS
FOR 50 AND 60 PERCENT SULPHUR ELIMINATION ROASTS

Nominal Degree of Roast (% of Feed Sulphur Eliminated)	50	60
Bottom Buildup (cm)	15	25
Power Consumption (kWh/tonne DSC)	328	347
Coke Consumption (% on feed)	3	3.9
Slag Magnetite Content	5	∿3
"Black Top" Condition (% of time)	100	70
Recycle Converter Slag (% on concentrate)	42	36
Electric Furnace Matte Grade (%Ni+%Cu+%Co)	25.8	31.3
Matte Sulphur Content (%)	26.9	26.0
Slag Assay - % Ni	.12	.11
- % Cu	.18	.20
- % Co	.11	.10

Converting. Selected data are presented in Table VII to show the effect of increasing the degree of roast on converter operations. The quantity of furnace matte treated or product matte per blowing hour increased significantly. This resulted from the higher furnace matte grade at a degree of roast of 60%. This reduction in converting time, in combination with the lower throughput rate, has allowed us to reduce the number of active converters to two. Thus, the second benefit of a reduced converting load, with an increased degree of roast, was realized.

TABLE VII

CONVERTER OPERATING DATA FOR
50 AND 60 PERCENT SULPHUR ELIMINATION ROASTS

Nominal Degree of Roast (% of Feed Sulphur Eliminated)	50	60
Furnace Matte Converted (Tonne/Blowing Hr.)	23	27.6
Product Matte (Tonne/Blowing Hr.)	8.2	11
No. of Active Converters	3	2

Process Modifications and Performance

Process changes and equipment modifications, which led to major improvements in the performance of the operation, are summarized in this section.

Concentrate Slurry Handling

Rail Tank Car Plug Valve. The bottom outlet on the rail slurry tank cars were changed from the spring-loaded doors to a 44 cm diameter opening sealed from the inside with a rubber plug. An overhead hoist pulls the plug at the unloading station allowing the slurry to discharge into the sump. The new plug has eliminated transit slurry losses and injury to employees opening the doors. The plugs, as well, have been virtually maintenance free.

A two man crew unloads 25 cars in an eight hour shift.

Secondary Repulpers. The second stage of repulping, after filtration for density control, was eliminated when a completely homogenous slurry and simplified density control was obtained with a single stage.

Roaster Operation

Roaster Feed Pumps. The hydraulic oil makeup and bleed-off system was improved to compensate for excessive transfer of oil from one diaphragm cavity to the other. This extended the diaphragm life to approximately 2000 hours. Check valves were replaced with standard Canadian-made ball type check valves. Inlet accumulators to the pump were redesigned to overcome vibration and material buildup in pumps and lines.

These variable speed positive displacement pumps now provide a reliable metered feed to the roasters and are used in the tonnage determination for plant feed.

Roaster Gas Cleaning

Cooler Sprays. Accumulations of dust on the walls of the off-gas cooler were eliminated by changing from sonic to air atomized sprays. A dedicated high pressure air and water supply to the spray nozzles through galvanized piping systems was provided to avoid fluctuations in availability from plant supply.

Electrostatic Precipitators. To minimize sulphate buildup on the discharge electrodes and structural members and to prevent discharge electrode wire breakage the following steps were taken:

a) More continuous roaster operation and higher roaster bed temperature at start-up (640°C) reduce sulphate formation conditions in the system.

b) Replacement of mild steel support brackets for collecting screens with stainless steel.

c) Replacement of ceramic high tension frame vibrator support shafts with fibreglass to improve the life.

Roaster Off-Gas Fans. Structural changes to the fan impellers were required shortly after start-up to improve the fan reliability by eliminating fan vibration and subsequent failure.

Off-Gas Flues. All bellows-type expansion joints, between the roaster off-gas fan and the acid plant, were replaced with slip-type joints designed at Falconbridge to overcome serious corrosion and thus leakage at this point.

All unlined mild steel flues from the roaster off-gas fan to the acid plant were lined to prevent corrosion. A gunned, acid-resistant refractory was used.

Performance. The combined operating time, for the two roasters in 1983, was 91.5% for the year - well above the initial design of 85%. There has been no roaster shutdown time attributed to defluidization of the roaster bed since start-up in 1978.

Acid Plant

Acid Plant Blower. Significant vibrations in the SO_2 blower occurred on several occasions. These were traced to movement of the inlet and outlet ducts. Both these ducts were isolated from the blower by installing rubber-base wrapped expansion joints. Vibration from this source has not recurred.

Performance. The high degree of plant reliability has been achieved since start-up in 1978 - with availability in excess of 90% each year reaching 93% in 1983.

Conversion efficiencies have been over 98% with tail gas well below the specified 0.3% SO_2.

The operating rate for 1983 approached 900 tonnes per day.

The typical analysis of the sulphuric acid for 1983 is shown in Table VIII.

TABLE VIII

TYPICAL 1983 ACID QUALITY

Parameter	Units	1983
Strength	% Wt. H_2SO_4	93.6
Colour	Hazen Units	20
Transparency	%	92
Particulate	Visible Suspended Matter	-
Analysis:		
SO_2	ppm	0.5
Fe	ppm	25
Cu	ppm	.2
Ni	ppm	.5
Zn	ppm	< .01
Pb	ppm	< .2
Hg	ppm	< .01
As	ppm	< .01
Cl	ppm	< .3
H_2O	ppm	< .5
Oxides of Nitrate	ppm	< .2

Electric Furnace

Calcine Transfer. The major source of smelter downtime in the first two years of operation, was attributed to failure in the calcine handling system.

To improve the reliability of this system, the following positive steps were taken:

a) The flexibility for transferring calcine to either side of the electric furnace feeding system was eliminated. Poor calcine flow characteristics prevented reliable operation of drag conveyors, transfer points and valves.

b) Failure of the drag conveyors was reduced by improving the preventive maintenance program, reinforcing drag links to prevent bending, increasing overload protection (shear pins), installing a chromium carbide wear strip on the riding rails and changing link connecting pins to stainless steel.

Reductant. The addition of petroleum coke as a reductant, which had been demonstrated in the pilot plant investigation, was added shortly after start-up to the electric furnace. The use of reductant improved the flow properties of the calcine to aid in the formation of a calcine cover of the molten bath to the electrodes, thus reducing off-gas temperatures.

Major metallurgical benefits, of operating in the reducing mode, have been covered in detail in a previous section of this paper.

Furnace Off-Gas Fans and Ducts. High speed (1800 rpm) furnace off-gas fans were replaced with slow speed (1200 rpm) paddle wheel bladed fans to prevent wear and excessive vibration. Two furnace fans were equipped with variable speed drives to provide for more positive control of the freeboard pressure and allow for maintenance. This replaced the original vane damper control.

The source of temperating air was relocated from the fan inlet to the furnace building. This reduced the potential for cold, damp air contributing to dust buildup and corrosion in the system. The flues, between the furnace cyclones and the fan inlet, were refractory-lined, or replaced with 316 stainless steel to overcome corrosion evident after the first year of operation. Bellows-type expansion joints were changed to slip-type joints. Only minor repairs have been required over the last four years.

Slag Tapping. The size of the water-cooled slag tapping blocks was increased to afford greater protection for the refractories in the slag end wall.

Water-cooled slag launders were installed, extending service life from one month to over one year. Hydraulically operating clay guns, similar to those on the matte end, replaced a water-cooled pneumatic stopper, which was not sufficiently reliable in stopping slag flow.

Matte Tapping. Water-cooled copper billets were added to the matte tapping blocks to provide additional protection to the refractory in this area. This made replacement of tap blocks less risky and reduced matte penetration into the refractory tap blocks.

Furnace Refractory. Round copper billet coolers were installed in the matte end wall directly below the converter slag return launder to reduce refractory consumption at this point. Regular thermoscanning of the electric furnace wall temperatures provide information in determining the extent and location of reduced refractory thickness. After six years service, there is no evidence of side wall attack on No. 2 electric furnace.

Furnace Power. In 1981, a computer program was developed to manage Falconbridge area peak power control. When the furnace power is under computer control, the electrode position in the slag bath and hence furnace power is automatically regulated. The three furnace phases are adjusted every 30 seconds, if required, to maintain preset power. For power control, megawatt rates are reduced automatically by raising furnace electrodes whenever the area forecast indicates a peak. Power shed, during peak load periods, is automatically picked up during low energy demand periods such as night shift.

Performance. No. 2 electric furnace has been in operation continuously since start-up in mid-1978 (six years) without being emptied of molten material. The operating availability in 1983, for receipt of roaster calcine, was over 99%.

The physical integrity of the furnaces has been maintained with side walls remaining essentially vertical over the operating period. This is attributed to the continuous operation, as well as dedication to furnace tie rod adjustments to maintain consistent pressure on the furnace refractory.

The installation of the copper cooling billets, in the matte tapping area of the furnace, has prevented run-out of molten material from this area and allowed for prolonged furnace operation. As yet, refractory measurement on the furnace side walls does not indicate the need for billet cooler installation.

Environmental Improvements

Falconbridge has been instrumental in reducing the impact of its operating facilities on the environment over the years. Significant reductions have been made in particulate and sulphur dioxide emissions, as well as in the working environment with the operation of the new smelting process.

Sulphur Dioxide Emissions

Reduction of sulphur dioxide emissions has been achieved by pyrrhotite rejection in the milling circuit and sulphur fixation from roaster off-gas to produce sulphuric acid. In 1983, the acid plant recovered 53% of the sulphur entering the smelter or 21.8% of the sulphur in the ore. A total of 58.5% of the sulphur in the ore is discarded with the mill tailings to the effluent treatment system. The sulphur distribution is shown in Table IX.

TABLE IX

1983 SULPHUR DISTRIBUTION

	Tonnes Sulphur	%
Rejected in Milling	196,800	58.5
Slag	3,900	1.2
Ni-Cu Matte Product	16,000	4.8
Sulphuric Acid	73,200	21.8
Cu Conc. Inventory, Miscellaneous	8,300	2.4
Emission	37,900	11.3
Total Ore and Other Feed	336,100	100.0

The overall reduction in sulphur emission since 1953 is shown in Figure 7. These emissions have decreased to 12% of the sulphur in the ore by the end of 1983.

With the increase, in degree of roast from 50 to 60%, emissions from the smelter decreased from 39% in 1980 to 28% in 1983, Figure 8.

FIGURE 7

FIGURE 8

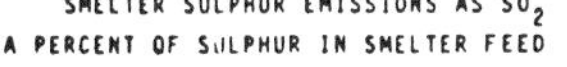

Particulate Emission

A comparison of the particulate emission, before start-up of the new smelter and that of 1983, are shown in Table X.

TABLE X

PARTICULATE EMISSIONS (TONNES/DAY)

	1974	1983	% Reduction
Sinter Stack	10.7	--	--
Smelter Stack	4.9	2.0	--
Total:	15.6	2.0	87
Ni	0.91	.031	96
Cu	0.69	.042	94
Fe	4.80	.35	93
Pb	0.18	.044	75

1983 particulate emissions were only 13% of those in 1974. The major factor in this reduction was the replacement of the sinter plant and blast furnaces with the roasting and electric furnace operation in the new smelting process.

Working Environment

The working environment in the smelter was significantly improved with the new smelter complex. Further improvements have been made to eliminate upset conditions over the last six years, which have reduced the dust exposure in the work place by approximately 50% from those of the old smelter.

Comparative dust exposures for equivalent jobs are shown in Table XI.

TABLE XI

AVERAGE SMELTER EMPLOYEE DUST EXPOSURE COMPARISON

1975-1977 Old Smelter	Mg/M^3	1983 New Smelter	Mg/M^3	% Decrease
Pellet Plant Operator	7.0	Roaster Operator	1.1	84
Blast Furnace Feederman	10.1	Fettling Conveyorman	2.3	76
Settler Tapper	2.6	Furnace Tapper	1.2	54
Settler Slagman	2.2	Furnace Slagman	1.2	45
Converter Skimmer	2.5	Converter Skimmer	1.1	56
Converter Puncher	2.3	Converter Puncher	0.9	61

The maximum allowable concentration for the majority of smelter dusts is 5 mg/m^3.

The main contributor to this improvement is the operation of major gas handling systems under slightly negative pressure at source, the installation of local hooding at source and the provision for sufficient air movement to remove fugitive emissions and excess heat.

This improvement in working environment provides the new smelting complex with a virtually dust-free atmosphere with only occasional detection of SO_2 in some work areas.

Process Development Focus

Higher Degree of Roast

At current production rates, capacity is available in the roasters and acid plant to further increase the sulphur elimination in the roasting step. Therefore, the extension of the work described in this paper to a degree of roast of 65% sulphur elimination is being considered at Falconbridge. The development approach, taken to raise the degree of roast to 60%, has been successful in combining laboratory, pilot scale and plant testwork to define the key metallurgical and operation parameters. Although the benefits such as reduced sulphur dioxide emissions and lower converting loads become greater with higher degree of roast, the burden placed on the electric furnace smelting stage particularly, becomes more severe. As the furnace operation is extended towards full capacity, with lower slag losses, greater understanding of the smelting mechanism, coke reaction and the behaviour of recycled converter slag will be required.

Converter Slag Cleaning

The second area for development is reduced slag losses. It was pointed out that slag loss is a major factor in the smelting of nickel-copper-cobalt concentrates. This subject must be tacked on three separate fronts:

1) slag cleaning in the electric smelting furnace;
2) control of metal oxidation into converter slag during converting;
3) separate cleaning of converter slag.

The combined efforts in these areas are necessary to significantly impact on metal recoveries, particularly that of cobalt.

Pyrrhotite Rejection

The third development area is concentrate feed upgrading with more efficient rejection of sulphur in the form of pyrrhotite from the milling circuits. The negative impact on nickel recovery, with current technology, makes this option extremely uneconomical as a means of reducing sulphur emission.

Summary

The process changes which have been developed and instituted since the start-up of the new smelting process in 1978, have provided major improvements to the environment both inside and out of the smelter complex. These have been achieved without seriously impacting on the economic metal recoveries in the process at current production levels.

Acknowledgement

The authors wish to thank Falconbridge Limited for the opportunity to present this paper. Recognition must be given to the Falconbridge staff who were instrumental in the implementation of process and equipment changes, which provided the basis for this presentation.

References

1. Mahant, R.P.; McKague, A.L.; Michelutti, R.E.; Norman, G.E.; "The Development and Operation of the Smelting and Environmental Improvement Process, Falconbridge Nickel Mines - Sudbury Operations, CIM Bulletin, June 1984.

2. Jackson, J.F.; McKague, A.L.; Norman, G.E.; "Falconbridge Nickel Mines Limited - New Smelting Process, CIM Bulletin, June 1980.

3. Nagamori, M.; Metallurgical Transactions, Volume 5, March 1974.

P.T. INCO'S INDONESIAN NICKEL PROJECT : AN UPDATE

by J.D. Guiry and A.D. Dalvi

President and Managing Director
Senior Technical Specialist

P.T. INCO
Soroako, S. Sulawesi, Indonesia

ABSTRACT

P.T. INCO constructed a smelter between 1974 and 1978 to produce more than 35,000 tonnes of nickel per year as a nickel matte from laterite ores on the island of Sulawesi in Indonesia. Since then significant changes have occurred in both mining and processing operations.

Original relatively high grade (2.4 % Ni) but acidic (SiO_2/MgO = 2.4) feed to the plant is now blended with a lower grade (1.8 % Ni) but more basic (SiO_2/MgO = 1.6) feed to give a plant feed of 2.0 % Ni at SiO_2/MgO ratio of 1.9. Several ore bodies from a different ore zone had to be developed to obtain this feed which reduced corrosive attack of slag on electric furnace sidewall refractory.

Improvements were made in the mechanical and materials handling systems around dryers and in the dryer internals. Installation of a modified dust handling system has resulted in more efficient materials handling. Changes in reduction kiln operation have reduced fuel requirements by over 30 percent. A part of the kiln oil has been replaced by coal. The electric furnace sidewall design was modified to cope with corrosive slags. Furnace power was steadily increased from 36 to 43 MW. Converter productivity has been substantially increased. A Pierce-Smith Converter has replaced one of three original rotary converters. The original furnace slag disposal system of casting onto ground, cooling and ripping was changed to granulation and then to molten slag disposal.

Intensive manpower training programs have substantially reduced both total and expatriate manpower. Improvements made over the last several years have increased energy efficiency about 30 % and productivity over 100 %, resulting in significant reduction in costs.

Paper presented at the
International Seminar on Laterite
Tokyo, October 14-17, 1985

INTRODUCTION

P.T. INCO is a fully integrated nickel mining and smelting company located on the island of Sulawesi in Indonesia. It has an annual capacity of about 35,000 tonnes of nickel in matte. The project area encompasses some 200,000 hectares with certain facilities near the sea port of Malili and the major installations around the minesite at Soroako, some 60 km inland (Figure 1). Exploration to date has outlined several hundreds of millions of tonnes of nickel bearing minerals which could support the project well into the next century. The present proven and probable ore reserves are 70,000,000 tonnes containing 1,270,000 tonnes of nickel. The company employs about 2,700 Indonesians : managers, engineers, technicians, tradesmen and operators. The expatriate group of experts and specialists is reducing rapidly and it numbers some 40 people at the present time. Essentially, all of the product in 1985 will be shipped to Japan with P.T. INCO now supplying almost 20 % of that market.

In 1979, at the International Laterite Symposium in New Orleans, a paper was presented with the title "P.T. INCO's Indonesian Nickel Project" (1). In this paper we would like to discuss technical developments and changes since 1979, which have substantially improved productivity and cost performance.

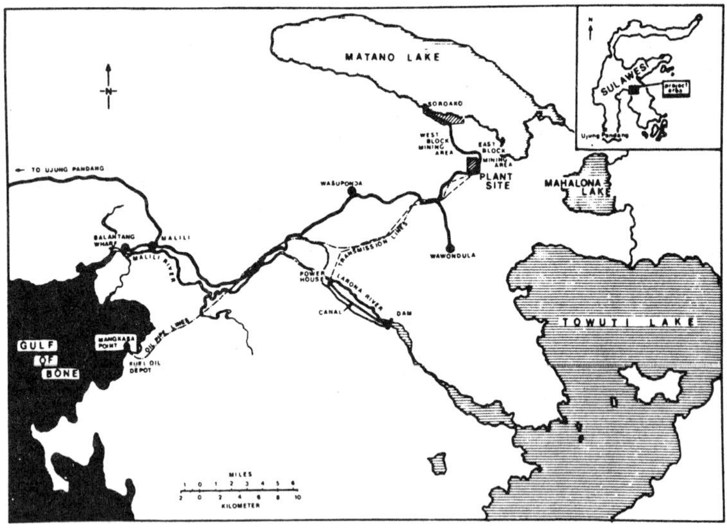

Fig.1. Location Of P.T. INCO's Soroako Project

Although several modifications are described in the following pages, the most significant change was the development of another major section of the ore body, known as the East Block. This provided lower acidic material for blending with the original West Block material. The blending of the feed to the smelter reduced the corrosive nature of the resultant slags produced in the Electric Furnaces. However, this also lowered the nickel grade of the plant feed and necessitated the restating of the Process Plant capacity from an initial rate in the early years of 45,000 tonnes to the present average rate of 35,000 tonnes per annum of nickel in matte form.

A second major corrective step taken to deal with the corrosive slags has been the development of an improved system for cooling the sidewalls of the electric furnaces which, in combination with the blended feed, protects the furnace refractories and permits sustained dependable operation. Most recently, one of the three Top Blown Kaldo-Type Rotary Converters (TBRC) was replaced by a conventional Pierce-Smith converter.

These and other less major but important improvements are described in more detail below.

MINING

The mine is fully developed and has demonstrated that it can readily produce the required quantity and quality of plant feed. It operates two shifts per day, five days per week throughout the year. In 1985 run-of-mine material deliveries to the Screening Stations averaged 15,500 wet tonnes per day and total mine material movement averaged 61,800 wet tonnes per day.

The mineral deposits fall into two geological types as described previously (2,3). These are referred to as the West Block and the East Block deposits. The West Block material is generally made up of unserpentinized peridotite boulders surrounded by weathered saprolite in which the nickel mineral values occur. The hard boulders are very low grade. The East Block material is generally made up of weathered products of serpentinized peridotite with more uniform distribution of mineralization through the matrix, and friable boulders.

Composition of the mine material falls into the ranges given in Table 1. Averages are based on production from 1975 to 1985.

Since the mine must provide a blended feed to the Process Plant, several production areas are operated concurrently to balance out the sometimes widely varying compositions.

In 1985 the mine is being operated to provide a plant feed with a mix of 2 parts West Block to 3 parts East Block. Present production

Table 1. Composition Of West And East Block Mine Material (wt.%)

West Block - Screened Material : - 50 mm

Assay	Low	High	Avg.
Ni	1.60	2.60	2.30
Fe	17.0	25.0	18.6
SiO_2	32.0	45.0	39.0
MgO	12.0	19.0	17.0
SiO_2/MgO*	1.9	3.0	2.3

East Block - Total Material

Assay	Low	High	Avg.
Ni	1.50	2.10	1.89
Fe	14.0	24.0	15.9
SiO_2	26.0	39.0	35.0
MgO	17.0	28.0	24.0
SiO_2/MgO*	1.3	2.1	1.45

*Ratio

requires three mining shovels in the East Block and two in the West Block. There are normally four stripping shovels working.

The overburden, which usually ranges in thickness from 5 m to 15 m, is moved by truck to waste dumps. Medium grade limonite, when it is encountered between the overburden and the run-of-mine material, is stockpiled for future use. The run-of-mine material is excavated down to grade cutoff and trucked to the Screening Stations. Shovels work on 8 m benches. Backhoes are used to recover deep roots of higher grade materials.

At the Screening Stations + 500 mm boulders are scalped off first. Then - 500 mm + 150 mm rock is screened out. In the West Block, these boulders and rocks are very low grade and are used for ballast and road building. In the East Block, rocks are of good grade and are crushed to - 150 mm and added to the Screening Station Product which constitures the Process Plant feed.

A degree of blending takes place as material from different mining faces is mixed at the Screening Stations. Usually the Screening

Station Product is blended further by placing it in intermediate storage piles which are maintained separately for East and West Block materials. The intermediate stockpiles also serve to improve the handling characteristics during rainy periods by providing time for the thixotropic phenomenon to "thicken" the material. When plant requirements and weather conditions permit, the Screening Station Product by-passes the intermediate stockpiles and is delivered directly to the Dryer Feed Hoppers.

Table 2 shows changes in mine quantities and improvements in productivity since the project went into operation.

Table 2. Mine Quantities And Productivity

Item	Original Design	1984 Actual	Current Design
Ni in Product, Million lbs	100	50	77
Ni in Product, thousand tonnes	45	23	35
Plant Feed Grade, % Ni	2.40	2.09	1.97
Ratio WB : EB in Dryer Product	1.0/0.0	1.0/0.7	1.0/1.5
Stripping, Million wet tonnes	5.24	5.10	6.55
Run-Of-Mine, Million wet tonnes	4.86	2.80	4.50
Moisture Content, %	27	31	31
Manpower	1400	1000	900
Heavy Equipment			
Shovels & draglines	27	15	17
Trucks	128	70	68
Dozers & Loaders	78	55	53
	233	140	138

PROCESSING

Several modifications have been made in the plant facilities since commissioning. The capacity of equipment has been increased and operating reliability has been improved.

The plant layout is shown in Figure 2 and the process flow sheet is shown in Figure 3.

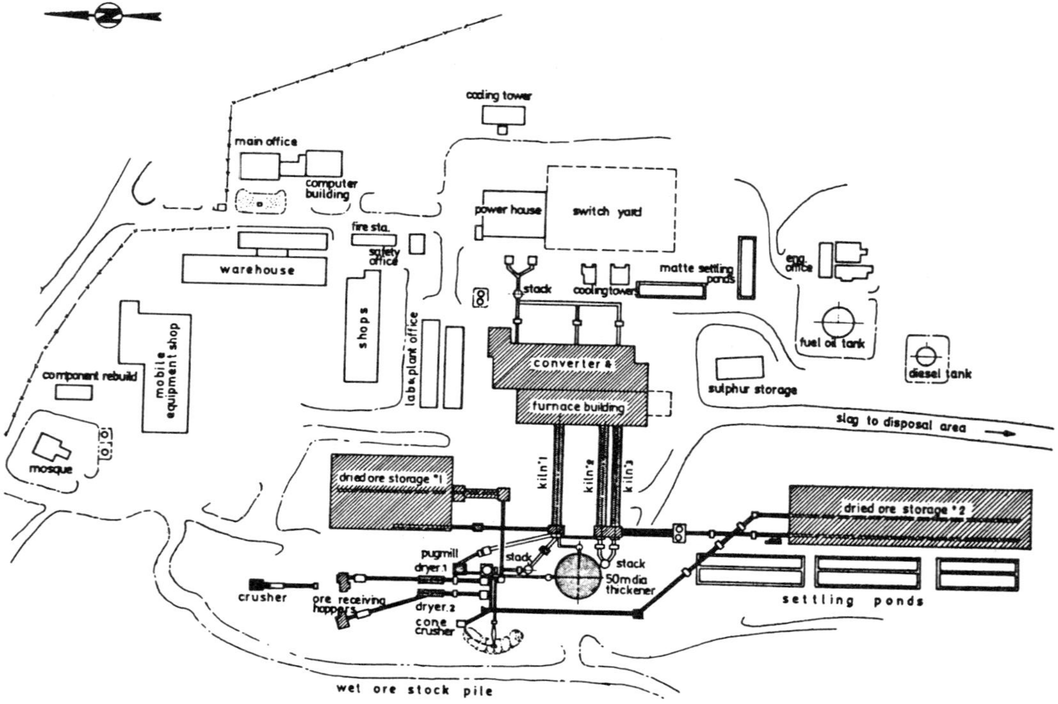

Fig. 2. Process Plant Layout

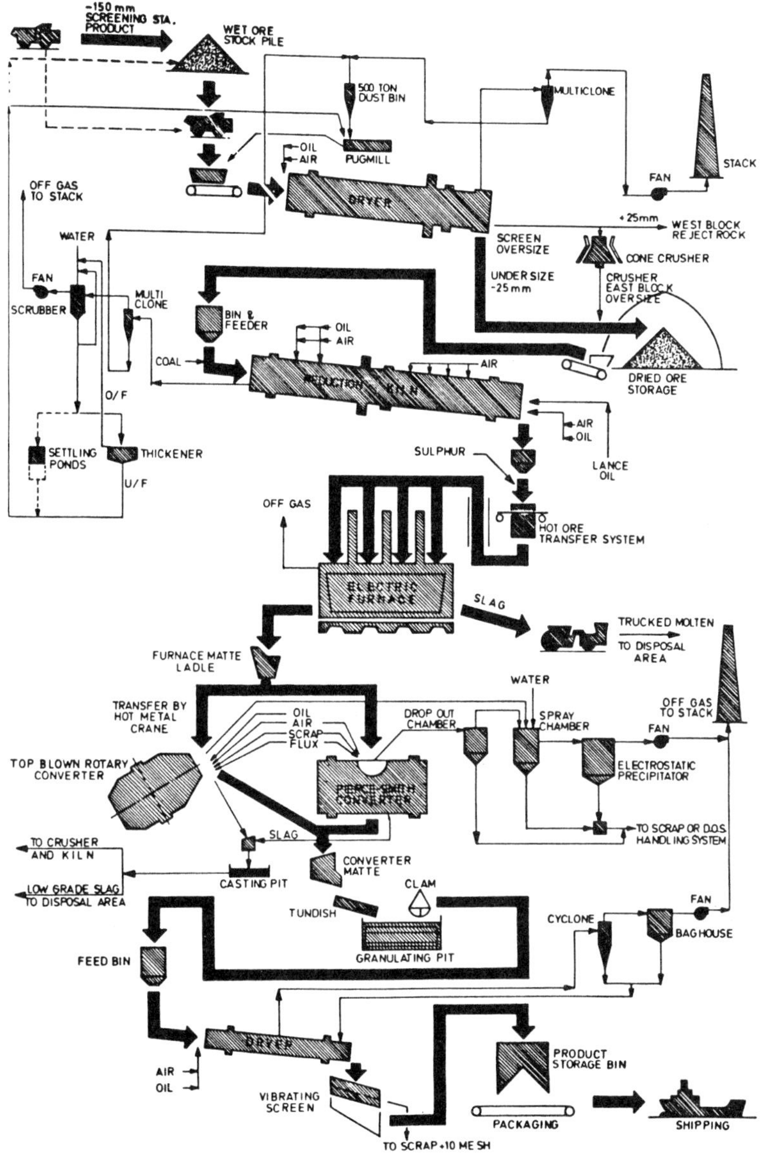

Fig.3. Simplified 1-Line Flowsheet

Drying

There are two dryers, one 5.0 m dia. x 50 m long, and the second 5.5 m dia. x 50 m long, which partially dry the Screening Station Product and screen out the + 25 mm rock. When handling West Block material the - 150 mm + 25 mm rock is rejected and used for building mine roads. East Block rock is crushed to - 25 mm and joins the Dryer Product. In 1985 the Dryer Product from new feed had the following composition:

Ni	2.00 %
Fe	19.3 %
SiO_2	33.3 %
MgO	17.4 %
SiO_2/MgO	1.9

Dust and miscellaneous reverts are mixed with the new feed to the Dryers. Crushed converter slag and a part of crushed scrap are mixed with Dryer Product.

Table 3 illustrates the changes in major operating parameters for the dryers.

To deal with changed condition and to improve the dryer operation, the following modifications have been made.

Originally bent plate lifters were used. More robust "tent type" and "bucket type" lifters are now installed and this has increased lifter life and improved dryer capacity and fuel efficiency.

The initial internal Dryer trommel screened out +50 mm rock and the undersize was conveyed to a screen house where +12 mm material was removed. The - 12 mm material was stockpiled in Dry Ore Storage buildings for subsequent feeding to the reduction kilns. It was necessary to dry to 18 % moisture to keep material flowing through the system. However, at 18 % moisture, fuel consumption was higher than design and dusting was a problem at times. Since then the screens have been bypassed and the area of the internal dryer trommel has been enlarged. The rocky material is now screened off at + 25 mm and the undersize is delivered directly to the Dry Ore Storage buildings. The Dryer system now operates at 20 % to 21 % moisture in the product with consequent savings in fuel, a significant increase in capacity and little or no dusting.

Table 3. Dryer Operating Parameters

Parameter	Original Design	Initial Operation	Current Conditions
Feed Moisture, %	28	28 - 31	28 - 31
Product Moisture, %	22*	18*	20 - 21
Discharge Product Size, mm	- 50**	- 50**	- 25
Product Rate***, dry tonnes/h	340 (WB)	230 (WB)	340 (WB) 530 (EB)
kg oil/dry tonne product	--	35	26

* 22% design. However, 18 % was required in initial operation to maintain material flow.

** The original plant employed a screen (now bypassed) to reject + 12 mm material.

*** WB : West Block; EB : East Block

Testwork during early plant operation showed good correlation between off-gas density and dryer product moisture. An optical dust density meter was installed and this increased the responsiveness of the burner control. Dust losses are lower, fuel efficiency is better, chute plugs from wet material have been reduced and the general workplace environment has been improved.

Dryer discharge conveyers are being upgraded to add flexibility to the system. Whereas originally the two dryers were intended to discharge to separate Ore Storage buildings it is advantageous to be able to send material from both dryers simultaneously to one or the other storage. Better utilization and maintenance flexibility will result.

A new permanent crushing system is being installed to handle the increased amount of East Block oversize from the dryer trommel.

Dry Ore Storage and Ore Blending

As stated above, blending begins in the mine and proceeds through screening plants, intermediate stockpiles and the dryers. However, materials from the West Block and the East Block are handled separately through to the Dry Ore Storage buildings. The actual proportioning of the materials for feeding to the kilns is done by front-end loaders by alternate recovery from the East Block and West Block stockpiles. Although the nominal proportion is 2 West Block to 3 East Block, minor adjustments are made to keep the composition of the electric furnace slag within the following limits.

SiO_2/MgO	1.85 - 1.95
% Fe	17 - 21

Reduction

There are three Reduction Kilns 5.5 m dia. and 100 m long, countercurrently fired. The internal arrangement is shown schematically in Figure 5. Throughputs and assays of certain key streams are shown as part of Table 4. Kiln Feed consists of new dried ore, recycled dust, crushed converter slag, scrap and other reverts.

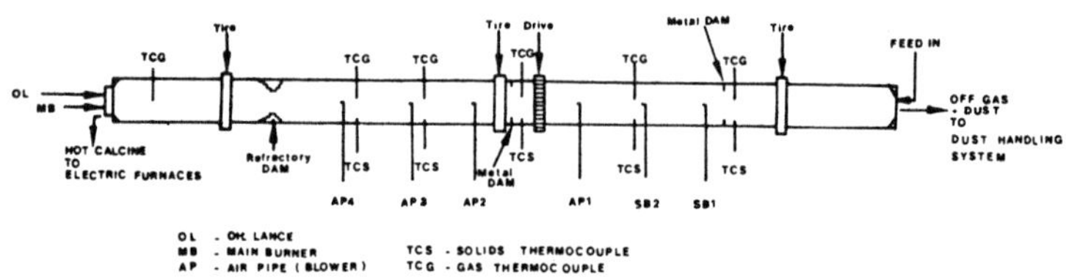

Fig.5. Schematic of Internal Arrangement of Reduction Kilns

The initial distribution of fuel plus air to the main burner and to the side burners and the air to the air pipes has been progressively altered to reduce fuel consumption while maintaining acceptable temperature profiles. The kiln off gas temperature has been lowered by some 150°C (150 K), from 350°C to 200°C (625 to 475 K). At the same time, because of the higher proportion of more refractory East Block ore, it has been possible to increase the kiln calcine discharge temperature by about 50°C (50 K) to 770°C (1045 K) without fusing the material. These changes have resulted in the following specific fuel consumption figures expressed as kg oil per tonne of calcine (Reduction Kiln Product):

1979	1984	Present
116	92	75

The Reduction Kilns have ample capacity to produce the necessary calcine for the Electric Furnaces and they operate in excess of 90 % availability.

Table 4. Process Plant Throughputs* and Assays of Key Streams

Process Streams	1000s dry tonnes/year	%					
		H_2O	Ni	Fe	S	SiO_2	MgO
Total Dryer Product	1910	20	2.0	19	-	36	20
Reduction Kiln Feed	2015	19	2.1	21	-	36	20
Electric Furnace Slag	1395		0.16	20	0.2	45	23.5
Electric Furnace Matte	108		32	57	10		
Matte Product	38		79	0.5	19.5		

* At 30000 tonnes Ni/year production rate

Off Gas and Dust Handling

The gas from the Dryers passes through multiclones to a stack and the dust is collected in a large bin. The gas from the Reduction Kilns first passes through multiclones and then through wet scrubbers before discharge to a stack. Initially the dry dust was combined and pneumatically conveyed to the feed end of the Dryers where it joined the wet feed. This resulted in increasing circulating dust loads. The scrubber slurry was originally placed in ponds for settling and the sediment was trucked back to the Dryer Feed Hopper. Both the dry dust and slurry systems needed improvement. In 1983 a thickener was installed in the scrubber circuit. The thickener underflow is mixed with dry dust in a pug mill and trucked to the Dryer Feed Hopper. It joins the new feed at this point and blends well with it.

The gas and dust handling system now requires less manpower and is more efficient. Energy savings have been effected and dust losses reduced.

Smelting

Smelting is carried out in three Electric Furnaces 18 m outside diameter and rated at 45 MVA. Based on pilot plant experience, there was a question as to the capability of the refractory sidewalls to withstand attack by the molten slag. The large furnace diameter selected and the relatively low power density were considered adequate to ensure slag freezing on the walls and protection of the refractories. However, very soon after start-up the highly corrosive nature of slags from the smelting of the West Block ore became evident. The refractory lining was rapidly eroded when operating at or near full power. Investigation showed that:

- The slag SiO_2/MgO ratio of 2.2 to 2.4 was too acidic for the basic MgO furnace refractory.
- The liquidus temperature of the slag was relatively low (see Figure 6).

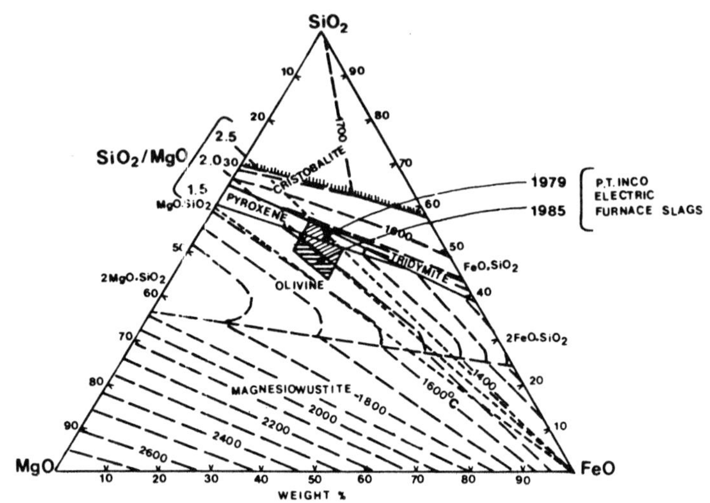

LIQUIDUS AT CONSTANT FeO SECTIONS

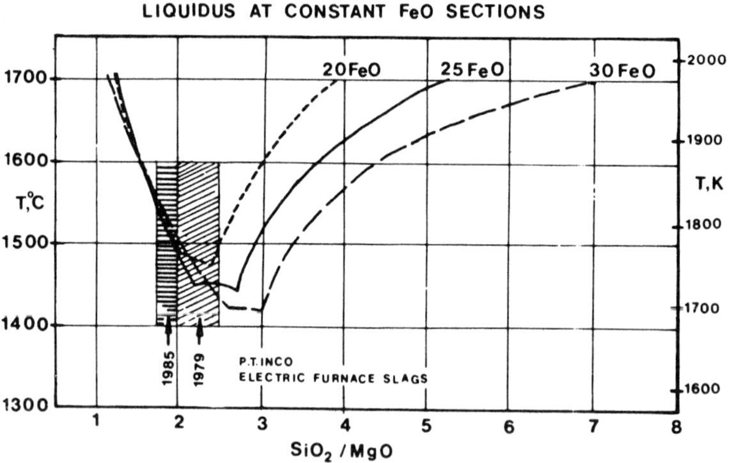

Fig. 6. P.T. INCO Electric Furnace Slag Composition Superimposed on the $FeO-MgO-SiO_2$ Phase Diagram

- A higher superheat was required to continually dissolve the coarse unaltered peridotite in the feed. This precluded the precipitation of a protective slag layer on the refractory walls.

Two major changes were made to ensure the establishment and maintenance of a protective slag coating on the side walls to prevent corrosion.

The acidity (SiO_2/MgO ratio) of the plant feed was reduced by opening the East Block mining area and blending increasing amounts of this more basic ore feed. Today the S/M ratio in the slag is held at about 1.9. The liquidus has increased some 50°C (50 K) and the coarse peridotite has been reduced (because of the smaller proportion of West Block ore) thereby lowering the slag superheat requirement. Skimming temperature has remained at about 1550°C (1825 K).

The falling film water cooling system on the outside of the steel sidewall was replaced in the slag zone with a system of water cooled copper fingers.

The greater proportion of East Block ore in the feed also improved electrical stability on the furnaces because the slag is now more conductive. The secondary current has been raised from 45 kA to 55 kA, on average.

Other improvements in furnace construction have included : additional cooling with copper blocks at the tapping modules and skimming points; reinforcement of the steel shell around the tapping modules; copper face plates at the blocks; protection of the underside of the water cooled roof beams with castable refractory to reduce roof heat losses; and spray water cooling of the furnace bottoms to provide a greater degree of operating flexibility.

The furnace matte tapping temperature has been increased from 1330°C to 1360°C (1600 to 1630 K) to reduce scrap generation and to lower fuel consumption in the converters.

When the first furnace went into operation, furnace slag was cast onto the ground, cooled, broken and trucked to disposal. An improved system using water jet granulation, with trucking to disposal, soon replaced the ground casting. When the plant was expanded to three production lines, a more efficient system was installed, using rubber tired vehicles to haul 17.5 m^3 ladles to disposal. Tilting tundishes were installed to permit continuous skimming.

Furnace power and energy consumption have been varied to optimize production rates and fuel costs. The partial substitution of oil with coal in the kilns has meant that more reduction has shifted to the furnaces. Power requirements have increased from 516-580 kWh/t noted in Reference 1 to 580-600 kWh/t today. The initial design provided for an average peak power of 36 MW. Today the furnaces average better than 40 MW.

Sulphur is introduced into the smelting process in two forms in order to obtain a nickel sulphide matte for shipment and sale. The high sulphur fuel oil used in the Reduction Kilns contributes a small proportion and the balance is added as elemental sulphur. Originally the elemental sulphur was introduced to the calcine at the kiln discharge. The escalating cost of sulphur and the relatively low pick-up efficiency have intensified the search for better methods of addition. A sulphur melter has been installed and a portion of the sulphur is now added in liquid form to the furnace matte during tapping. Pickup efficiency at this point is better than 80 %. Further improvements are being pursued to lower overall sulphur consumption.

Converting

The original three top blown rotary Kaldo-type converters (TBRC) were installed to provide maximum flexibility in selecting finished product forms. The top blow feature offered the possibility of providing a low sulphur final product. However, maintenance time and costs for refractory and mechanical running gear have been relatively high. With the nickel sulphide matte product well established as the desired product for the Japanese market, it was decided last year that a conventional side-blown Pierce-Smith converter (P.S.) offered operating advantages. In early 1985 one of the three TBRCs was removed and replaced with a P.S. converter.

The converter department is now capable of adequately and dependably handling current and foreseeable matte production requirements. The single Pierce-Smith converter, which went into operation in July can handle nearly all the furnace matte production. The two remaining TBRCs have been completely refurbished and are available for operation. One supplements the P.S. as required while the two on stream can handle the full production when the P.S. is being relined and is down for major maintenance. This arrangement ensures that converter costs will be lower and reliability will be higher.

Energy Utilization and Energy Costs

The improvements made to the Dryer and Reduction Kiln operations have resulted in significantly reduced consumption of fuel oil. Substitution of coal for fuel oil also reduced fuel oil consumption but increased energy consumption in the Electric Furnaces.

Improvements made in the converting practice resulted in reductions in fuel consumption for converting - particularly with the installation of the Pierce-Smith converter.

Improved mobile equipment utilization in the mine and support areas has also reduced diesel consumption.

The breakdown of energy consumption by source is given in Table 5 for the years 1980, 1984 and 1985 (first quarter). Table 4 also shows the energy utilization and cash costs for the same years.

Since energy is the single largest item of cost in the production of nickel at our plant the energy savings have led to significant unit cost reductions.

Table 5. Energy Utilization And Energy Costs

Year :	1980	1984	1985
(a) Distribution of Energy Usage (%)			
Hydro Electric Power*	25	30	32
Fuel Oil	73	64	57
Coal	2	6	11
Total	100	100	100
(b) Energy Utilization And Cost :			
Energy Utilization :			
GJ/MT Ni*	620	495	415
Index	100	80	67
Energy Cash Cost :			
$/kg Ni	1.74	1.50	1.25
Index	100	86	72

*Converting hydro electric energy at theoretical conversion of 3600 kJ/kWh.

Process Plant Throughput And Productivity

Process Plant throughput and manpower is summarized in Table 6. The projected increase in productivity is fourfold since 1979 and approximately twofold since 1983.

Table 6. Process Plant Throughput And Productivity*

Year :	1979	1983	1984	Current
Manpower	730	680	680	630
New Feed To Process Plant, Dry tonnes (1000's)	647	962	1327	1650
Product, 1000's tonnes Ni	9	18	23	30
Productivity :				
t Feed Processed/Man-year	880	1400	1960	2610
t Nickel produced/man year	12	27	34	48

*Annual basis.

CONCLUSION

Since 1979, when the P.T. INCO project was described in a paper at New Orleans, and when the project was in its first years of Operation, many significant changes have been made. As in most pioneering enterprises much had to be learned and many problems had to be solved. As this paper illurstrates, P.T. INCO has been no exception. Important progress has been made in the six years since commissioning.

The plant can produce consistently and dependably and the project is among the world's lowest cost nickel laterite projects. The workforce is compact and well trained, having been reduced from 3900 to 2750. Energy costs, with the benefit of the Larona Hydro Generating Station, are relatively low and have been reduced 30 %. Our infrastructure is complete and mature. Our contract with the Indonesian Government has been observed fully, and our relations with the Government are excellent. The Government is gradually taking over community services which had been provided by the Company. Our efforts are aimed at ensuring that P.T. INCO continues to provide benefits from this vast natural resource for our employees and the Indonesian economy, and with the expectation that the shareholders will receive a return on investment in the not too distant future. Customers here in Japan can look to P.T. INCO as a dependable long term supplier of nickel matte suitable for further refining and processing in Japan.

REFERENCES

(1) MUSU, R. and BELL, J.A.E. - P.T. INCO's Indonesian Nickel Project, International Laterite Symposium, New Orleans, Feb 19 - 21, 1979, AIME, 1979, pp 300 - 322.

(2) GOLIGHTLY, J.P. - Geology of Soroako Nickeliferous Laterite Deposits, International Laterite Symposium, New Orleans, Feb 19-21, 1979, AIME, 1979, pp 38-56.

(3) HARJU, H.O. - Exploration of P.T. INCO's Nickel Laterite Deposits in Sulawesi, Indonesia, International Laterite Symposium, New Orleans, Feb 19-21, 1979, AIME, 1979, pp. 292-299.

PROCESSING OF COMPLEX CONCENTRATES AND BY-PRODUCTS AT METALLURGIE HOBOKEN-OVERPELT

Robert Maes	Fons Lauwers	Luc Groothaert
MHO	MHO	MHO
Research Department	Raw Materials Purchase	Research Department
2710 Hoboken, Belgium	2710 Hoboken, Belgium	3583 Overpelt, Belgium

Abstract

Metallurgie Hoboken-Overpelt has evolved from the integration of three plants with several smelting and refining facilities. The processing of feed materials relies on the production circuits of three non-ferrous base metals - lead, copper and zinc - with major interflow of by-products containing several secondary metals. The flow-sheet of the company is reviewed, with particular emphasis on the interconnections between the different plants and on the possibilities of processing sulphide concentrates of varied origin.

For many years, a progressive diversification of the plant inputs has been taking place, impure and complex materials being introduced in increasing proportions, which has forced the company to recover a maximum of contained elements by developing the production of secondary metals. Production currently embraces more than twenty-five non-ferrous metals, of which precious metals occupy an outstanding place from an economic point of view.

Some perspectives of developments in the fields of smelting, refining of base metals and recovering of by-product metals are outlined.

This paper was previously published in the TMS Complex Sulfide Symposium on Complex Sulfides, Processing of Ores, Concentrates and By-Products, held at the TMS-AIME Fall Extractive Meeting, San Deigo, California, November 10-13, 1985.

Introduction

At the 1962 AIME Extractive Metallurgy Symposium a paper (1) was presented on the subject "The World's Most Complex Metallurgy (Copper, Lead and Zinc)". The possible complexity of the three base circuits of the lead-, copper- and zinc metallurgy was described and an integrated flow-sheet of the three circuits was proposed. It has been reproduced in fig. 1, which shows the main processing steps of the base circuits, with the possible locations for the extraction of some traditional by-product metals, recycling points to another circuit and their possible recovery in commercial form.

The aim of the present paper is to explain how, a quarter of a century later, Metallurgie Hoboken-Overpelt has succeeded in developing such an integrated flow-sheet on an industrial scale and how this enables us to process complex concentrates and by-products of varied origin.

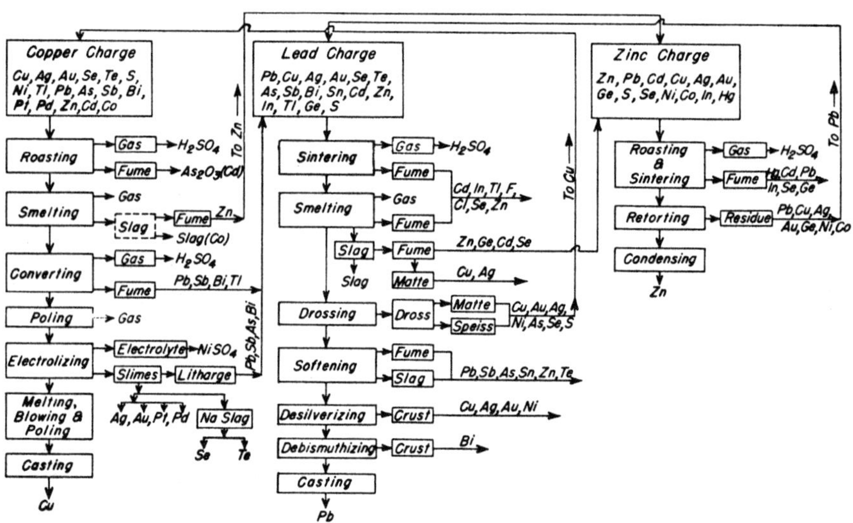

Fig. 1 Integrated copper, lead and zinc circuits (1)

Flow-sheet of Metallurgie Hoboken-Overpelt

Metallurgie Hoboken-Overpelt has evolved from the integration of three plants processing their feed materials following the production circuits of lead, copper and zinc, with an important interflow of by-products. Its flow-sheet is represented in fig. 2.

The Hoboken plant is a complex smelter devoted to a mixed Pb-Cu metallurgy complicated by the presence of several secondary metals. Some principles of this metallurgy have been described elsewhere (2).

The Olen plant is based on four main productions : Cu, Co, Ge and Ni. The Cu-circuit, which is the most important in tonnage, refines the company's blister together with various foreign blisters.

The Overpelt plant is mainly oriented towards the processing of zinc-bearing materials and scrap. The traditional hydrometallurgical circuit for blende treatment - roasting, leaching, electrowinning - constitutes an important part of the operations.

The Hoboken plant

The feed materials of the Hoboken plant consist mainly of lead-, copper- and precious metal-bearing concentrates and by-products. In addition to the base metals, these materials also contain a variety of associated metals, such as nickel, tin, zinc, bismuth, cobalt, arsenic, antimony, indium, cadmium, selenium, tellurium, etc.

Processing methods rely on a combination of the known processes used in lead and copper extraction and consist of the sequence "sinter roasting - reduction smelting - converting". Partial descriptions of the operations have been given in earlier publications (3, 4, 5, 6, 7, 8).

The fine sulphur-bearing materials - concentrates and lead sulphate residues - are introduced at the roasting and sintering stages. Sinter is smelted and reduced in blast-furnaces, together with various other materials, mostly non-sulfidic. Due to the complexity of the charges and the number of metals which they contain, smelting regularly produces four liquid phases, which can be separated from each other : slag, matte, speiss and lead bullion. The slag is discarded after being processed in a slag cleaning furnace. The matte is blown in siphon converters in order to produce blister and a slag which is returned to the blast furnaces. Speiss is processed by hydrometallurgy in order to extract the valuable metals.

Lead bullion is forwarded to the lead refinery which operates a Harris process plant. Precious metals, which are concentrated in various by-products of the different plants of the company, are separated in a refinery operating a nitric acid cycle. Several refining plants for special metals produce elements such as selenium, tellurium, arsenic, antimony, bismuth and indium, either in elemental form, as salts or, possibly, as saleable intermediate products.

The Olen plant

The Cu-circuit of the Olen plant is a refinery operating on blister and copper-bearing scrap. It comprises the steps of anode casting, electrorefining and casting of commercial shapes. Descriptions of these operations have been given elsewhere (9, 10, 11, 12).

The Ni-circuit processes the various nickel-bearing by-products of the company by hydrometallurgy and extracts the nickel as a sulphate.

The Co-circuit processes cobalt-bearing materials, scrap and by-products by hydrometallurgy and produces this metal in various commercial forms : oxides, salts and metal powders. A description of some of these operations has been given earlier (13).

A Ge-circuit extracts and refines this element from germanium-bearing materials and scrap.

The Overpelt plant

The operations of the Overpelt plant consist principally of a circuit for processing blende by roasting, leaching and electrowinning. These operations have been described in earlier papers (14, 15). The flow-sheet also includes a cadmium electrowinning plant.

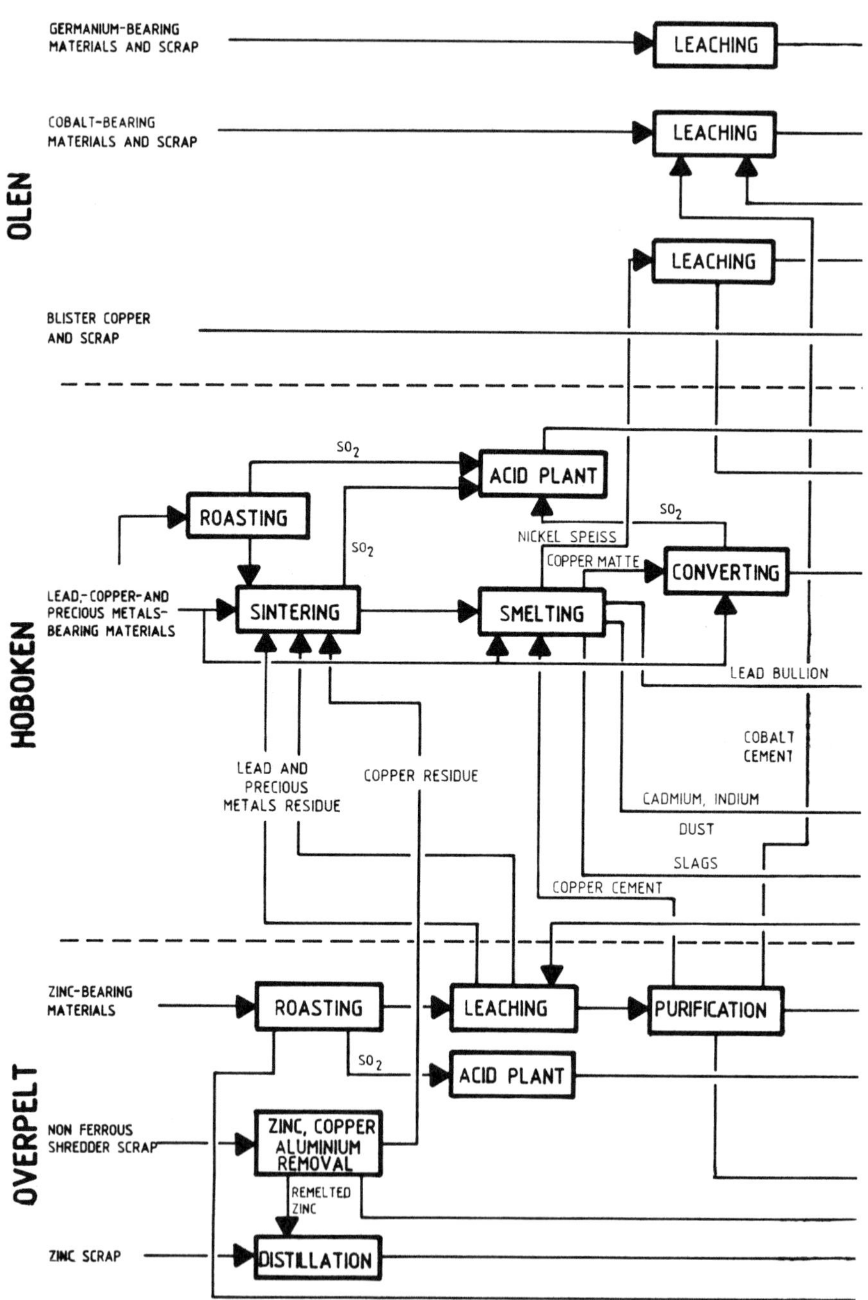

Fig. 2 Flow-sheet of Metallurgie Hoboken-Overpelt

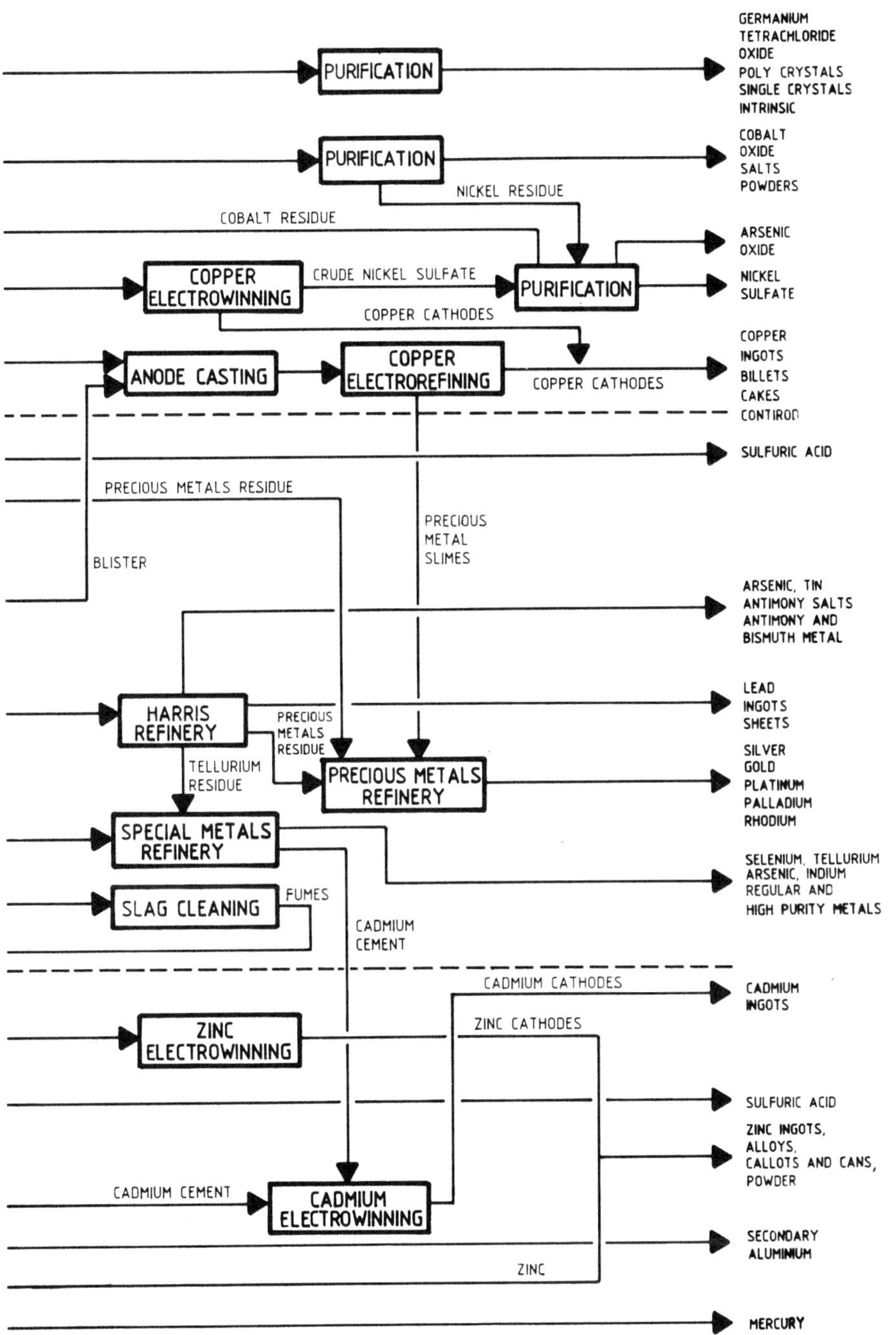

PURIFICATION
GERMANIUM
TETRACHLORIDE
OXIDE
POLY CRYSTALS
SINGLE CRYSTALS
INTRINSIC
PURIFICATION
COBALT
OXIDE
SALTS
POWDERS
NICKEL RESIDUE
COBALT RESIDUE
ARSENIC
OXIDE
COPPER ELECTROWINNING
CRUDE NICKEL SULFATE
PURIFICATION
NICKEL
SULFATE
COPPER CATHODES
ANODE CASTING
COPPER ELECTROREFINING
COPPER CATHODES
COPPER
INGOTS
BILLETS
CAKES
CONTIROD
SULFURIC ACID
PRECIOUS METALS RESIDUE
PRECIOUS
METAL
SLIMES
BLISTER
ARSENIC, TIN
ANTIMONY SALTS
ANTIMONY AND
BISMUTH METAL
HARRIS REFINERY
LEAD
INGOTS
SHEETS
PRECIOUS
METALS
RESIDUE
TELLURIUM
RESIDUE
PRECIOUS METALS REFINERY
SILVER
GOLD
PLATINUM
PALLADIUM
RHODIUM
SPECIAL METALS REFINERY
SELENIUM, TELLURIUM
ARSENIC, INDIUM
REGULAR AND
HIGH PURITY METALS
SLAG CLEANING
FUMES
CADMIUM
CEMENT
CADMIUM CATHODES
CADMIUM
INGOTS
ZINC ELECTROWINNING
ZINC CATHODES
SULFURIC ACID
ZINC INGOTS,
ALLOYS,
CALLOTS AND CANS,
POWDER
CADMIUM CEMENT
CADMIUM ELECTROWINNING
SECONDARY
ALUMINIUM
ZINC
MERCURY

Another part of the plant feed materials consists of non-ferrous shredder scrap, which is separated into four fractions collecting, respectively, copper, aluminium, stainless steel and zinc. The latter is fed to refining distillation columns, together with other zinc-bearing materials.

In addition to zinc, the plant produces cadmium, mercury and clean aluminium scrap, as saleable by-product metals.

Interflow of by-products

It is hardly possible to give an exhaustive list of materials interflowing between the three plants. The most important circulations are shown in the flow-sheet of fig. 2 and can be summarized as follows :

- most of the complex by-products from the Olen and Overpelt plants are returned to one of the feed points of the Hoboken smelter
- all the precious metal concentrates - leaching residues and tankhouse slimes - are sent to the precious metals refinery at Hoboken
- nickel-, cobalt- and germanium-bearing materials are sent to the corresponding circuits of the Olen plant
- zinc-bearing materials (particularly the slag cleaning fumes of Hoboken) and cadmium- and mercury-bearing products are sent to the Overpelt plant
- lead sulphate residues from the Overpelt electrolytic plant are returned to the Hoboken sinter plant
- the Hoboken blister is refined at Olen and the Olen refining slags returned to Hoboken.

Range of metals produced and processing flexibility

If the sulphur recovered as sulphuric acid is added to the other elements produced by the company in one or another commercial form - metal, alloy, oxide or salt - a total of 27 recovered elements results. They are indexed in the Periodic Table of fig. 3.

This variety, together with the complexity of the flow-sheet, provides flexibility to process the most varied materials, as well as intermediate smelter by-products similar to those of our own flow-sheet as primary materials of complex composition. Some examples will illustrate this in the following paragraphs for the Hoboken smelter and the Overpelt plant.

Feed materials of the Hoboken smelter

The Hoboken smelter processes about 300,000 tonnes of raw materials per year. Only one third of these are concentrates, whose compositions vary over a wide range. In general, Hoboken does not process simple Pb-concentrates, but rather more complex Pb Cu-concentrates containing precious metals, As, Sb, Bi, etc.; the As and Sb contents can reach 5 and 15 % respectively. One of the important features of the smelter is thus that it produces a very impure lead bullion which is further processed in the Harris refinery, capable of removing up to 500 tonnes of As + Sb + Sn per month.

An average composition of the feed concentrates is the following :

Ag :	3,000 g/t	Sb :	2 %
Au :	10 g/t	Bi :	0.2 %
Pb :	40 %	Zn :	5 %
Cu :	4 %	Fe :	10 %
As :	1 %	S :	23 %

IA	IIA	IIIB	IVB	VB	VIB	VIIB		VIII		IB	IIB	IIIA	IVA	VA	VIA	VIIA	INERT GASES
H																	He
Li	Be											B	C	N	O	F	Ne
Na	Mg											Al	Si	P	S	Cl	A
K	Ca	Sc	Ti	V	Cr	Mn	Fe	Co	Ni	Cu	Zn	Ga	Ge	As	Se	Br	Kr
Rb	Sr	Y	Zr	Cb	Mo	Tc	Ru	Rh	Pd	Ag	Cd	In	Sn	Sb	Te	I	Xe
Cs	Ba	La	Hf	Ta	W	Re	Os	Ir	Pt	Au	Hg	Tl	Pb	Bi	Po	At	Rn
Fr	Ra																
			Ce	Pr	Nd	Pm	Sm	Eu	Gd	Tb	Dy	Ho	Er	Tm	Yb	Lu	
			Th	Pa	U	Np	Pu	Am	Cm	Bk	Cf	Es	Fm	Md	No	Lw	

Fig. 3 Range of metals produced by the Company

In addition to these concentrates, a large proportion of by-product materials is processed, most of which originate from foreign non-ferrous operations :

- Pb smelter by-products : - dross
  - Pb Cu matte
  - speiss
  - slags
  - Pb bullion
- Cu plant by-products : - black Cu
  - tankhouse slimes
  - Cu slags
- Electrolytic Zn plant residues : - Pb sulphate residues
  - Cu cements

Another important feature of the smelter feed consists in the large proportion of precious metal-bearing materials, which can be efficiently processed owing to the availability of a Pb-circuit, a Cu-circuit and a fast circuit for materials with higher P.M.-content. In addition to P.M.-bearing concentrates and tankhouse slimes, the smelter processes large quantities of sweeps, bullions, catalysts, P.M.-scrap, film ashes and complex residues containing Se, Te and In.

The percentage distribution of seven elements in the various types of materials over the last five years is given in the following table :

	Ag	Au	Pb	Cu	As	Sb	Bi
Concentrates	20	7	35	13	35	52	47
Pb-bullions	4	-	23	-	1	5	26
Black Cu	-	-	-	5	-	-	-
Cu-scrap	2	7	-	7	-	-	-
Pb Cu-matte	3	5	10	22	14	7	2
Dross	3	-	16	7	10	9	4
Speiss	-	-	-	2	12	3	1
Tankhouse slimes	9	16	-	1	2	4	3
Slags	-	-	1	10	4	6	2
PM-Cu base bullions	8	23	-	6	-	-	-
Sweeps	4	14	-	-	-	-	-
Cu-cement	-	-	-	7	2	-	-
Pb-sulphate	4	1	7	1	3	6	5
Doré and coins	13	10	-	-	-	-	-
From Olen plant	13	11	1	13	5	3	2
From Overpelt plant	3	-	4	4	7	2	2
Miscellaneous	14	6	3	2	5	3	6
	100	100	100	100	100	100	100

Feed materials of the Overpelt plant

The zinc-bearing feed of the Overpelt plant also consists of a wide range of raw materials and by-products, amongst which a distinction can be made between sulphidic, oxidized and metallic materials.

Owing to the characteristics of the roasting process developed at Overpelt, which enables zinc-bearing concentrates containing up to 10 % Pb and 5 % Cu to be fed, some quantities of less traditional concentrates can be processed, as illustrated by the following table :

	Zn, %	Pb, %	Cu, %	Ag, g/t	Hg, g/t
Traditional blende	50	2	0,25	120	
Less traditional Zn concentrates	40-50	2-4	0,2-3	300-2000	
Bulk concentrates	30	18	5	800-1200	1700

By-products of oxidized type can be fed either into the roasting furnaces or directly into the hydrometallurgical circuit, according to their degree of purity and, more particularly, their halogen content. The following table gives some examples of products which we are attempting to introduce in increasing proportions in the feeds :

	Zn, %	Pb, %	Cu, %	Cd, %	Sn, %	Ag, g/t	Cl, %
Slag cleaning fumes	54	16	0,15	0,2	0,8	40	0,6
Smelting fumes	30-50	8-16	3-7	0,4-0,9	0,6-2	25-600	2-3
Zinc ashes	60-70						3

Metallic materials are introduced into the scrap processing plant and the distillation refining units. The following materials are, for instance, suitable for these circuits :

- non-ferrous shredder scrap : this is the non-ferrous fraction, containing about 40 to 50 % aluminium alloy and 20 to 25 % zinc, copper and stainless steel alloys
- zinc scrap such as old roofing zinc with 98 % Zn and 2 % various impurities
- shot separated from zinc ashes, with 85 to 90 % Zn.

In addition to these operations, the cadmium plant processes various foreign Cd-bearing residues, together with the cements and by-products produced in the company.

Ancillary sampling and assaying installations

It is worth emphasizing that the treatment of about 9,000 lots per year of often highly valuable materials of most varied origin has led Metallurgie Hoboken-Overpelt to expand considerably its assaying and sampling facilities. These at present employ about 400 people.

The most complex sampling problems are being surmounted. They are carefully examined, if necessary in co-operation with the supplier and possibly even with the assistance of our R & D departments.

Some perspectives of developments

The main lines of development of our productions are related, on the one hand, to the processes applied and, on the other, to the resulting products. For the latter, a permanent effort of diversification is maintained in order to meet demands from the consumer markets. Selenium, tellurium and arsenic, for instance, are produced in high purity grades for reprographic uses. Germanium is produced in special shapes for infrared optics and in a 12 nines grade for radiation detectors. Zinc illustrates for instance the case of a base metal whose production is diversified into ingots, alloys, amalgamated powders, slugs and cans.

As regards the extraction and refining processes, our efforts are concentrated on their continual adaptation to the increasing complexity of the feed materials and on the investigation of new, more efficient processes. Without attempting to be exhaustive, here are some development directions on which our attention will be focused in the following years :

- the investigation of modern smelting processes, such as autogenous smelting processes for concentrates
- the development of continuous and automated processes for the various steps of copper refining, in the traditions of the Olen refinery
- the adaptation of the zinc electrolytic process to increasingly complex feed materials and the development of new pyro- as well as hydro-metallurgical processes for treating zinc-bearing materials
- the extension of the precious metals refining capacity and the development of new, more efficient techniques for concentrating and separating these metals
- the development of a processing circuit for complex cobalt-bearing materials such as by-products and scrap.

Acknowledgement

The authors gratefully acknowledge the General Management of Metallurgie Hoboken-Overpelt for permission to publish this paper. The Belgian Institute for the Promotion of Scientific Research in Industry and Agriculture (IWONL) is thanked for financial support of several developments described in this paper.

References

1. Phillips A.J., "The World's Most Complex Metallurgy", Transactions of the Metallurgical Society of AIME, vol. 224, August 1962, pp. 657-68

2. Fontainas L., Coussement M. and Maes R., "Some Metallurgical Principles in the Smelting of Complex Materials", Proceedings of the Symposium on Complex Metallurgy, The Institution of Mining and Metallurgy, Bad Harzburg, September 1978, pp. 13-23

3. Leroy J.L., Lenoir P.J. and Escoyez L.E., "Lead Smelter Operation at N.V. Metallurgie Hoboken S.A.", Extractive Metallurgy of Lead and Zinc, Cotterill C.H. and Cigan J.M. eds. (New York, AIME 1970), pp. 824-52

4. Coekelbergs C. and Delvaux A.L., "Lead Sulfates Processing at Metallurgie Hoboken-Overpelt", 108th AIME Meeting, New Orleans, February 1979, AIME-TMS Paper n° A-79-33

5. Leroy J.L. and Lenoir P.J., "Hoboken Type of Copper Converter and its Operation", Proceedings of the Symposium on Advances in Extractive Metallurgy, The Institution of Mining and Metallurgy, London, April 1967, pp. 333-43

6. De Keyser J. and Jespers W., "The Harris Refinery of Metallurgie Hoboken-Overpelt", 110the AIME Meeting, Chicago, February 1981, AIME-TMS Paper n° A-81-10

7. Tougarinoff B., Van Goetsenhoven F. and Dewulf A., "Recovery by a Nitric Acid Cycle of Gold and Platinum Metals from the Anode Slimes Arising from the Electrolysis of Doré Metal", Proceedings of the Symposium on Advances in Extractive Metallurgy, The Institution of Mining and Metallurgy, London, April 1967, pp. 741-58

8. Maes R., "Fusion en Deux Etapes de Matières Plombifères Complexes", Metallurgie, vol. 22, n° 4, 1982, pp. 211-16

9. Lecointe G., "La Raffinerie Electrolytique de Cuivre d'Olen", Cuivre, Laitons, Alliages, N° 1, 1971, pp. 10-14

10. Dompas J.M., "Continuous Casting of Copper Products at the Olen Refinery", 107th AIME Meeting, Denver, February 1978, AIME-TMS Paper n° A-78-20

11. Dompas J.M., "Continuous Casting of Anodes with Integral Lugs for Electrolytic Copper Refining", 109th AIME Meeting, Las Vegas, February 1980, AIME-TMS Paper n° A-80-36

12. Dompas J., Govaerts M. and Hens K., "Anode and Tankhouse Developments at MHO Olen Works", Journal of Metals, vol. 37, Januari 1985, pp. 44-46

13. Van Peteghem A., Willekens M. and Tougarinoff B., "Chemische Verfahren zur Verarbeitung von Nickel- und Kobalthaltigen Vorstoffen auf Chloridischer Weg", Proceedings of the Nickel Symposium, Gesellschaft Deutscher Metallhütten- und Bergleute, Wiesbaden, September 1970, pp. 147-54

14. Denoiseux R., Winand R., Willekens H. and Vos L., "Metallurgie Hoboken-Overpelt Process for Roasting Zinc Concentrates in a Fluid Bed", Lead-Zinc-Tin '80, Cigan J.M., Mackey T.S. and O'Keefe T.J. eds (Las Vegas, AIME 1980), pp. 69-84

15. Van den Neste E., "Metallurgie Hoboken-Overpelt's Zinc Electrowinning Plant", 6th Annual CIM Hydrometallurgical Meeting, CIM Bulletin, August 1977, pp. 173-85

EXTRACTION PROCESSES FOR METAL RECOVERY FROM COMPLEX ORES AND BY-PRODUCTS AT BOLIDEN METALL

Arne A. Björnberg, Göran Lindkvist and Åke Holmström

Boliden Metall AB, 932 00 Skelleftehamn, Sweden

ABSTRACT

The complex nature of the sulphidic ores of the Skellefte field has made the capacity to treat a large variety of raw materials essential for the Boliden Rönnskär smelter. The important components of the plant's processes are:

Fluidized bed roasting for As and Sb elimination/reduction.
Electric smelting furnace (19 MVA) giving maximum independence on energy content and composition of raw materials.
Two TBRC converters, one for sulphidic and oxidic copper raw materials and one for lead dusts from the zinc clinker and copper PS converter furnaces.
Zinc fuming furnace for the recovery of Zn from copper and lead slags.
A copper electrorefining process well tuned to the use of anodes with high precious metal and impurity content.

The different process steps will be presented and discussed.

INTRODUCTION

The Rönnskär Smelter located in the northern part of Sweden has during the last five years strongly increased its production of copper and lead as well as by-products. This has been done in the existing plant without any increased smelting capacity.

The production has been as follows (ton)

	1980	1981	1984
Copper incl. blister	56 500	74 000	103 000
Lead incl. crude	47 000	22 000	66 000
Gold tot.	4.5	4.6	7.6
Silver tot.	242	213	305
Selenium	51	45	68
Zinc clinker	18 000	24 000	29 400
Sulphuric acid	129 000	167 000	244 000
Liquid SO_2	48 000	53 000	67 000

This means an increased production of about 50% since 1980/81.

RAW MATERIALS

The production of copper is based on three types of raw materials: domestic complex copper concentrates from Boliden's own mines, imported copper concentrates with precious metals and impurities like arsenic and finally secondary materials such as ashes and scrap.

The domestic concentrates are produced in eight ore dressing plants in the middle and northern part of Sweden as showed in figure 1. These concentrates are transported by rail-way to the smelter and mixed together with imported concentrates prior to processing. Normally the mix consists of twenty to thirty different concentrates.

Scrap and ashes are mainly imported from western European countries and North America. The concentrates and secondary materials differ very widely in analyses but a typical charge is presented in table I.

Table I. Composition of charge to the copper plant.

	% Cu	% Fe	% Zn	% Pb	% As	% S	g/t Ag
Concentrate	20	22	3-5	1-3	0.4-3.0	30	500
Secondary materials	18	13	18-20	1-3	0.5	<5	500

Compared to most other copper smelter's charges the contents of zinc, lead and arsenic are considerably higher in the Rönnskär charge.

BASIC PROCESSES USED IN THE COPPER PLANT

The basic flowsheet for the copper plant is shown in figure 2.

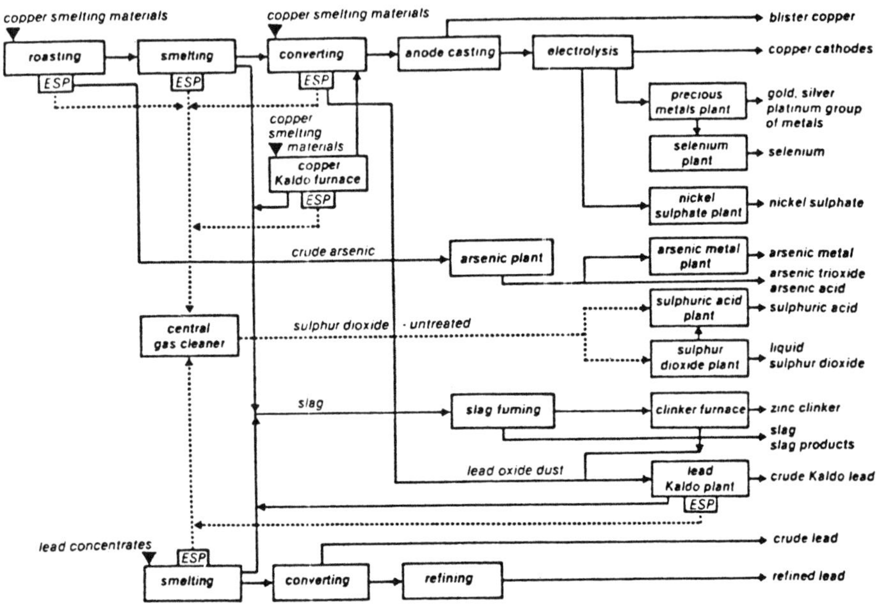

The main production line consists of roasting of concentrate and drying of secondary materials prior to matte smelting in the electric furnace. Besides that, matte is also produced in the TBRC plant by autogenous smelting of concentrates and scrap with oxygen. All copper matte is converted to blister copper in two conventional PS-converters and reduced to anode copper by liquid ammonia in the anode furnace.

Slag from the converters are recirculated to the electric furnace and all smelting slags are treated in the fuming plant for the recovery of zinc and lead.

When the two matte production lines, the roaster - electric furnace line and the TBRC plant are compared, the following can be concluded: the roasting and electric smelting operation are most suitable for the treatment of very dirty arsenic and antimony rich concentrates and secondary materials while the TBRC plant is most suitable for the treatment of fine grained and lead-zinc rich concentrates.

TREATMENT OF COMPLEX RAW MATERIALS

The main impurities in the copper plant are arsenic, antimony, bismuth, lead and nickel, and the most important by-products are gold, silver, zinc and selenium. To optimize the operations a computor model calculates on a ten day basis the mixture of the roaster and the dryer charges considering actual plant capacities and raw material stock and supply.

Roasting and smelting

The roasting is done in a Dorr Oliver fluidized bed roaster charged with copper concentrates containing various amounts of impurities. During the roasting operation arsenic, partly antimony and sulphur are removed in order to produce a clean calcine for matte smelting.

In the dryers all kinds of copper and zinc containing oxides, residues and scrap are dried prior to smelting with calcine in the electric furnace. Here a matte phase with 45% Cu is produced together with a zinc rich fayalite slag. Table II gives the roasting and smelting operations including the TBRC plant.

Table II. Production data for the roaster, the electric furnace and the TBRC plant

	Prod.	Analyses							
		% Cu	% Fe	% As	% S	% Pb	% Zn	% CaO	% SiO_2
Roaster	35 t/h	24	26	0.20	20	1.1	3.2	0.7	14
Dryer	15 t/h	18	13	0.5	3	2	19	3.7	16
Electric furnace	1100 t/d								
matte		45	18	0.25	22	3.8	4.2	-	-
slag		1.0	34	0.18	-	1.2	9.0	1.6	32
TBRC	450 t/d								
matte		50	15	0.14	23	3.3	4.2	-	-
slag		1.0	33	0.12	-	1.0	9.4	1.5	32

Converting

The matte is converted to blister copper in conventional PS-converters. Besides matte, the converters are also charged with large amounts of scrap mainly burned cable, electronic and telcom. scrap containing copper and precious metals. The burning of scrap is done in the lead TBRC-plant prior to charging to the electric furnace and converters.

Due to high amounts of lead and antimony the blister copper is overblown in order to oxidize the impurities. This also reduces the sulphur to a very low level. Blister copper is transferred to the anode furnaces and reduced by liquid ammonia to anode copper. Table III shows the composition of converter slag and blister copper produced in the converter.

Table III. Converter products.

	% Cu	% Pb	% Zn	% Fe	% As	% Sb	ppm Bi	ppm Ag	ppm Au
Converter slag	4.3	7.0	7.4	35.7	0.07	0.14	-	-	-
Blister copper	98.5	0.11	-	-	0.20	0.04	180	3 000	70

The anode copper is cast into anodes weighing 305 kg which are refined in the electrolytic refinery for the production of high purity cathodes. The precious metal containing slime is used as raw material for the gold and silver production.

Copper refining

The blister anodes are refined in a conventional tankhouse with 14 groups of tanks DC and 4 groups PRC. The total refining capacity is 63 000 tons. Anode spacing is 120 mm and DC current density 240 A/m^2.

Slag treatment

Slag from the electric furnace and the TBRC contain 8-10% Zn and 1-3% Pb. All the slag is treated in 100 ton charges in the fuming plant for the recovery of zinc and lead. This is done by reduction of the slag with coal injection and vaporization of zinc and lead fume. The fume is oxidized and the oxides are precipitated in the electric precipitators. The oxides are further treated in a rotary kiln, the clinker kiln, for the vaporization of halogens and lead oxide.

After fuming the slag is tapped to a settling furnace and then granulated and dumped. A copper matte is tapped intermittently from the settler thus increasing the recoveries of copper and precious metals. Table IV shows the analyses of the dump slag, matte and zinc oxide produced in the fuming and clinker plant.

Table IV. Products from the fuming plant.

	% Cu	% Pb	% Zn	% Fe	% SiO_2	% As	% Sb
Dump slag	0.56	0.03	1.3	35.2	37	0.01	0.01
Matte	33.5	0.15	0.95	30.4	-	2.0	1.0
Zinc oxide	0.10	6.0	73.0	0.34	0.8	0.2	0.3

Dust treatment

The processing of complex raw materials creates large amounts of intermediate products and dusts. The arsenic rich dust collected at low temperature from the roaster gas is treated for production of arsenictrioxide and arsenic metal.

During the converting of matte a lead and zinc rich dust is recovered in the low temperature electric precipitators after the converters. When the crude zinc oxide from the fuming plant is deleaded and dehalogenized in the clinker kiln a very contaminated oxide high in lead is recovered in the precipitator. The dusts from the copper converters and the clinker plant are pelletized and used as raw material in the lead TBRC plant for crude lead production. Table V shows the composition of the dusts utilized for the production of crude lead.

Table V. Lead TBRC: Raw materials and products.

	% Cu	% Zn	% Pb	% As	% Bi	% Sb
Converter dust	0.6	11.1	43.1	3.3	0.65	0.22
Clinker plant dust	0.1	10.4	57.0	4.2	0.10	0.26
Crude lead	0.07	-	97.6	0.13	0.60	0.75

Pretreatment of scrap

As secondary materials play an increasingly more important role as a raw material source for the smelter the pretreatment of scrap containing organic materials is an important process.

If organic materials are treated directly in the copper plant hazardous explosions can occur and the sulphuric acid produced from the smelter gases will be coloured black. To avoid this the scrap containing organics such as plastic, rubber or oil is burnt in the lead TBRC. This is done in the way that the scrap first is heated and then burnt with oxygen. Hydrocarbons released are combusted in the hood and gas duct and hydrochloric acid is neutralized in the venturi scrubber. Hence no acid, hydrocarbons or dust are released to the atmosphere. The burnt scrap can after this treatment and screening be charged directly to the converters and the electric furnace depending on the particle size.

RÖNNSKÄR IN THE NEAR FUTURE

As the production of copper and therefore also by-products such as sulphur and precious metals has increased a great need exists to expand the capacity of sulphur fixation and copper refining. In the end of 1985 a double contact sulphuric acid plant will be installed and in the beginning of 1986 the extended copper refinery will be started up. The refined copper production will have increased by 50% from 63 000 to 92 000 tons of cathodes by the end of 1986. At the same time the precious metals plant will be modernized and expanded to meet the increased output of anode slime. The dryers have been exchanged 1985 and an extensive program is planned for the modernization of the electric furnace and fuming plant.

Intensive R & D activities are performed in the areas: pretreatment of concentrates, dust leaching and refining of matte and blister which can lead to plant modifications and additions within the next decade.

In combination with an increasingly more flexibel process in terms of possibilities for treatment of dirty and complex raw materials the Rönnskär Smelter will be prepared for the needs and demands in the 1990's.

RECENT DEVELOPMENTS IN THE BOLIDEN LEAD KALDO PLANT

L. Hedlund, L. Johansson, Boliden Metall AB, S-93200 Skelleftehamn, Sweden

Abstract

Boliden's Kaldo Plant was built in 1976 to treat dusts from the copper plant containing Pb, Zn, Ag, Cu, Sn, Sb, As, Cl and F. Since the start-up the process has been modified several times and adjusted successfully to overcome problems such as foaming slag, high consumption of bricks etc. Consumption of oil has also decreased. In the paper the flexibility of our Kaldo plant is illustrated.

At the start-up of the plant, large amounts of dusts were stored, but these stocks were consumed after a few years. New materials and processes have therefore been tested and developed. Different lead bearing materials such as leach residues, batteries and various concentrates can be treated in the plant. At the moment, the most interesting issue seems to be complex lead concentrates. Many of the complex lead concentrates can be smelted together with our lead dusts, thus saving oil. For a test 2000 tons of such a concentrate were smelted in October 1984.

A completely different application is to burn copper scrap in the lead Kaldo furnace. Material with a large proportion of organics and chlorine (halides) can be burned, with high recoveries of copper and precious metals.

The plant meets up to the stringent in-plant and environmental legislation of Sweden.

Introduction

The Rönnskär Smelter of Boliden Metall AB is situated in the north of Sweden, 800 km north of Stockholm. The operations began in 1930 with treating of the very complex ore from the Boliden mine. The development has led to treatment of complex concentrates with precious metals, lead, zinc clinker, arsenic, sulfuric acid and liquid SO_2 as important by-products. Since 1945 lead is also produced in a separate lead-line.

The copper plant consist of an electric furnace, fed by calcine from a fluid-bed roaster and dried secondary materials. The slag is treated for recovery of zinc, lead and copper in a slag-fuming furnace. The matte is converted in PS converters. A Copper Kaldo Plant, erected in 1978 also delivers matte to the converters. The tankhouse has a capacity of 63 000 tons a year, but it is under expansion to 92 000 tons a year.

The copper concentrates treated contain considerable amounts of lead and zinc and hence 12 000 tons of lead-bearing by-products are produced every year. In order to treat these very complex materials a Lead Kaldo Plant was erected in 1976.

Smelting of lead concentrates started up in a separate lead-line 1945. The production has increased significantly the last years, since the lead plant was modernized in 1981. The process was designed by Boliden in 1963 (1). The concenrates are fed directly into the furnace and oxidized by oxygen-enriched air (25% O_2). Additional heat is provided by four electrodes with a capacity of 8 MW. The slag with 4% Pb is transferred to the fuming plant, and the lead with 2% S has to be converted in small PS converters before cooling to refining temperatures.

Treatment of lead-bearing dusts

Lead-bearing dusts from the copper converters and the clinker furnace were in the 60's sold to other lead producers. Penalties for arsenic, antimony, cadmium and halides (table I) soon made the sales unprofitable.

Table I. Compositions of materials currently treated in the Lead Kaldo Plant

	Dust from copper converters %	Dust from zinc clinker plant %	Residue from zinc clinker leaching %
Pb	43,1	50,0	48,9
Zn	11,1	14,7	5,0
As	3,3	7,4	2,2
Sb	0,22	0,43	0,95
Sn	0,81	1,21	1,96
Cu	0,6	0,1	0,12
S	9,9	6,4	9,2
Cl	0,61	1,46	0,06
F	0,022	0,7	0,1
Cd	0,73	0,42	

It was necessary to develop a process of our own to treat these materials.

In 1972 the first trials were conducted of the Mefos 5-ton Kaldo furnace in Luleå. The tests indicated that the Kaldo furnace (a top-blown rotary converter) was a suitable unit for the smelting of these materials.

Plant description

The dusts were stored in a separate buildning and via daybin and chargebin charged to the furnace by drag chains. Fluxes, iron and coke are charged from separate bins above the furnace.

The furnace is totally enclosed by a ventilated hood. Inside diameter of the furnace is 3,6 m and the length is 6,5 m. During operation the furnace inclination was first 22 degreees, later on changed to 28 degrees to give larger working volume. Maximum rotation speed is now 15 rpm.

Heat is supplied by an oxygen/oil burner in a watercooled lance.

The offgas hood is water cooled and reinforced by a basic castable.

For the gas cleaning a venturi scrubber is used, operating at 2 000 mm pressure drop. Venturi water flows to a thickener for settling. The sludge is filtrated, and the filter cake dried in a oil heated drum dryer and then returned to the process.

Lead is cooled in ladles before pouring to kettles and cast into 2,5 ton blocks from the kettles. (Figure 1.)

After several test-campaigns the full scale process was designed as follows (figure 2).

Pelletized dust, granulated slag from the fuming furnace, soda, lime and dross from the lead kettles were charged batchwise to the furnace. Heat is supplied by an oxygen/oil burner. During the melting the lead sulphates decomposes to lead oxide and SO_2. After melting coke was charged continously for the reduction of the lead oxide in the slag. When a lead content of less than 1% was reached, the slag was skimmed into ladles and transferred to the fuming plant. The lead was collected in a kettle. When 60-90 tons of lead were collected in the kettle, the lead was pumped to the Kaldo furnace for refining. In pilot scale it was easy to get a selective oxidation of tin to a dry dross that was pulled out of the furnace before the oxidation was continued to produce a fluid arsenic-antimony dross.

Process modifications

The time and labour consuming oxidizing refining method was discontinued quite soon. Instead iron was added to form a iron-arsenic-speiss. The fluid speiss was easy to skim off to a ladle and could be discarded after solidification. Instead of collecting lead in a kettle and pumping it back, it was collected in the furnace and refined after 2-3 heats, then tapped in a ladle.

Minor modifications of the slag composition made it possible to operate without the expensive soda. (Table II)

Table II. Products from the Lead Kaldo Plant.

	Slag	Crude Lead	Speiss
Pb	1,5	96-97	3,5
Zn	15		
Fe	22		50
As	0,5	<0,1	30
Cu	<0,1	<0,1	12
Sb	<0,1	0,7	
Bi	<0,1	0,3-1,5	
SiO_2	25		
CaO	23		

The pneumatic feed of coke for the reduction was unreliable, and batchwise charging of coke was introduced.

The most serious remaining problem was foaming slag. The reaction between coke and lead oxide evolves gas, and if the viscosity of the slag is too high, bubbles will be trapped in the slag and cause foaming. During melting the fayalite slag from the fuming plant oxidized to magnetite. This was avoided by adding the slag at the end of the melting period.

Now the process ran smoothly, except for the wet gas cleaning system. Variations in the raw materials caused variations in the dust and thus the pH in venturi water was unstable. At the end settling conditions in the thickener and properties of slurry were almost out of control. The rotary drumfilter could not take this, and large amounts of lead were lost with filtrate and overflow water. A change to a press filter was a big improvement. It does not operate continuously, but it always gives a good separation and a good filter cake.

During mid -82 the refractory wear increased. This coincided with higher lead content in the dust treated. Tests were done to prereduce the charge to bring down the lead content of the slag. Quite surprising it was found out that all the coke could be added at the start of the charge. The reduction would start before melting, and when the melt was fluid, it was also reduced. This method was also a lot faster than the previous 2-step process.

The theory was that the melting should be oxidizing to avoid PbS-evaporation, which had been experienced during the pilot plant tests. When one old truth was found to be false, others were to be tested. How should the burner be run? Oxidizing, neutral or reducing? The answer was unexpected: Oxidizing burner is best for reduction. The melting time became shorter and oil consumption lower without increased coke consumption.

The refractory wear was caused by the refining step. At the beginning sponge iron was used, but later it was exchanged for cheaper iron cuttings with only 70% metallic iron, the rest oxide. The metal reacts very fast with the arsenic in lead, and forms a fluid speiss at temperatures below 1000°C. On the other hand the iron oxide, mixed with some remaining slag, needed very high temperatures to melt. This overheating was very harmful for the lining. Fortunately, it was not necessary to melt all the contents of the furnace. It was sufficient to add the iron and rotate the furnace while heating for a few minutes. The speiss could be skimmed and the unmolten slag was left in the furnace.

In this way the refining time was reduced from 90 minutes to 30 minutes, also saving a lot of oil and oxygen. In 1982 processing of leach residues from the leaching of zinc clinker was started. No major changes in the process were needed. Smaller amounts of lead batteries were also smelted without any preparation of the batteries, except for emptying the acid.

Smelting of lead concentrates

After some years of operation most of the stored lead dusts were consumed. Melting lead concentrates was a natural option. During the pilot plant tests at Mefos prior to the erection of the Lead Kaldo Plant, a process for smelting lead concentrates was demonstrated. In 1981 some full scale tests with concentrates were conducted and in 1982 several thousand tons were smelted in a bigger campaign.

The process has two steps: melting and reduction. The melting takes place when dry concentrate is fed through a lance to a nozzle where it is mixed with oxygen and air. The heat of oxidation is in most cases sufficient to melt the concentrate and slag added. In order to get low sulphur content in the lead some lead must be oxidized to the slag, resulting in about 30% lead in the slag. In the case of Boliden own "Laisvall" concentrate, the sulphur content is only 13% and more lead must be oxidized to provide enough heat for autogenous melting.

After melting the lead in the slag is reduced. At lead contents of 30-50% in the slag, there is the risk of foaming slag. This is avoided be finishing the melting period by adding concentrate without oxygen supply. The heat is provided by oxygen/oil burner. After that the reduction is continued by adding coke.

In 1983-1984 the technique of smelting concentrates was further developed. Concentrates with 50% lead and 25% pyrite were smelted successfully. A process for smelting lead-carbonate concentrates was also demonstrated. On basis of these experiences a process for smelting complex lead concentrates high in Cu, As, Sb etc. was developed.

Processing of dusts from the copper plant together with complex lead concentrate in the Lead Kaldo Plant

Melting

By mixing the dusts (unpelletized) from the copper plant with a lead concentrate or several concentrates which are high in energy, e.g. high in sulphur etc., this mix can be melted autogenously.

The concentrates, the dusts and recycled dust are dried to less than 0,5% of water in the drier. The dried feed is fed pneumatically into the furnace where the feed is efficiently distributed into flash suspension by air and oxygen.

In order to get a good combustion, it is important that the feed is well mixed because of varying properties of the materials. If it is not completely autogenous, the flash smelting can be supported by an oxygen/oil burner.

By having as high oxygen enrichment as possible the dust content in the feed mix can be maximized. The fluxes ought to be charged continously through a special material lance into the furnace. That is because of the often high amount of fluxes per ton of mix, which can disturb the flash. The fluxes can also be added batchwise during the melting period.

The dusts from the copper plant are complex i.e. they contain a lot of antimony, arsenic, bismuth, zinc, etc. For economic reasons the concentrates then also ought to be complex.

The main task is still to process the dusts from copper plant but by mixing suitable concentrates in the dusts we can get better economy of the process. The capacity of the melting process is dependent on various parameters, e.g. percentage of dusts in the mix, concentrate analysis, gas capacity, etc.

Reduction

After melting, the slag is reduced with coke. A lead-phase is then formed under the slag-phase. The coke can be charged either batchwise or continuously through a lance. With a coke lance the control of the reduction process can be more exact.

During reduction, energy is added to the furnace from an oxygen/oil burner to cover the heat losses and the endothermic reduction processes.

By having excess of coke in the furnace, this can be used to cover a part of the necessary energy that is needed. The coke is then burnt with an excess of oxygen coming from the oxygen/oil burner. This combustion of coke is preferable in a way that the energy is developed where it best is needed i.e. in the coke layer where the endothermic process is going on. This increases the reduction speed. Because of the rotation of the furnace, the slag is well mixed and therefore zinc fuming is avoided. When the lead content in the slag is lower than about 2 percent the slag is tapped. Since the slag is often high in zinc (about 15% zinc) the slag is delivered to our fuming plant where the zinc and the remaining lead in the slag is fumed off.

Melting and reduction is repeated until the furnace contains about 25 tons of bullion.

The process gas is treated all over in the usual way i.e. the gases are washed in the venturi scrubber. The dust is precipitated in the venturi and recirculated.

Refining of bullion

The further treatment of the bullion is depending on especially the copper/arsenic ratio.

Low copper/arsenic ratio. In this case the bullion is treated in the same way as in the ordinary pellets processing. Here iron cuttings are charged to the bullion and a speiss phase is formed. This speiss is discarded because it contains very little of valuable elements. The refined bullion is tapped in a ladle and is then cast after cooling and drossing. All formed drosses are recycled back to the Lead Kaldo Furnace.

High copper/arsenic ratio. The bullion is tapped from the furnace into a ladle. The bullion is cooled down to about 500°C. During this cooling period a dross, so called ladle-dross, is formed. The cooled bullion is tapped in a kettle. The ladle-dross is cut away from the ladle and is then stored or returned to the furnace, depending on the copper loading.

The drosses formed in the kettles are returned back to the Kaldo Furnace. After kettle drossing the bullion is cast.

The stored copper-rich dross (ladle-dross) is treated in a special campaign. The ladle-dross is then remelted in the Lead Kaldo Furnace together with coke and some fluxes. The melting is performed by the oxygen/oil burner. After melting, the slag is tapped and after that, iron cuttings are added to the bullion. After a time of rotation of the furnace one or two phases have been formed over the lead phase.

The phases over the lead phase are tapped in a ladle to separate and be solidified. The ladle is emptied and the different phases can then be mechanically separated and they are treated as follows:

The lead phase is returned back to the Kaldo Furnace. The matte is transferred to the copper plant. The speiss is discarded.

The refined bullion in the Lead Kaldo Furnace is tapped and treated in the same way as mentioned above.

Burning of copper scrap

Boliden Metall has traditionally always processed all domestic low-grade copper scrap. Organic material has to be removed from cables and electronic scrap before further processing.

In the early 70's the scrap was burned in a stationary furnace. However, the removal of organics was not perfect. In addition the dust and gas emissions were too large. The authorities demanded new equipment with lower emissions to be built.

In this situation it was decided to test burning of scrap in the Lead Kaldo Plant. The results were very encouraging. The rotation of the furnace makes a perfect combustion possible. A part of the organics is vaporized and is burnt with secondary air in the water cooled offgas hood at a high temperature. This makes it possible to dispose of the large amount of heat evolved. The highly effective venturi takes care of the chlorine and dust and thus the emissions are well below the legislated values.

In 1982 the furnace hood was extended to accomodate a fully enclosed skip hoist for charging of scrap to the furnace.

Burning of scrap was first tested in 1980 and today it contributes to half of the production time.

Summary

After the start-up of the Lead Kaldo Plant in 1976 this entirely new process has been modified several times with doubling of the production capacity as a result.

The Kaldo Furnace is also suitable for treating of complex lead concentrates.

In addition copper scrap can be burned efficiently.

The plant meets up to the Swedish legislation of less than 0,1 mg Pb per cubic metre of air in working areas. In the last years no workers have been transferred due to high blood-lead levels (>500 µg/l). The emissions to the environment are also well below Swedish legislation.

References

1. P.-L. Nystedt; Operation of the Boliden Lead Kaldo Plant for treatment of lead-bearing dust, CIM-Bulletin May 1980 pp. 131-136.

2. H.I. Elvander; The Boliden Lead Process, Symposium on "Pyrometallurgical Processes in Non-Ferrous Metallurgy", Pittsburgh 1965.

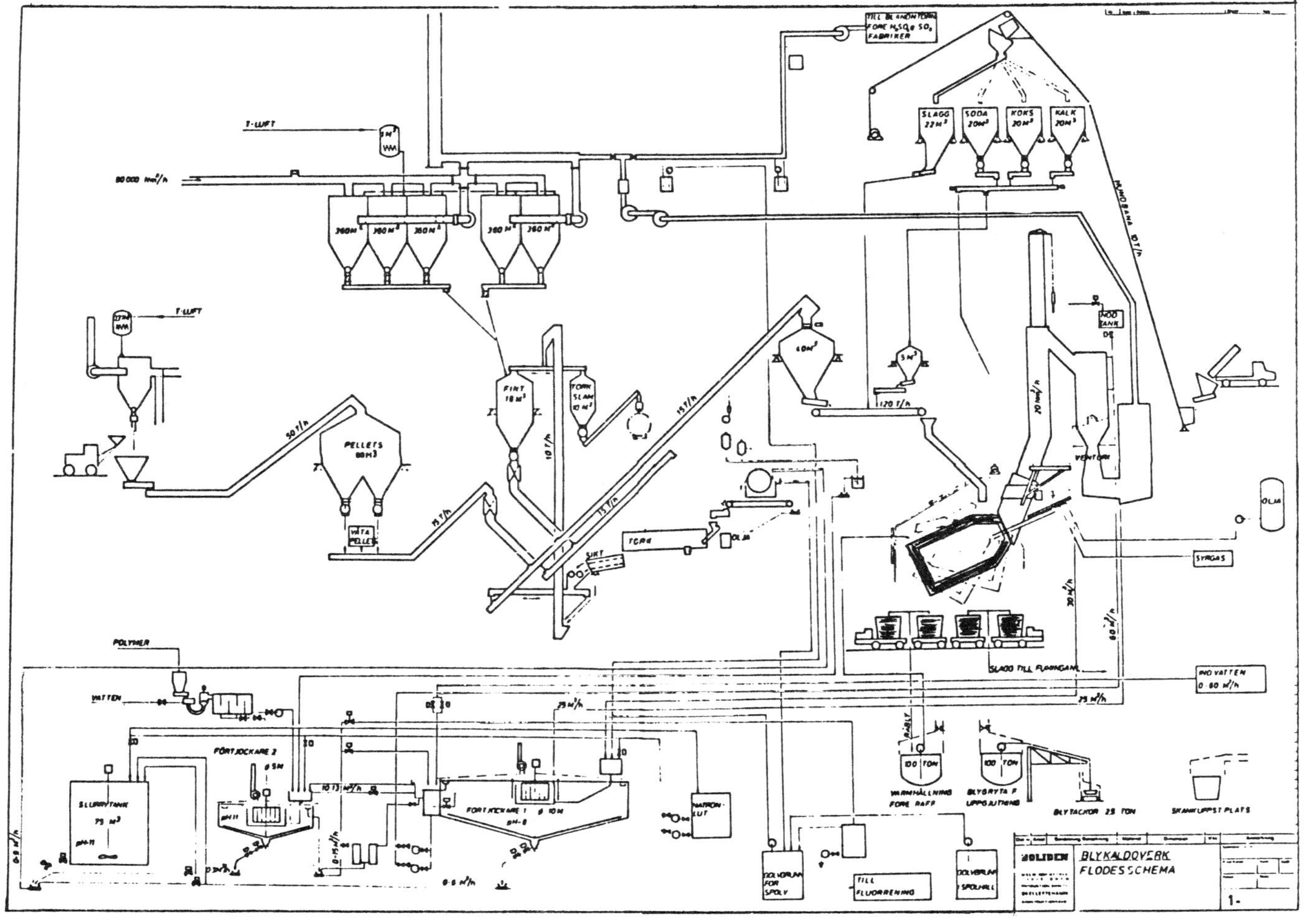
SLAGG 22 M³
SODA 20 M³
KOKS 20 M³
KALK 20 M³
PELLETS
FINT 18 M³
VÅTA PELLETS
VENTURI
OLJA
SYRGAS
SLAGG TILL FUMINGANL.
POLYMER
VATTEN
FÖRTJOCKARE 2
FÖRTJOCKARE 1
SLURRYTANK 75 M³
NATRONLUT
TILL FLUORRENING
VÄRMHÅLLNING FÖRE RAFF
BLYGRYTA F UPPGJUTNING
BLYTACKOR 25 TON
SKÄNKUPPST PLATS
BOLIDEN
BLYKALDOVERK
FLODESSCHEMA

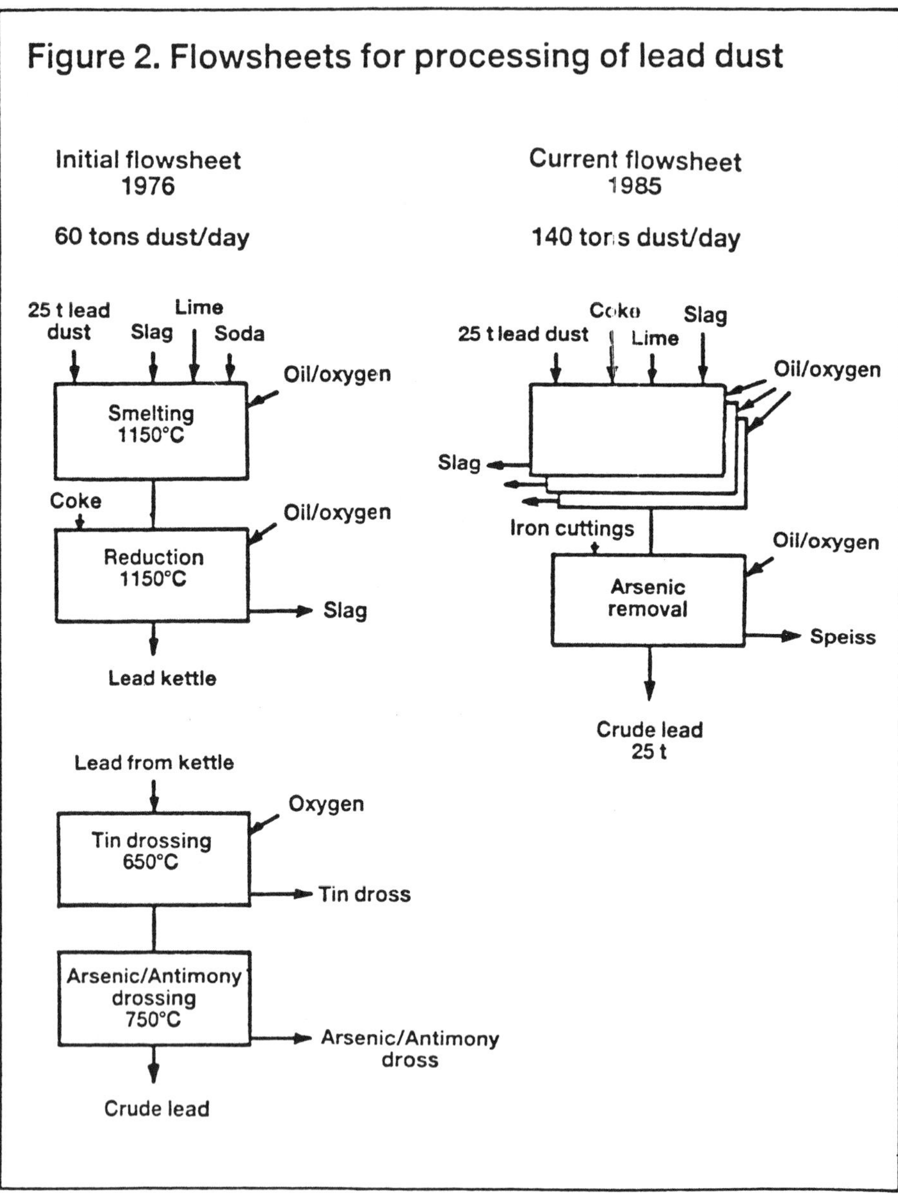
Figure 2. Flowsheets for processing of lead dust
Initial flowsheet
1976
60 tons dust/day
25 t lead dust
Slag
Lime
Soda
Oil/oxygen
Smelting
1150°C
Coke
Oil/oxygen
Reduction
1150°C
Slag
Lead kettle
Lead from kettle
Oxygen
Tin drossing
650°C
Tin dross
Arsenic/Antimony drossing
750°C
Arsenic/Antimony dross
Crude lead
Current flowsheet
1985
140 tons dust/day
25 t lead dust
Coke
Lime
Slag
Oil/oxygen
Slag
Iron cuttings
Oil/oxygen
Arsenic removal
Speiss
Crude lead
25 t

ST. JOE RESOURCES COMPANY

NATIONAL ZINC DIVISION

The following summary of operations has been prepared by the Editor from the brochure supplied with the survey reply.

INTRODUCTION

The new plant of National Zinc was the first modern plant to be built in the United States in over thirty years, that is, since ASARCO's Corpus Christi plant was built in 1941. Bartlesville is about 80 km north of Tulsa, Oklahoma. It may not be common knowledge that Tulsa as well as being an important center in the oil industry is a port on the Arkansas-Mississippi River system giving direct connection with the port of New Orleans.

Zinc was first produced in Bartlesville in 1907. The discovery of lead and zinc ores in the Tri-State District towards the end of the last century, together with the abundance of natural gas from shallow wells, prompted the building of at least ten horizontal gas-fired retort smelters within a 100 kilometer radius of the Bartlesville, Oklahoma area. When National Zinc's furnaces were phased out in the first half of 1976, it was the last remaining horizontal retort zinc plant in the United States.

After extensive feasibility studies, it was decided to maintain the existing Dorr fluo-solids roasting system and the relatively new (only nine years old) Leonard Monsanto sulphuric acid plant and to build only a new modern electrolytic plant with the most practical residue treatment system allowing optimum disposal characteristics.

Existing offices, machine and maintenance shops, concentrate storage and handling facilities and all associated service facilities were left intact to serve the new electrolytic plant. Thus, the capacity of the acid plant became the determining factor for the capacity of the tankhouse and the new plant. This is 51,000 MT of slab zinc per annum.

The layout of the new plant was designed in such a way that a second 51,000 ton per annum tankhouse and expanded residue treatment and purification facilities can be added at any later date; that is, as soon as the availability of additional concentrates within the area of influence of the Bartlesville plant will justify such an expansion. The area of influence for raw materials to supply the Bartlesville plant reaches well into Canada in the north and into Mexico in the south.

Process know-how and design details were supplied by Vieille-Montagne who had responsibility for all process engineering. The plant utilizes the Vieille-Montagne goethite residue treatment process and purification technology.

Construction of the new plant was started in December 1974 and completed by July 1976. It was started up during the fourth quarter of 1976 and the first cathodes were produced early in December 1976.

Concentrate Roasting and Acid Plant

With the horizontal retort plant, NZ had two Dorr fluo-solid roasters designed to produce calcine for sintering feed with a relatively high sulfide sulfur content at a capacity to satisfy the horizontal retort zinc smelter. These two 5.5 meter diameter roasters plus associated facilities did not have adequate capacity to produce sufficient quantity of low sulfide sulfur calcine for the new 51,000 ton electrolytic plant. A third Dorr type fluo-solid roaster with a 6.7 meter diameter grate was therefore built. Above the grate, it has a combustion chamber of 7.3 meter diameter, 7.3 meter height. All three roasters are slurry fed with Moyno pumps. With these three units, NZ has ample roasting capacity to produce the 51,000 tons of slabs per year.

The sulfide sulfur content of the calcine is not as low as we had hoped for (Table I). It is about 0.6%. If the plant is enlarged one day or a new acid plant is built we shall build new VM-Lurgi fluo bed roasters, mainly to obtain a consistently low sulfide sulfur in the calcine.

The Bartlesville roasters have no waste heat boilers. In the old plant and with our own natural gas, no one thought of recovering heat in those days. NZ has its own gas wells and pipeline system from the gas fields into the plant.

TABLE I

Typical Concentrate and Calcine Mix Analysis

Elements	Concentrates	Calcine
Zn	55.3	62.1
SO_4	-	1.6
SS	-	0.6
S Total	31.36	2.2
Pb	1.74	2.06
Fe	5.84	6.53
Cu	0.541	0.62
Cd	0.544	0.62
Co	0.028	0.03
MgO	0.286	0.33
Mn	0.179	0.20
SiO_2	1.20	1.36
Al_2O_3	0.347	0.39
INSOL	2.406	2.56
Ag g/t	258.7	276.8
Au g/t	1.29	1.89

Calcine Handling

All of the roasted concentrates produced are ground in a 2 meter by 2 meter dry ball mill in a non-recirculating system. The ground calcine is elevated and distributed to six 350 ton storage bins, providing approximately eight days of feed storage for the electrolytic plant. From these storage bins the calcine is weighed and transported by bottom hopper rail cars to the leach plant where the calcine is sampled and elevated by a drag chain type inclined conveyor to two 250-tons conical storage bins.

a) One bin is used for feeding calcine to the neutral leach system and the other bin provides feed for the neutralization and hydrolysis to precipitate goethite. All bins are equipped with vibrating bin discharges. To maintain the desired rate of calcine feed for the neutral leach, a manually controlled variable speed screw conveyor is installed between bin discharge and the first leach tank. The calcine feed is kept constant, while the volume of spent acid is regulated and controlled by the pH in the first of the five leaching tanks (pH is kept between 4.6 and 4.8 in the fifth tank).

b) The calcine fed to neutralization is regulated by one variable speed screw conveyor automatically controlled by the pH in the neutralization tank.

c) The calcine fed to hydrolysis is conveyed to the first three of a total of four hydrolysis reactors by a "side-pull horizontal recirculating continuous conveyor" equipped with rubber hoses with pinch valve hoses discharging to each hydrolysis reactor. In the fourth tank the pH is adjusted by adding spent acid. The pinch valves are intermittently opened by a timer actuated from the pH controllers. Table I gives typical assay of NZ's concentrates and calcines.

The typical minor element analysis of the concentrate and calcine mix is given in Table II.

TABLE II

Minor Element Analysis

Elements	Concentrates	Calcine
As	0.041	0.045
Sb	0.124	0.14
Bi	0.005	-
Cl	0.03	0.015
Ni	0.044	0.05
CaO	0.819	0.93
F	0.039	0.008
In	0.003	Nil
Tl	0.002	Nil
Te	0.005	Nil
Sn	0.009	0.01
Hg	0.0004	Nil
Se	0.0002	Nil

Melting and Casting

Primary zinc production is 51,500 MTPY of slab zinc. Cathodes are melted in a 1800 KW Ajax furnace equipped with four air cooled inductors. Holding and alloying is carried out using two 200 KW tilting Ajax furnaces and a third 125 KW unit is used for the production of zinc dust required in purification of the neutral leach solution.

Zinc is cast in various shapes using a Morganite continuous slab casting machine with automatic stacking. Cast zinc, most of which is debased with lead and aluminum, is in the form of 26 kg slabs, 272 kg slabs, or 1090 kg "logs".

The 26 kg slabs are automatically stacked with four leg castings and 48 regular slabs per stack. The stacks are cooled overnight, straightened in a hydromatic press and then banded. The banded stacks will weigh 1.36 MT and are ready for shipment to market by rail.

RESPONDENTS TO THE SURVEY

The Pyrometallurgical Committee of The Metallurgical Society of AIME acknowledges with thanks the cooperation of the individuals and companies who took time to respond to the questionnaire. Listed below are the addresses of the respondents in the event that the reader wishes further information or to make contact with those who participated in this survey.

COMPANY	CONTACT	PLANT ADDRESS
NORTH AND CENTRAL AMERICA		
AMAX Lead Company	K. Buckley	Route KK, Boss, MO 65540, USA
AMAX Zinc Co. Inc.	J.T. Krieg	P.O. Box 2347 East St. Louis, IL 62202, USA
Brunswick Mining & Smelting Corp. Ltd.	P. G. Evans	Belledune, N.B., CANADA EOB 1GO Telex: 014-24583
Canadian Electrolytic Zinc Ltd.	M. J. Agnew	860 boul. Cadieux Valleyfield, Quebec CANADA J6S 4W2
Chino Mines Company	B.K. Wichers	Chino Smelter Hurley, NM 88043, USA
Cominco Ltd.	A. W. Toop	Trail, B.C., CANADA V1R 4L8 Telex: 041-4426
Compania Minera de Cananea S.A.	J. Corona M.	Cananea, Sonora MEXICO 84620 Telex: 54706
Falconbridge Ltd.	R.R. Hoffman	Falconbridge Smelter Falconbridge, Ontario CANADA POM 1SO Telex: 06-77194
	R.J. St. Eloi	Kidd Creek Smelter Bag 2002, Timmins, Ontario CANADA P4N 7K1 Telex: 06-781553
Falconbridge Dominicana c.por A.	L.G. Tejada	Maximo Gomez 30 Apartedo 1343, Santo Domingo DOMINICAN REPUBLIC Telex: ITT 346-0356
Hudson Bay Mining and Smelting	B. Barlin	Box 1500, Flin Flon Manitoba, CANADA R8A 1N9
Inco Limited	T.N. Antonioni	Copper Cliff Smelter Copper Cliff, Ontario CANADA POM 1NO
	R. G. Orr	Thompson Smelter Thompson, Manitoba CANADA R8N 1C1

COMPANY	CONTACT	PLANT ADDRESS
NORTH AND CENTRAL AMERICA (cont'd)		
Inspiration Consolidated Copper Co.	R. Prescott	Inspiration Smelter Box 4444, Claypool Arizona 85532, USA
Jersey Minière Zinc Co.	D. A. Rice	Clarksville, TN, USA
Kennecott Corporation	T. Weddick	Utah Smelter Box 329, Magna, UT 84047, USA
	D.J. Nelson	Ray Mines Division Hayden, AZ 85235, USA
Mexicana de Cobre, S.A.		La Caridad Smelter Nacozani, Sonora, MEXICO
McLaren Forest Products Gaspé Mines Division	D.G. Pannell	Gaspé Smelter Murdochville, Quebec, CANADA
Noranda Minerals Inc.	G.Kochaniwsky	Horne Division Box 4000, Noranda, Quebec CANADA J9X 5B6 Telex: 051-3897
Phelps Dodge Corp.	W. L. Cage	Douglas Smelter P.O. Drawer E Douglas, AZ 85608-0019, USA
	W. J. Chen	Hidalgo Smelter Box 67, Playas, NM 88009, USA
St. Joe Lead Company	J.T. O'Reilly	Herculaneum Smelter 881 Main St., Herculaneum Missouri 63048, USA
St. Joe Resources Co.	J. M. Buxau	National Zinc Division Box 5791, Bartlesville OK 74005, USA
	R.L. Williams	Monaca Smelter 300 Frankfort Road Monaca, PA 15061-2295, USA
CHILE		
Empresa Nacional de Mineria (ENAMI)	J.S. Lamarca	Fundicion y Refineria Ventanas Camino a Puchuncav s/n Comuma de Quintero V Region CHILE
		Fundicion Hernan Videla Lira (Ex-Paipote), Paipote, III Region, Copiapo, CHILE
Corporación Nacional del Cobre del Chile	G.V. Bustos	Division El Teniente Millán 1040 Rancagua, CHILE Telex: 251001 TTE CL

COMPANY	CONTACT	PLANT ADDRESS
UNITED KINGDOM, SCANDINAVIA AND EUROPE		
Associated Lead Manufactuers Ltd.	Dr. H. Forrest	Elswick Lead Works Newcastle-Upon-Tyne ENGLAND NE99 1GE Telex: 53668
Austria Metall Aktiengesellschaft	J. Wallner	Montanwerke Brixlegg P. O. Box 19 A-6230 Brixlegg, AUSTRIA Telex: 5125126
BCL Limited	K. Robinson	P. O. Box 3 Selebi Phikwe, BOTSWANA Telex: 2286
Berzelius Metallhütten GmbH	R. Kola	Duisburg ISF and Tin Smelter Richard-Seiffert Strasse 20 Postfach 28 11 80 D-4100 Duisburg 28, F.R. GERMANY
	Günter Fuchs	Bleihütte Binsfeldhammer Binsfeldhammer 14, Postfach 1160 D-5190 Stolberg, F.R. GERMANY
Boliden Metall AB	B. Rudling	Rönnskär Smelter S-932 00 Skelleftehamn, SWEDEN
Commonwealth Smelting Limited	A.C. Perrins	Kingsweston Lane Avonmouth, Bristol ENGLAND BS11 8HT Telex: 449178
Copper Mining and Smelting Group	W. Grabowski	Legnica Copper Smelter 59-220 Legnica ul. Zlotoryjska, POLAND Telex: 034264
		Glogow Copper Smelters 67-231 Zukowice, POLAND Telex: 0787541
Espanola del Zinc SA	V. Gonsalez-Pumeriage Diaz	Pasao de la Castellana, 126 Madrid 16, SPAIN
Hüttenwerke Kayser Aktiengesellschaft	Dr. Bussmann	Kupferstrasse D-4670 Luenen, F. R. GERMANY
La Metalli Industrial s.p.a.	S. Zarone	Borgo Pinti, 99 50121 Firenze, ITALY
Metallurgie Hoboken-Overpelt	A.Franckaerts	Hoboken Smelter A. Greinerstraat 2710 Hoboken, BELGIUM
Mining and Smelting Combine Bor	C. Knezevic	R.T.B. Bor Topionica Bor M. Titay 29 19210 Bor, YUGOSLAVIA
Norddeutsche Affinerie Aktiengesellschaft	Dipl.-Ing. Kopke	Alsterterrasse 2, 2000 Hamburg 26, F. R. GERMANY

COMPANY	CONTACT	PLANT ADDRESS
UNITED KINGDOM, SCANDINAVIA AND EUROPE (cont'd)		
Norzink AS		5751 Odda, NORWAY
Outokumpu Oy	J. Makinen	Harjavalta Smelter 29200 Harjavalta, FINLAND Telex: 66379
Preussag-Boliden-Bloi GmbH	P. Klotz	Nordenham Lead Smelter Johannastrasse 2 D-2890 Nordenham 16, F.R. GERMANY
Preussag AG Metall	W. Blum	Hüttenwerk Harz P.O. Box 2240 D-3380, Goslar 1, F. R. GERMANY
Preussag-Weser-Zink GmbH	Dr. Riecke	Zinkelektrolyse Nordenham Postfach 17/18 D-2890 Nordenham, F. R. GERMANY
Quimica de Portugal, EP	I. Oliveira	2830 Barreiro, PORTUGAL Telex: 44282
Rio Tinto Minera, SA	P. Barrios	P. O. Box 110, Huelva, SPAIN Telex: 75516
AFRICA AND TURKEY		
Black Sea Copper	Works Director	K.B.I.A.S. Murgul Copper Works Goktas, TURKEY Telex: 83323
BSR Limited	G.C.P.Donoghue	Bindura Smelter Box 35, Bindura, ZIMBABWE Telex: 0907
Cinko Kursun Metal Sanaysii, A.S.	Works Director	Cinkur P.K. 184 Kayseri, TURKEY
Ergani Copper Plant	M. Azmi Ayhan	E.B.I. Müessesesi Müdürlügü Maden, Elazig, TURKEY
Palabora Mining Corp. Ltd. (Pty.)	P.D. Immelman	P. O. Box 65, Phalaborwa 1390 Transvaal, SOUTH AFRICA Telex: 350045
Tsumeb Corporation	Dr. R. Haegele	P.O. Box 40, Tsumeb SOUTH WEST AFRICA Telex: 908-680
Zambia Consolidated Copper Mines Ltd.	P. Mambo	Luanshya Smelter Box 90456, Luanshya, ZAMBIA Telex: 71270
	G. Cushing	Mufulira Smelter Box 40067, Mufulira, ZAMBIA
	D. I. Hughes	Nkana Smelter Box 22000, Kitwe, ZAMBIA Telex: 53140
	D. Littleford	Kabwe Smelter Box 80045, Kabwe, ZAMBIA

COMPANY	CONTACT	PLANT ADDRESS
AUSTRALIA, SOUTH EAST ASIA AND JAPAN		
Dowa Mining Co. Ltd.	C. Hayashi	Kosaka Smelter 29 Otarube, KosakaMachi Akita, JAPAN
Electrolytic Refining & Smelting Co. of Australia Limited	P. Owen	Military Road, Box 42 Port Kembla, NSW, AUSTRALIA Telex: 29020
Furukawa Mining Co. Ltd.		Ashio Smelter 1-1, Honzan, Ashio-cho Kamitsuga-gun, Tochigi Pref. JAPAN
Hibi Kyodo Smelting Co. Ltd.		Tamano Smelter 6-1-1, Tamano, Okayama Pref. JAPAN
Korea Mining & Smelting Co. Ltd.	C. Y. Lee	Onsan Smelter 70 Daejung, Onsan Uljoo, Kyong-nam, KOREA Telex: K28598
Mitsui Mining & Smelting Co. Ltd.		Miike Smelter 1-3, Asamuta-cho, Omuta City Fukoka Pref., JAPAN
	N. Ogata	Kamioka Smelter Shikama, Kamioka-cho Yoshiki-gun, Gifu-ken, JAPAN
	Y. Miyaji	Takehara Refinery 1436-3 Takehara-cho, Takehara-shi Hiroshima Pref., JAPAN
Mitsubishi Metal Corporation	E. Oshima	Naoshima Smelter Naoshima-cho, Kagawa-gun Kagawa-ken, JAPAN Telex: 2226533
Mount Isa Mines Ltd.	J. Pritchard & K. Ramus	Mount Isa, Queensland AUSTRALIA 4825 Telex: 49959
Nikko Zinc Co. Ltd.	T. Okura	Mikkaichi Smelter 8, Tenjinshin, Kurobe-shi Toyama, JAPAN 938
Nippon Mining Co. Ltd.	T. Okura	Saganoseki Smelter 10-1, Toranomon, 2-chome Minato-ku, Tokyo, JAPAN
Onahama Smelting & Refining Co. Ltd.	Y. Sugawara	Onahama Smelter 1-1, Nagisa, Onahama Iwaki City, Fukushima, JAPAN
Philippine Associated Smelting & Refining Corp.	Jan H. Reimers & Assoc. Inc.	PASAR Smelter Lide, Isabel, Leyte, PHILIPPINES Telex: 63758
P.T. International Nickel Indonesia	A. D. Dalvi	Soroako, South Sulawesi INDONESIA Telex: 46806

COMPANY	CONTACT	PLANT ADDRESS
AUSTRALIA, SOUTH EAST ASIA AND JAPAN (cont'd)		
Sumitomo Metal Mining Co. Ltd.		Toyo Smelter 145-1, Shinchi Funaya Saijo City, Ehime Pref., JAPAN
Western Mining Corporation Ltd.	D. Tillotson	Kalgoorlie Nickel Smelter P.O. Box 448, Kalgoorlie WESTERN AUSTRALIA 6430

BIBLIOGRAPHY AND ADDITIONAL REFERENCES

1. The Electrorefining and Winning of Copper, edited by J. E. Hoffmann, R.G. Bautista, V.A. Ettel, V. Kudryk, R.J. Wesely, The Metallurgical Society (TMS), 420 Commonwealth Drive, Warrendale, PA 15086, USA, February 1987.

2. Proceedings - Nickel Metallurgy, Volume I, Extraction and Refining of Nickel; 25th Annual Conference of Metallurgists, CIMM, 1130 Sherbrooke Street West, Montreal, Quebec H3A 2M8, August 1986.

3. Extraction Metallurgy '85, Institution of Mining and Metallurgy, 44 Portland Place, London W1, England, September 1985.

4. Mineral Processing Handbook, S.W. Mudd Memorial Series, edited by N.L. Weiss, SME-AIME, New York, NY, 1985.

5. Extractive Metallurgy Symposium - Melbourne, the Australian Institute of Mining and Metallurgy, 191 Royal Parade, Parkville, Vic., Australia 3052, November 1984.

6. Mineral Processing and Extractive Metallurgy, Kunming Conference, op cit (3), October 1984.

7. Complex Sulfides, edited by A. D. Zunkel, R. S. Boorman, A. E. Morris and R. J. Wesely, TMS-AIME, op cit (1), November 1983.

8. Copper Smelting - An Update, edited by D.B. George and J.C. Taylor, TMS-AIME, op cit (1), February 1982.

9. Mining and Metallurgical Practices in Australia, The Sir Maurice Mawby Memorial Volume, op cit (5), 1980.

10. Lead-Zinc-Tin '80, edited by J.M. Cigan, T.S. Mackey and T.J. O'Keefe, TMS-AIME, op cit (1), February 1980.

11. International Laterite Symposium, edited by D.J.I. Evans, R.S. Shoemaker and H. Veltman, op cit (4), February 1979.

12. Copper and Nickel Converters, edited by R.E. Johnson, TMS-AIME, op cit (1), February 1979.

13. Extractive Metallurgy of Copper, Volume I, edited by J.C. Yannopoulos and J.C. Agarwal, TMS-AIME, op cit (1), 1976.

14. Extractive Metallurgy of Copper, A.K. Biswas and W.G. Davenport, Pergamon International Library, Pergamon Press Inc., Maxwell House, Fairview Park, Elmsford, New York 10523, U.S.A., 1976.

15. The Future of Copper Metallurgy, edited by Carlos Diaz, Proceedings of the Chilean Institute of Mining Engineers, Santiago, Chile, 1973.

16. Lead and Zinc, Volume II, edited by C.H. Cotterhill and J.M. Cigan, AIME World Symposium, TMS-SME-AIME, op cit (4), 1970.

17. Copper Metallurgy, edited by R.P. Ehrlich, Extractive Metallurgy Symposium, TMS-AIME, op cit (1), February 1970.

18. The Winning of Nickel, edited by J.R. Boldt and P. Queneau, Meuthuen & Co. Ltd., 11 New Fetter Lane, London EC4, 1967.

19. Pyrometallurgical Processes in Nonferrous Metallurgy, Metallurgical Society Conference Volume 39, edited by J.N. Anderson and P.E. Queneau, TMS-AIME, op cit (1), 1965.

20. Sintering Symposium, Port Pirie, South Australia, op cit (5), September 1958.

SUBJECT INDEX

AUTHOR INDEX